Giovanni Giudice

The Sea Urchin Embryo

A Developmental Biological System

With 53 Figures

Springer-Verlag Berlin Heidelberg New York Tokyo

Professor Dr. Giovanni Giudice
Università di Palermo
Dipartimento di Biologia Cellulare
e dello Sviluppo
Via Archirafi, 22
I-90123 Palermo

The figure on the cover was reproduced by permission of the University of South Carolina
Press from J. B. Morrill and L. L. Santos, (1985) "A Scanning Electron Microscopical
Overview of Cellular and Extracellular Patterns During Blastulation and Gastrulation in the
Sea Urchin *Lytechinus variegatus*," in *The Cellular and Molecular Biology of Invertebrate
Development*, R. H. Sawyer and R. M. Showman, eds.

ISBN 3-540-15353-5 Springer-Verlag Berlin Heidelberg New York Tokyo
ISBN 0-387-15353-5 Springer-Verlag New York Heidelberg Berlin Tokyo

Library of Congress Cataloging in Publication Data:
Giudice, G. (Giovanni), 1933 — The sea urchin embryo. A developmental biological system.
Includes index. 1. Sea urchin embryo. I. Giudice, G. (Giovanni), 1933. II. Title. QL958.G5
1985 593.9′5 85-2819
ISBN 0-387-15353-5 (U.S.)

Printing: Sala-Druck, Berlin. Bookbinding: Lüderitz & Bauer, Berlin.
2131/3020-543210

Preface

This book should be regarded as the continuation to my previous book *Developmental Biology of the Sea Urchin Embryo*, edited by the Academic Press in 1973, rather than as a new edition.

Due to the exceedingly high rate of development in this field (something like 2000 papers have been published on this subject in these last 10 years), I preferred, in fact, not to describe again in detail the enormous amount of the old literature, as was attempted in my previous book, but to briefly summarize the state of the art in each problem and to describe in some detail the experiments performed in the last 12 years. In doing so, more emphasis was given to the more recent ones and to those which can be considered as corner stones in each subject. Care was, however, taken to mention the reviews or key papers in which the reader can find a source of the details of the older literature, besides refering him to my previous book. This detail may still be of relevance when problems of classical experimental embryology are to be investigated, because it is not unfrequent to discover, on reading the classical experiments, that we have an oversimplified idea of them, as they have been schematised in the course of transfer of information from one author to the next, so that the last one of the chain, for the sake of simplicity or in order to define a rule, may present as a proved fact what was in the beginning only a proposed theory.

The importance of going into the older literature in detail applies much less to the chapters on nucleic acid and proteins because the technologies developed in the last 10 years have so much improved the possibility of investigation that the old data can usually be considered obsolete.

More than giving a complete list of all the literature of the last 12 years (an almost impossible task) I have tried to discuss problems in the light of more recent knowledge. I was, however, often unable to resist the temptation of also quoting, for the sake of completeness, papers which appeared of minor interest to me. This, however, offers the advantage that the reader will not miss works that I have perhaps underestimated. This, of course, does not free me of the responsibility of providing the reader with my judgement on the papers discussed which I have done at least every time I felt able to do so.

Therefore I think the reader will also find in the book a fairly complete key to the current literature on the subject; I apologize, however, to the authors of those papers that may have escaped my attention, and also to those, I hope in both cases not too many, whose papers I have not quoted accurately.

Since, as I have already stated, the studies of sea urchin developmental biology are advancing at an exceedingly rapid pace, I found myself continuously rewriting entire chapters, so that I decided to stop and then to write an addendum while correcting the proofs. The reader is therefore asked to take particular notice of this if he wants an updated view of the subject.

I am very indebted to Drs. I. Albanese, D. Epel, M. Hoshi, D. Mazia, A. Monroy, E. Nakano, C. Petzelt, G. Sardet, G. Schatten, B. M. Shapiro, G. Spinelli, for reading and commenting single chapters, and to Mrs. A. Cascino and Mr. D. Cascino for their invaluable technical help.

This book is dedicated to Anna, Silvia and Lisa in exchange for the time it took me away from them.

Palermo, October 1985   G. Giudice

Table of Contents

VIII

Part I

Development

Chapter 1

Fertilization

1.1 The Egg Surface

1.1.1 The Jelly Coat

The sea urchin egg is surrounded by a gelatinous layer usually referred to as the jelly coat, which Lillie (1914) called "fertilizin". Lillie's original observation was that the jellycoat, dissolved in sea water, is able to cause a transitory activation of sperm motility and also sperm agglutination. From this the author inferred that fertilizin might be the molecule that causes sperm-egg interaction, because it binds the sperm surface from one side and the egg surface from the other.

The observation of the effects of the jelly coat on sperm motility and agglutination still stands, but its supposed role in fertilization should be modified (see Metz, 1978, for one of the latest reviews). As we shall describe later, specific egg-binding substances have in fact been isolated directly from sperm, and sperm receptors directly from eggs. Yoshida and Aketa (1978), moreover, by means of an immunofluorescent probe, failed to demonstrate the presence of the sperm-binding factor of the egg in the jelly coat. An important role played by the jelly coat may be that of causing the so-called "acrosomal reaction" of the sperm.

Beside this effect, the jelly coat has other effects on the sperm physiology; in fact, during the transit of the sperm through the jelly, the sperm mitochondrion becomes round, the internal pH is transiently alkalinized and then reacidified, and respiration becomes uncoupled and rapidly decreases, after an initial stimulation, which is lower than that caused by the simple dilution in sea water, Ca^{2+} is rapidly taken up (Schackmann and Shapiro, 1981; Collins and Epel, 1977; Tilney et al., 1978) and enzymatic activities like protease (Levine et al., 1978; Yamada et al., 1982) protein kinase (Garbers et al., 1980); phospholipase (Conway and Metz, 1976; SeGall and Lennarz 1981) increase or are released in the surrounding sea water (see Christen et al., 1983a for one of the latest studies on this subject, and later in this chapter for a discussion on the acrosome reaction and sperm motility).

How is all this brought about? We do not know yet, but an important event might be that of the dephosphorylation of one major protein of the sperm flagellum membrane of about 150,000 molecular weight (Ward and Vacquier, 1983).

The jelly coat is made of large glycoproteins, about 300,000 in molecular weight, containing SO_4 groups (Tyler, 1949, 1956; Vasseur, 1952; Nakano and Ohashi, 1954; Minganti and Vasseur, 1959). Early fractionation of the jelly coats of *Anthocidaris crassispina* and *Pseudocentrotus depressus* (Isaka et al., 1969, 1970) on hydroxylapatite columns suggested that all fractions are represented by sialopolysaccharides, rich in sialic acid, and poor in protein and sulfate, and that they contain L-fucose-4 sulfate (Ishihara et al., 1973). At least five antigenic fractions have been described by Lorenzi and Hedrick (1973) by means of immunoelectrophoresis. A more recent fractionation on Sepharose 4B (see Gall and Lennarz, 1978, 1979) of the jelly coats of several species of sea urchins has yielded a high molecular weight component composed of two fractions, one, 20% of the total by weight, represented by a sialo-protein, and the other by a fucose sulfate polymer. The latter is entirely responsible for the sperm acrosomal reaction. This is species-specific in some instances but not in others. SeGall and Lennarz (1981) have been able to show that species specificity may reside not merely in the composition but also in the structure of the jelly coat polysaccharides, as shown, e.g., by NMR studies.

Uno and Hoshi (1978) have fractionated the jelly coat of starfish by means of Sephadex G-100 columns into three fractions, J1, J2 and J3. Only J1 induces the acrosome reaction and only J2 induces sperm agglutination. The J1 fraction was further fractionated on Sepharose 4B into two subfractions consisting of glycoproteins rich in methylpentose and hexose respectively. J2 was further purified to a substance called agglutinin T1, that appears to correspond to asterosaponin A, i.e., a molecule made by a sulfated steroidal ring and by a sugar moiety containing two moles each of quinovose and fucose.

Hansbrough and Garber (1981a) have isolated from the jelly coat of *Strongylo-centrotus* and *Lytechinus* a factor, FS1, which is composed in a major proportion by a polymer containing fucose and sulfate, and is able to stimulate acrosome reaction, Ca^{2+} uptake and elevation of the cyclic AMP in the sperm (see also Kopf and Garbers, 1979). The same authors have purified from the jelly coat a peptide, named speract, which in sea water at acidic pH stimulates sperm movement and respiration, and causes an increase in the content of cyclic AMP and GMP. Suzuki *et al.*, (1981) have also purified from the jelly coat of *Hemicentrotus pulcherrimus* two peptides able to stimulate sperm respiration whose sequence is Gly-Phe-Asp-Leu-Tre-Gly-Gly-Gly-Va, and Gly-Phe-Asp-Leu-Asn-Gly-Gly-Gly-Val-Gly respectively.

The latter peptide corresponds to speract. Repaske and Garbers (1983) have recently shown that speract and analogs stimulate sperm respiration by causing a H^+ efflux (see Sect. 1.3.1). The mechanism through which these peptides affect the cyclic nucleotide concentration of the sperm is not yet known; Randany *et al.*, (1983) have, however, purified a particulate guanylate cyclase from the sperm of *Strongylocentrotus purpuratus*, which may represent an important step in the study of such a mechanism.

An important point raised by Holland and N. L. Cross (1983) is that the pH of the jelly coat is not acidic, as commonly supposed, but close to that of the sea water; these authors therefore question the role of speract in nature, because speract acts only at acidic pH's. Repaske and Garbers (1983) have recently shown, however, that speract stimulates sperm respiration also at the pH of sea water.

What can we conclude about the role of jelly coat in fertilization? We can easily understand its importance for the initiation of the acrosomal reaction (although, as we shall see later, this role has been questioned) and its role in the stimulation of sperm motility. We do not have a clear explanation for the role of sperm agglutination, which is a widespread phenomenon in nature, and which indicates that processes resembling antigen-antibody interactions occur in the jelly coat-sperm interaction. These processes may thus initiate what will be the ultimate interaction between the surfaces of the sperm and of the egg, which again is a species-specific process (Kinsey *et al.*, 1980a).

It has long been known that the treatment of the eggs with "egg water" of a different sea urchin species, i.e., water where these eggs have been kept and which therefore contains some jelly coat, may increase the chances of interspecific hybridization (see Giudice, 1973, for a review). More recent data obtained on Japanese sea urchins (Osanai and Kyozuka, 1979) confirm the older observations and suggest that the egg water acts not by eliciting the homologous sperm acrosomal reaction, but by modifying the heterologous egg surface.

An important point to bear in mind is that the stimulatory effects of the jelly are transient and the sperm rapidly loses fertility after jelly addition. One may speculate that this is of some relevance to the block of polyspermy.

1.1.2 The Vitelline Layer

Underlying the jelly coat there is a layer superimposed on the egg plasma membrane; that is the so-called vitelline layer, formerly called vitelline membrane, (see Moser, 1939, and Runnström, 1966, for early reviews). Loenning (1967a, b, c) and Millonig (1969) saw it at the electron microscope as a fluffy amorphous layer 3.5 nm thick (Lönning) or 10 nm (Millonig) which surrounds the plasma membrane. Chandler and Heuser (1979) after quick-freezing, freeze-fracturing and deep-etching, saw it as a network of fibers separated from the egg plasmamembrane by a distance of 20 nm.

Significantly, this layer is the first surface that the sperm meets on its way toward the egg, and it is therefore a good candidate to contain a specific sperm receptor. That this is the case has been suggested by Glabe and Vacquier, (1977a), who have developed a method for isolating the vitelline layer from eggs of *Strongylocentrotus purpuratus*. In a thorough analysis of the vitelline layer, they have demonstrated that it is composed of several proteins, ranging from 25,000 to 213,000 in molecular weight. They contain 3.5% sugar (fucose, mannose, galactose, glucose, xylose, glucosamine, galactosamine and sialic acid).

Only two high molecular weight proteins appear to be exposed on the outer side of the egg, based on their iodinability by the lactoperoxidase method. The isolated vitelline layers are able to bind sperms in species-specific fashion. Significantly, the sperms bind only to the outer side of the vitelline layer, which can be recognized under the electron microscope by the absence of protruding knobs which are characteristic of the inner surface.

Moreover (Glabe and Vacquier, 1977b, 1978), the eggs of *Strongylocentrotus purpuratus*, if parthenogenetically activated in the presence of low concentrations of protease inhibitors, release a glycoprotein into the sea water. This glycoprotein specifically binds the sperm surface protein called bindin that, as we shall see later, binds specifically to the egg surface. This is in line with the idea that a protease released from the egg upon fertilization displaces and then destroys the sperm receptors present on the surface of the egg, thereby preventing further sperm attachment. Furthermore, fertilization or parthenogenetic activation in the presence of low concentrations of a protease inhibitor may release the sperm receptor without destroying it. That the released receptor comes from the egg surface is also suggested by the fact that it can be labeled with ^{125}I with procedures that label only the proteins on the outer surface of the entire egg.

Additionally, Aketa and coworkers, who have long been involved in the search for a specific sperm receptor of the egg surface (Aketa, 1967a, 1973; Aketa and Tsuzuki,

1968; Aketa and Onitake, 1969; Aketa *et al.*, 1968, 1972, 1979; Onitake *et al.*, 1972; Tsuzuki *et al.*, 1977; Yoshida and Aketa, 1983), have extracted a 225,000 molecular weight glycoprotein from the isolated vitelline membranes of several Japanese sea urchin eggs. This glycoprotein is able to specifically interact with homologous sperm, thus completely inhibiting their fertilizing ability, without reducing their motility.

Treatment of the egg with antibodies against this protein inhibits sperm attachment in a species-specific fashion and sperm penetration in a non-species-specific way. Interestingly, sperm specifically bind to agarose beads covered with this glycoprotein (Aketa *et al.*, 1979, Yoshida and Aketa, 1979).

Using immunological methods Yoshida and Aketa (1982) have partially purified the sperm-binding factor from eggs of *Anthocidaris crossispina*, and found that it consists of three proteins of molecular weight 225,000; 87,000 and 80,000 respectively; the first and the third of these proteins contain carbohydrates.

The group of Vacquier has also initiated an immunological approach to understand the molecular structure of the vitelline layer. Gache *et al.*, (1983) have in fact produced 31 monoclonal antibodies against the vitelline layer of *Strongylocentrotus purpuratus* eggs. Eight of them bind the vitelline layer in a species-specific way, and inhibit fertilization without causing any egg wrinkling or cortical reaction, or other signs of egg activation.

A third description of a sperm receptor on the surface of the eggs of *Arbacia punctulata* comes from Schmell *et al.*, (1977). Here again a glycoprotein factor was extracted from isolated egg "ghosts" that were shown to contain surface proteins by the iodination method. This glycoproteic factor contains seven major proteins of 80,000 to more than 150,000 in molecular weight, as determined by electrophoretic analysis. This factor also binds to sperm and inhibits their fertilizing ability, in a species-specific way. Finally Kinsey and Lennarz (1981) have, by treatment with pronase, isolated from *Arbacia punctulata* egg surface a glycopeptide of 6000 molecular weight that binds acrosomal reacted sperm, thus inhibiting fertilization.

It is therefore quite evident that the vitelline layer contains glycoproteins representing species-specific sperm receptors (see also Glabe *et al.*, 1981). The question then arises whether these represent the only sperm receptors on the egg surface, or if the underlying plasma membrane also contains sperm receptors. This question has been investigated in experiments in which the vitelline layer has been removed by treatment with proteases (E. J. Carroll *et al.*, 1975, 1977; Tegner, 1974). The fertilizability of the egg invariably decreases, although the egg could still be activated parthenogenetically. Although sperm seem to attach to the microvilli of the underlying plasma membrane in protease-treated eggs, there is increased polyspermy and a decreased species specificity. This is in line with earlier observations on the possibility of obtaining cross-fertilization by proteolytic treatment of the egg (Hultin, 1948a, b; Bohus-Jensen, 1953; Tyler and Metz, 1955). These experiments, while again stressing the role of the vitelline layer as a site of sperm receptor, do not exclude the existence of other receptors in the plasma membrane (see also H. Schatten and G. Schatten, 1979). Interestingly, Longo (1982) has reported failure to inhibit fertilization by treating the eggs with concanavalin A, which binds to egg plasma membrane.

1.1.3 The Plasma Membrane

Although the egg plasma membrane has the usual lipoprotein composition, further analysis has long been hampered by the lack of adequate isolation methods. Whereas methods have been developed to purify the plasma membrane from cells of sea urchin embryos, purification of the egg plasma membrane is rendered more difficult by the presence of the vitelline layer. Methods aimed at isolating the egg plasma membrane are generally preceded by a treatment with dithiothreitol and proteases in an effort to remove all of the vitelline layer without damaging the plasma membrane. Aketa (1967b) has presented a simple method for isolating the plasma membrane of fertilized eggs, after removal of the fertilization layer. However, no mention is made of the hyaline layer, which may stick to it. Later analyses of the lipid composition of unfertilized and fertilized egg plasma membranes have been performed on preparations of "ghosts", which also contain the vitelline membrane and potentially some egg "cortex" (Barber and Mead, 1973; 1975; Barber and Foy, 1973). Preliminary analyses of the iodinable (and therefore exposed) proteins of the plasma membrane have been reported by J. L. Grainger and Barrett (1973), by Johnson et al. (1974), and by Dunbar et al. (1974).

Kinsey et al. (1980b) eventually elaborated a satisfactory method for isolating the plasma membranes of *Strongylocentrotus purpuratus* eggs; under the electron microscope these appear free of cortical granules, as markers of cortex contamination, although still covered here and there on the external surfaces with a fluff, which may represent a residue of hyaline layer. The latter, however, can be completely removed by extraction with KCl; moreover an enzymatic analysis shows an enrichment in Na^+/K^+-dependent and oubaine-sensitive ATPase, characteristic of cell membranes, and absence of protoesterase and ovoperoxidase, characteristic of the cortical granules. An electrophoretic analysis of the porteins of these membranes shows many bands, some of which are Schiff-positive; their lipidic composition expressed in nmol μg^{-1} protein is 61 phosphatilyl ethanolamine, 25.3 phosphatidylinositol, 16.2 phosphatydylcholine, 86.0 cholesterol, plus minor lipids (see also Decker and Kinsey, 1983 for the fatty acid analysis). More recently Ribot et al. (1983) have presented a new method to obtain purified membranes from fertilized eggs of *Lytechinus variegatus*, essentially based on the centrifugation of the membrane pellet through several sucrose gradients, the last of which do not contain Ca^{2+} and Mg^{2+}, in order to remove the hyaline layer. It is interesting that several proteins of the membranes prepared with this method from unfertilized eggs are lost following fertilization.

An in situ analysis of the plasma membrane sterol distribution has been carried out by Carron and Longo (1983) by use of the drug filipin, which binds to the membrane sterols, causing the formation of 20 to 25 nm protuberances visible in freeze-fracture replicas. These are numerous in the plasma membrane of *Arbacia punctulata* unfertilized eggs, but are fewer in the membrane delimiting the cortical granules (see next paragraph). When these, following fertilization, fuse with the plasma membrane, they acquire a higher number of sterol-rich sites.

Attempts to measure the lateral mobility of the lipidic and proteic components of the egg plasma membrane have been made both before and after fertilization by the use of a variety of probes, which are incorporated into the membrane and whose translational diffusion is thereafter measured (Wolf et al. 1979; Campisi and Scandella, 1980a, b; R. Peters and Richter, 1981). The most important conclusion appears to us that of Wolf *et al.* (1981), who after measuring the fluorescence photobleaching recovery of the diffusion of a series of lipid probes, found that this changes depending upon the probe structure, probably because different probes localize in different microenvironments or domains, and therefore it is inappropriate to discuss "bulk fluidity or viscosity" of the egg membrane.

It has long been known that the unfertilized egg plasma membrane protrudes into many short microvilli (Endo, 1961; Tyler and Tyler, 1966; P. Harris, 1968; Kane and Stephens, 1969; Vacquier and Mazia, 1968). An elegant description of them, based on observations with the scanning electromicroscope (Fig. 1.1), has been given by Schroeder (1978, 1979). He has estimated their length to be about 0.35 μm in the unfertilized egg of *Strongylocentrotus purpuratus*. They elongate to about 1.0 μm soon after fertilization and then shorten again to about 0.5 μm. Interestingly,

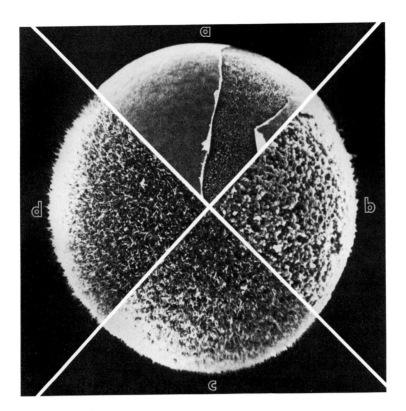

Fig. 1.1a–c. A composite of SEM images of plasma membrane surfaces of eggs at selected stages: **a** before fertilization with the plasma membrane revealed where the vitelline layer is torn away; **b** 1 min after fertilization, when the surface is obscured by globules from the cortical vesicles; **c** 5 min after fertilization, following removal of the hyaline layer precursor; and **d** 13 min after fertilization, when the presence of a few long microvilli give a distinctive "fuzzy" appearance. The two early eggs (*a* and *b*) were fixed from sea water; later eggs (*c* and *d*) were fixed from calcium-free sea water after treatment (2200). (T. E. Schroeder 1979)

this elongation can occur without the contribution of the cortical granule surface (see later) (E. Spiegel and H. Spiegel, 1977b, H. Spiegel and E. Spiegel 1978a). The extension of the microvilli is inhibited by cytochalasin (Eddy and Shapiro, 1976), which is in agreement with the presence of actin filaments within microvilli (Spudich and Amos, 1979; Burgess and Schroeder, 1977). These actin filaments have been more recently described by Tilney and Jaffe (1980) as growing in the cortical cytoplasm between 1 and 2 min after fertilization and as organizing bundles along the microvillar axes between 3 and 5 min in *Arbacia punctulata*. A beautiful photographic documentation of the process of microvilli elongation in *Strongylocentrotus purpuratus* has been presented by Chandler and Heuser (1979) after quick-freezing, freeze-fracturing and deep etching. Here again the role of microfilaments is stressed, which form a network within a system of ruffles which arise to cover the egg surface within 5 min after fertilization, while microvilli grow to about 0.7 μm in groups of 2–4 from mound-like formations and contain bundles of microfilaments parallel to the microvillar axis. According to Carron and Longo (1980a, b, 1982), the liberation of Ca^{2+} which follows fertilization is necessary for microvillar elongation, but, as suggested by experiments of egg activation in the presence of the Ca^{2+} ionophore A 23187 in the absence of external Na^+, cytoplasmic alkalinization of the *Arbacia punctulata* eggs in needed for the organization of the microvillar microfilaments into bundles (see also Begg *et al.* 1982).

Contrary to earlier reports, the microvilli are completely covered with the vitelline layer in the unfertilized egg (see, e.g. Lönning, 1967a, b, c; Veron *et al.*, 1977), although this layer is not strictly needed for sperm attachment and engulfment and microvillar elongation. These processes in fact occur as well in eggs denuded of the vitelline layer, the first because of the residual receptors on the plasma membrane and the other two, cytochalasin-sensitive, because of the recruitment of the actin of the egg cortex into filament bundles (H. Schatten and G. Schatten 1979; 1980).

1.1.4 The Cortical Granules and the Formation of the Fertilization Membrane

The cortical granules are highly important structures which are present immediately adjacent to the inner surface of the egg plasma membrane. We shall describe them here because of their central role in the process of forming the fertilization membrane (see fig. 1.4). They were first described by E. N. Harvey (1911), who already recognized their function. Since then they have received a great deal of attention (see Runnström, 1966, for an early review). A characteristic appearance of the cortical granules under the electron microscope is depicted in Fig. 1.2. They are spherical bodies, limited by a membrane 20–30 nm thick which encloses an electron-dense core, with electron-dense granules hanging from its internal surface. Other internal morphologies are observed, such as concentrically arranged lamellar structures at he interior of the granules in *Paracentrotus lividus* or in *Hemicentrotus pulcherrimus* (Takashima, 1960).

The mechanism through which the cortical granules contribute to the formation of the fertilization membrane is essentially that proposed by Endo (1961). Upon fertilization or parthenogenetic activation of the egg, the cortical granules become

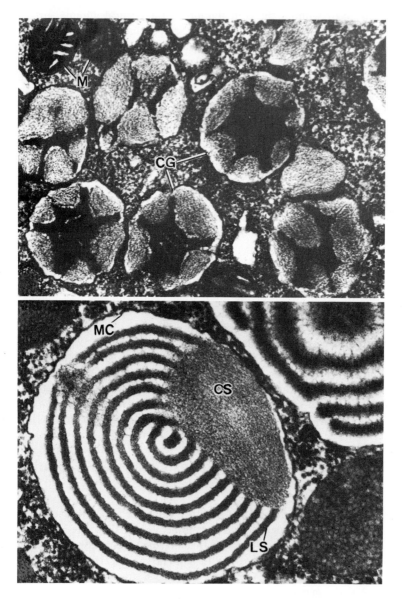

Fig. 1.2. a A tangential section through the peripheral ooplasm of a mature egg illustrates cortical granules (*CG*) and mitochondria (*M*). (× 12,000). **b** A section of a late oocyte of *Strongylocentrotus purpuratus* illustrates portions of two cortical granules, one of which shows a compact structure (*CS*) associated with lamellar units (*LS*). The lamellar units are associated with each other by fine filaments. Note the unit membrane (*MC*) of the cortical granule (× 90,000). (E. Anderson, 1968)

attached to the egg plasma membrane, with which they fuse through a process of membrane vesiculation (Millonig, 1969; Chandler and Heuser, 1979). They discharge their amorphous contents into the space between the plasma membrane and the vitelline layer, improperly called the perivitelline space, and contribute to the elevation, formation and hardening of the fertilization membrane (Fig. 1,3), and of the so-called hyaline layer (Hylander and Summers, 1980), which will be described later. Longo (1980), using the freeze-facture replica method, has described a lower concentration of intramembranous particles in the areas of the plasmamembrane overlying the cortical granules. The cortical granule dehiscence is linked to the formation of microfilaments in the egg cortex following fertilization or parthenogentic activation, as indicated by its sensitivity to cytochalasin B (Banzhaf et al. 1980).

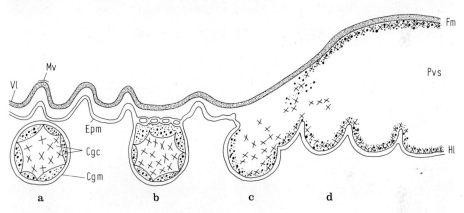

Fig. 1.3a–d. Scheme representing the processes of cortical granule reaction and of elevation of the fertilization membrane. **a** before fertilization. **b** the cortical granule membrane (*C.G.M.*) fuses with the egg plasma membrane (*E.P.M.*). **c** the cortical granule content (*C.G.C.*) is discharged into the perivitelline space (*P.V.S.*). **d** the fertilization membrane (*F.M.*) and the hyaline layer (*H.L.*) are formed with the contribution of the cortical granule content and of the vitelline layer (*V.L.*). Microvilli = (*M.V.*)

Further information about the so-called cortex of the egg will be provided in Chapter 2, when the mechanism of egg cleavage will be discussed.

The process of cortical granule dehiscence is believed to occur in a self-propagating wave (see Giudice, 1973, for a review), starting from the point of sperm penetration and traveling along the egg cortex (Uehara, 1971). Chambers and Hinkley (1979a, b), however, have demonstrated that it is possible to cause a localized granule discharge by touching only a portion of the egg surface with the Ca-ionophore A 23187.

Several experiments have been performed to elucidate the biochemical mechanisms by which the cortical granules contribute to the formation of the fertilization membrane. One important contribution to the understanding of such a mechanism comes from studies in which the cortical granules have been isolated and their contents analyzed. The criteria for identifying the cortical granules in cell fractionation experiments and for determining the purity of the fractions have been based on (1) the recognition within the fraction of substances (e.g enzymes) that are normally extruded from the eggs into the surrounding sea water following fertilization, i.e., in the so-called

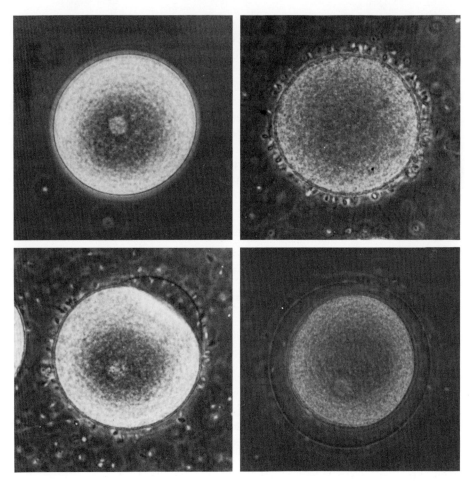

Fig. 1.4. a unfertilized egg of *Paracentrotus lividus*; **b** inseminated egg; **c** and **d** elevation of the fertilization membrane (× 400). (Courtesy of Prof. A. Monroy)

"fertilization product"; or (2) on the recognition of enzymes cytochemically known to be in the cortical granules, and, eventually, (3) on ultrastructural criteria. The cortical granules isolated by the methods of Schuel *et al.* (1972a, b) have a high content of β-1,3 glucanohydrolase, of arlyesterase, and of a trypsin-like protease (Schuel *et al.*, 1972a, b, 1973). These authors have also demonstrated the presence of sulfated mucopolysaccharides in the cortical granules by cytochemistry (Schuel *et al.*, 1974). The glycoprotein content of the fertilization product of eggs deprived of the vitelline membrane has been analyzed by Ishihara (1968a, b). An elegant and simple way of isolating the cortical granules has been described by Vacquier (Vacquier, 1974, 1975a, b) and more recently scaled up in order to permit isolation of the cortical granules in milligram amounts (Kopf *et al.* 1982). Eggs are caused to adhere to a protamine-coated slide and then briskly washed off with a jet of artificial sea water. The vitelline layer, the plasma membrane, and cortical granules remain attached to the slide. The granules retain some biological activities, such as undergoing dehiscence upon addition of Ca^{2+} ions, in a way that is inhibited by procaine. This dehiscence is accompanied by the expected ejection of Ca^{2+}-activated proteases and causes the formation of a protein gel. The excitability by Ca^{2+} ions, as well as the role of ATP, (Bryan, 1970; Baker and Whitaker, 1978), and the inhibition by procaine (Schuel and Kesner, 1977; B. S. Harris et al., 1977) are all well-known features regulating cortical granules dehiscence.

Haggerty and Jackson (1983), by adapting a procedure previously described by Sasaki, have been able to follow the decrease in turbidity of isolated egg cortices, which accompanies the Ca^{2+} excited cortical granule discharge, and to confirm the inhibitory effect of sulphydryl-modifying reagents and phenothiazine drugs. Similar experiments have been carried out by Sasaki and Epel (1983), who concluded that 6 µM calcium is needed for one-half maximal discharge of the cortical granules of isolated cortices of *Strongylocentrotus purpuratus* eggs, and that of the two media for isolation of cortices tested, the one that most closely simulates the in vivo conditions is that containing primarily potassium gluconate (250 mM) and glycine (500 mM).

A method for isolating the cortical granules still in association with the plasma membrane has been described by Detering et al. (1977), who have also reported the electrophoretic analysis of the proteins of such a complex.

Decker and Kinsey (1983) have recently succeeded into releasing the cortical granules from the complex isolated by Detering et al. (1977), by centrifuging them through 1 M sucrose, and have analyzed their fatty acid composition, finding a very high content of arachidonic acid, and a level of cholesterol higher than that of the plasmamembrane. The high content in glycoproteins of the cortical granules was confirmed by these authors, who by electrophoresis resolved at least 10 major PAS-positive proteins ranging in molecular weight from over 300,000 to 32,000.

Hylander et al. (1979) described another convenient method for the isolation of cortical granules in bulk. This is based on the observation that local anesthetics or ammonia prevent cortical granule attachment to the plasma membrane. If the fertilized egg of *Strongylocentrotus* or *Arbacia* is therefore centrifuged, its cortical granules are free to sediment to the bottom of the egg, and following homogenation can easily be purified by centrifugation through a sucrose density gradient. A method for collecting the cortical granule content has been described by Baginski et al. (1982), based on treatment with Ca- and Mg-free sea water of the freshly fertilized eggs, which causes the elevation of a fertilization membrane permeable to the cortical granule exudate (E. J. Carroll and Endress, 1982).

Based on immunological evidence Villacorta-Moeller and E. J. Carroll (1982) proposed that the soluble proteins of the fertilization membrane derive from the cortical granules.

The process of cortical granule exocytosis can therefore be viewed as a secretory process, according to Elhai and Scandella (1983). The granules in fact derive from the Golgi apparatus (E. Anderson, 1968) and under the effect of Ca^{2+} fuse with the plasma membrane and discharge their proteinaceus content outside the cell (beneath the vitelline layer). In confirmation of this idea, these authors also reported that arachidonic acid and some related fatty acids inhibit the elevation of the fertilization membrane.

Table 1.1. Fatty Acid and Cholesterol Composition of the Cortical Vesicle and Plasma Membrane

	Cortical vesicle (mol %)	Plasma membrane (mol %)
Myristate (C14:0)	4.7 (1.1)	4.8 (1.0)
Pentadecanoate (C15:0)	7.6 (1.4)	7.6 (1.6)
Palmitate (C16:0)	27.5 (3.2)	30.1 (2.9)
Heptadecanoate (C17:0)	8.1 (3.4)	9.8 (3.1)
Stearate (C18:0)	13.2 (1.8)	13.1 (1.7)
Oleate (C18:1)	2.2 (1.4)	0.8 (1.1)
Linoleate (C18:2)	1.9 (1.0)	2.5 (1.3)
Arachidate (C20:0)	1.5 (1.7)	1.9 (1.2)
Eicosenoate (C20:1)/ linolenate (C18:3)	7.3 (1.4)	5.7 (1.7)
Eicosadienoate (C20:2)	0.6 (1.0)	0.1 (1.0)
Arachidonate (C20:4)	17.7 (6.8)	14.7 (6.1)
Lignocerate (C24:0)	8.1 (1.5)	8.9 (1.7)
	mole/mole Lipid $PO_4^=$	mole/mole Lipid $PO_4^=$
Cholesterol	1.33 (0.12)[a]	0.55 (0.14)[a]

Note. Lipid extracts from six cortical vesicle and plasma membrane preparations were analyzed as described under Materials and Methods. Each sample was analyzed in duplicate and the values averaged. The numbers in parentheses represent the standard deviation.

[a] Ten samples of cortical vesicle and plasma membrane were analyzed for cholesterol.

Evidence has been presented that the cortical granules, in addition to the other enzymes, contain a peroxidase (Katsura and Tominaga, 1974; Foerder et al., 1978; Hall, 1978). What is the role of the enzymes contained in the cortical granules? One of the functions of the proteases is certainly that of establishing the block against polyspermy by removing the excess sperm receptors from the egg surface after the attachment of the first sperm (Vacquier et al., 1972a, b; Longo and Schuel, 1973; Schuel et al., 1973, 1976a, b, Longo et al., 1974; Troll et al., 1974, 1975; Byrd and Collins, 1975a). This idea is essentially based on the observation that protease inhibitors such as the soybean trypsin inhibitor, or inhibitors of the cortical granule dehiscence, e.g., procaine, cause polyspermy. Another important role of these proteases is that of helping the detachment of the vitelline layer from the plasma membrane in the process of elevation of the fertilization layer, as shown by the fact that protease inhibitors or inhibition of cortical granule breakdown cause incomplete elevation of the fertilization membrane, which appears blebbed because it is still attached to the plasma membrane at discrete points (Vacquier et al., 1972a; Longo and Schuel, 1973; Byrd, 1975a). By subfractionating the proteases of the fertilization product E. J. Carrol and Epel (1975a) have been able to separate two groups of proteases which, according to their functions, were called sperm receptor hydrolases and vitelline delaminase respectively.

A role for the peroxidase in the prevention of polyspermy has also been proposed

(Boldt *et al.*, 1980), which is in agreement with the observation that H_2O_2 inhibits sperm motility and that catalase, which destroys H_2O_2, causes polyspermy (Coburn *et al.*, 1981). The most important role of the peroxidase of the cortical granules, or at least the role that has been most thoroughly described, is, however, that played in the hardening of the fertilization membrane, as will be shown in detail later on.

The mechanism of action of the β-glucanase in the process of formation of the fertilization layer is still unknown. That it plays a role in the elevation of the latter is, however, suggested by the observation that among 13 sea urchin species studied, it is absent only in the egg of *Echinometra vanbrunti*, which are the only ones which form a fertilization membrane which does not elevate (Vacquier, 1975c). One may therefore suggest that it helps dissolve polysaccharides contained in the cortical granules. That such a polysaccharide dissolution occurs is suggested by the fact that concanavalin A inhibits the dispersion of cortical granule content of sand dollar (*Dendraster excentricus*) eggs, and that α-methyl-D-mannoside prevents this effect. Unfortunately, this sand dollar does not contain glucanase, whose effect might, however, be substituted for by another carbohydrase (Vacquier and O'Dell, 1975).

Schmidt and Epel (1983), confirming old experiments, have inhibited cortical granule dehiscence by subjecting inseminated eggs of *Strongylocentrotus purpuratus* to high hydrostatic pressure (4000–6000 p.s.i.). Under these conditions Ca^{2+} release occurs, as expected, but acid release is inhibited.

Once elevated, the fertilization membrane becomes quickly hardened, as proved, for example, by its loss of solubility in S—S reducing agents. There has been much discussion about the mechanism of this hardening process (see, e.g., Monroy, 1949; Motomura, 1957; Lallier, 1970, 1971, 1974b). The problem seems to have received a satisfactory answer following the more recent work of Foerder *et al.*, and of Hall (Foerder and Shapiro, 1977; Foerder *et al.*, 1977, 1978; Showman and Foerder, 1979; Hall, 1978). These authors have demonstrated that the peroxydase contained within the cortical granules causes the hardening of the fertilization membrane by catalyzing the formation of cross-links between the tyrosine residues of a glycoprotein of the fertilization layer. The proportion of dityrosines generated is one per 55,000 in the molecular weight of protein. Ca^{2+} and Mg^{2+} ions are required for such a process (E. J. Carroll *et al.*, 1979; Kay *et al.*, 1980; Cariello *et al.*, 1980). The only result difficult to reconcile with this hypothesis is that of E. J. Carroll and Epel (1975b), who found lack of effect of the fertilization product on membrane hardening. As Perry and Epel (1981) have demonstrated, part of the H_2O_2 produced by Ca^{2+} stimulation, and which may contribute to the hardening of the fertilization membrane, derives, at least in *Arbacia* eggs, from the oxidation of a naphtoquinone pigment, echinochrome A.

Sato *et al.*, (1973) reported the presence within the hardened fertilization layer of fibers departing from the surface of the egg. In a recent study on the mechanical properties of such "perivitelline fibers", Oshima (1983) has concluded that they are stiff enough to perform the supposed function of supporting the egg in the center of the perivitelline space.

The process of cortical granule exocytosis brings about nearly a doubling of the egg membrane, which is partially utilized for the microvillar elongation, and is partially reabsorbed through a rapid uptake into "coated vesicles" (Fisher and Rebhun, 1983).

1.1.5 The Block to Polyspermy

Although it seems quite clear that the ejection of proteases with the fertilization product contributes to the block to polyspermy, it does not appear justified to oversimplify such a mechanism. As reviewed by Epel (1978b), the block to polyspermy can be divided into two main steps. There is a fast step (Rothschild and Swann, 1952), which occurs within seconds after the entrance of the first sperm, and which has been attributed by Jaffe (1976) to a quick depolarization (within 30 s) of the egg membrane (see also Whitaker and Steinhardt, 1983), and which can be experimentally abolished by means of microelectrodes. This fast block to polyspermy is Na^+-dependent according to Schuel and Schuel (1981), since eggs fertilized in the presence of low Na^+ concentrations become polyspermic. As suggested by experiments of cross-species fertilization it appears to act on sperms in which the height of the experimentally applied voltage needed to block sperm penetration is that characteristic of the paternal species (Jaffe et al., 1982).

Secondly, there is a late block to polyspermy which is due to the cortical granule dehiscence. Byrd and Collins (1975a, b) questioned the existence of the fast block to polyspermy on the basis of the observation that polyspermy is obtained with high sperm concentration and if the jelly coat has been previously removed. This necessitated a re-evaluation of the role for the jelly coat in reducing the number of sperm which in nature reach the egg surface, and which one can suppose to be usually much lower than that found under experimental conditions and which has been questioned by De Felice and Dale (1979), who found that several sperm can penetrate the eggs within the first 13 s after the first successful sperm collision (see Dale and Monroy, 1981, and later in this chapter for a further discussion).

Relevant to this question is the work of G. Schatten and Hülser (1983), who have carefully measured the timing of the early events following fertilization by means of a special apparatus which superimposes by electronic video mixing the bioelectric recordings of impaled eggs to the microscopic differential contrast image of the egg. While confirming that the unfertilized egg potential membrane is about 15 mV in the impaled eggs, these authors found that the impaled eggs become polyspermic at the same sperm concentration that made the other, nonimpaled, eggs mono-spermic; moreover, if an egg is impaled and the electrode withdrawn before fertilization, then monospermy is the rule. The authors therefore suggest that the impalement, as suggested by Jaffe and Robinson (1978), Chambers and De Armendi (1979) and Shen and Steinhardt (1979), causes an ionic leakage, which results in a reduction of the membrane potential and interferes with the establishment of a fast block to polyspermy.

The data of Whitaker and Steinhardt (1983), also favor the hypothesis of the existence of a fast block to polyspermy based on the observation of an extracellular action current corresponding to the rising phase of an intracellular action potential, which is indicative of a membrane potential of -50 mV in the unfertilized egg of *Echinus esculentus* and *Psammechinus miliaris*.

Dunham et al., (1982) have found that a number of nonspecific proteins are able to reactivate aged sperms of *Arbacia punctulata* and also to induce polyspermy, which suggests that a block to polyspermy can be brought about in nature by sperm-inactivating substances. It is interesting to observe that G. Schatten (1977) has suggested a third block to polyspermy, which arises after the elevation of the fertilization membrane and which may be initiated by internal calcium release (Epel, 1975).

1.1.6 The Hyaline Layer

Another important egg envelope that is formed following the cortical granules breakdown is the so-called hyaline layer. This is found immediately adjacent to the outer surface of the membrane of the fertilized egg and has received a great deal of attention, as it has been hypothesized that it has the function of holding the blastomeres together (Osanai, 1960; K. Dan, 1960; Vacquier and Mazia, 1968b). In the scheme proposed by Endo (1961) this layer was thought to originate from material extruded from the cortical granules. This idea was mainly based on the observation that agents which inhibit the cortical granule breakdown also inhibit hyaline formation. This had never been criticized until McBlain and E. J. Carroll (1977) reported that it is possible to inhibit cortical granule breakdown by means of erytrosine and still observe the formation of an apparently normal hyaline layer in eggs of *Strongylocentrotus purpuratus*. More recently, however, Hylander and Summers (1980, 1982a, b) were able to show in the electron microscope the presence of antigens of the hyaline layer within the cortical granules.

The most reasonable conclusion seems to us that the hyaline layer, or at least good part of it, is produced at fertilization by the discharge into the perivitelline space of the cortical granule content, but that the egg is able to synthesize more of it and/or contains a further store of hyaline besides that of the cortical granules (see also Chap. 2.4.2).

The hyaline layer consists, as viewed in the electron microscope after ruthenium staining (Lundgren, 1973), of three distinct layers, of which the most internal is the thickest. Its chemical composition has been studied after isolation by several authors (Vacquier, 1969; Kane, 1970, 1973; Kondo, 1973; Citkowitz, 1971, 1972). It consists of 95 % protein and 2.5 % sugar, among which are fucose, xylose, mannose, galactose, glucose, N-acetylglucosamine, N-acetylgalactosamine and N-acetylneuraminic acid. The proteic part is precipitable by Ca^{2+} ions to form a gel, and can be electrophoretically resolved into at least three fractions. On the basic of electron microscopic and biochemical observations E. Spiegel and M. Spiegel (1979) have recently concluded that it also contains collagen.

1.1.7 Ultrastructure and Composition of the Fertilization Membrane

The fertilization membrane was first studied in situ by electron microscopy and found to be composed of three layers which, in total, are 50 nm thick (Ito *et al.*, 1967). An intermediate amorphous layer is enclosed between two layers made of tubular structures running antiparallel to each other. On comparison of the electron microscopic pictures taken before and after complete fusion of the cortical grabuke contents with the vitelline layer, S. Inoué *et al.*, (1967, 1970) concluded that tight rolls of an elongated sheet, which has a crossed-grid network of about 15×24-nm mesh, unroll from the cortical granules and adhere to the vitelline membrane.

Further ultrastructural (S. Inoué *et al.*, 1971, S. Inoué and Hardy, 1971, Chandler and Heuser, 1979) and biochemical (Carrol and Baginski, 1977, 1978) analyses have been performed on isolated fertilization membrane. The latter studies have shown that it is possible to detect three glycoproteins with molecular weights of 91,600; 71,200; 53,000 and two other proteins of 32,600 and 18,200.

All these proteins appear to be on the outer surface of the fertilization membrane, as indicated by their accessibility to ^{125}I labeling. Which of them is derived from the cortical granules remains to be established; preliminary attempts to answer such a question have been reported by Moeller *et al.*, (1980), who found an immunological correlation between the proteins isolated and electrophoretically resolved, from cortical granules and fertilization membrane.

The macromolecules secreted into the perivitelline space seem to play a role in the observed contraction of the egg cortex wich follows fertilization; this contraction in fact does not occur if the fertilization membrane is made permeable to macromolecules by treatment with dithiotreitol (T. D. Green and Summers, 1979).

A preliminary analysis of the fertilization membrane isolated from *Arbacia punctulata* eggs has also been reported by Cariello *et al.*, (1980).

1.2 The Sperm Surface and the Acrosomal Reaction

As shown in Fig. 1.5, the sperm surface is represented by the plasma membrane, without the outer layers observed in the egg. However, the part which interacts with the egg, the sperm apical zone, has an important specialization developed during spermiogenesis, the so-called acrosome.

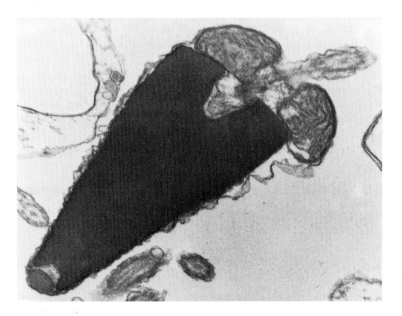

Fig. 1.5. Sperm of *Paracentrotus lividus* (× 49,000) (Courtesy of Dr. I. Salcher)

The overall sperm morphology can vary among echinoderms, as thoroughly reviewed by Chia *et al.*, (1975) and by Summers *et al.*, (1975): crinoids, ophiuiroids, asteroids, and many holothuroids have spherical sperm, whereas echinoids all have elongated sperm. However, the anatomy of the acrosomal region is strikingly similar in all the chinoderm classes studied so far (Summers *et al.*, 1971, 1975).

As first recognized by Popa (1927) and then thoroughly described by J. C. Dan (1952, 1954, 1956, 1960, 1967, 1970), the echinoderm spermatozoa, when in the presence of egg jelly in sea water, undergo a change in their apical region that is called the acrosomal reaction. As a consequence of this, and as shown in Fig. 1.6, the proteins which have to match the egg surface are exposed. The egg-binding protein(s) have been found in the acrosomal granule by Vacquier, who has described and partially purified a protein of 30,500 molecular weight which he has called bindin. Bindin meets the expected requirements that it is contained in the acrosomal granule, and binds to the egg vitelline layer glycoproteins in a species-specific way (Vacquier and Moy, 1977; Bellet *et al.*, 1977 a, b; Moy *et al.*, 1977; Glabe and Lennarz, 1979).

Bindin according to Glabe *et al.*, (1982), acts, by virtue of its lectin activity, as suggested by its ability to agglutinate eggs or also erythrocytes. This agglutinating activity is inhibited by an egg-surface polysaccharide and, in decreasing order, by fucoidin, by the jelly fucan, and xylan, thus suggesting that sperm-egg adhesion is mediated by a lectin-polysaccharide type of interaction.

Additionally, Aketa (Aketa, 1975; Aketa *et al.*, 1978, 1979) has extracted from sperms a factor composed of 95% carbohydrate and 5% amino acids. Because of

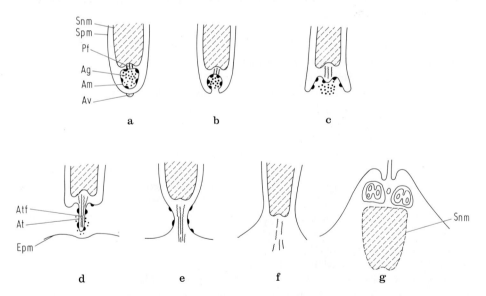

Fig. 1.6. Diagram illustrating the acrosomal reaction and sperm penetration. Abbreviations: *s.n.m.* sperm nuclear membrane; *s.p.m.* sperm plasma membrane; *p.f.* precursors of the acrosomal tubule fibrils; *a.g.* acrosomal granule; *a.m.* acrosomal membrane; *a.t.f.* acrosomal tubule fibrils; *a.t.* acrosomal tubule; *e.p.m.* egg plasma membrane; *a.v.* apical vericle. (Giudice, 1973)

its properties, it has been suggested by the author to be the counterpart to the sperm-binding factor of the homologous eggs.

Another important contribution to the analysis of the sperm-surface components involved in egg-surface recognition comes from the work of Metz and coworkers (reviewed by Metz, 1978). These authors have also purified a sperm glycoprotein antigen, which is able to block the antifertilizing activity of univalent antisperm antibodies (Cordle and Metz, 1973; Maitra and Metz, 1974). Lopo and Vacquier (1979, 1980a, b) have also purified from sperm of *Strongylocentrotus purpuratus* two glycoproteins of molecular weight 84,000 and 64,000 respectively, whose localization on sperm surface is demonstrated by radio-iodination, and by the lactoperoxidase reaction under the electron microscope. Univalent antibodies against the 84,000 molecular weight protein inhibit the acrosome reaction and fertilization. Species-specific sialosphingolipids have been described in the surface of sea urchin spermatozoa (Nagai and Hoshi, 1975; Hoshi and Nagai, 1975; Ohsawa and Nagai, 1975).

Cross (1983) has succeeded in isolating the plasma membranes from *Strongylocentrotus purpuratus* sperm. An electrophoresis of the proteins extracted from these membranes shows the same banding pattern as that obtained after lactoperoxidase-catalyzed external radio-iodination of entire sperm. This will permit further studies on the sperm plasma membrane function.

A great deal of work has been dedicated to answering the question of what causes the acrosome reaction. From what we have said until now the obvious answer seems to be "the jelly coat", but we have to recall that in dejellied eggs it is the egg surface that induces the acrosomal reaction. If the egg surface has been damaged by trypsin treatment, fertilization can occur only with sperm that have already undergone the acrosomal reaction (Kimura-Furukawa *et al.*, 1978). Moreover, Aketa and Ohta (1977) raise some doubts that the jelly has ever had a role in triggering such a reaction, because sperm can pass through isolated jelly coats without undergoing the reaction, and because their electron microscope studies show that the sperm undergoes the acrosomal reaction not when it passes through the jelly layer, but only when it reaches and touches the cell surface. The fact that many authors have consistently and repeatedly demonstrated that the egg jelly coat is able to induce the acrosomal reaction needs not be surprising because many physical or chemical agents have been described that are also able to induce the acrosomal reaction (see Metz, 1978, for a review). The jelly coat, however, seems to be the most specific candidate (Aketa and Ohata, 1979) for inducing the acrosomal reaction, so that one might suppose that the activating factor is continuously secreted by the egg surface and becomes loosely associated with the jelly coat. G. L. Decker and Lennarz (1979) have demonstrated that isolated vitelline layers containing the underlying plasma membranes and cortical granules elicit the acrosomal reaction only when they meet the sperm with their outer surfaces. Whatever induces the acrosomal reaction, its effect seems to be exerted through an increase in the influx of Ca^{2+} ions into the acrosome, as suggested by the negative effect of drugs inhibiting Ca^{2+} transport, by the removal of Ca^{2+} from sea water and by the positive effect of the Ca^{2+} ionophore A 23187, or of increasing the external Ca^{2+} concentration, Y. M. Takahashi and Sugiyama, 1973; G. L. Decker *et al.*, 1976; Collins and Epel, 1977; Schackmann *et al.*, 1977, 1978). An increase in the external pH seems to cause the acrosomal reaction through increasing the permeability to Ca^{2+}.

Acrosomal reaction is inhibited below pH 7.0 and 20 mM KCl (Christen et al., 1980). Schackman and Shapiro (1980), using various inhibitors of ionic flow, proposed that the ionic requirements and the permeability changes associated with the acrosome reaction of the *Strongylocentrotus purpuratus* sperm, stimulated by jelly coat, can be divided into four steps: Initial Ca^{2+}-dependent step; a second phase sensitive to inhibitors of Ca^{2+} and K^+ movements and to high K^+ concentrations; a third step implying Na^+ uptake and H^+ efflux, a fourth step, which follows filament extension and includes K^+ loss and mitochondrial Ca^{2+} accumulation. More recently Schackman et al., (1981), by measuring the concentration of weak bases and the excretion of weak acid, have shown that membrane potential depolarization and increase in the intracellular pH accompany the acrosomal reaction.

This increase has been measured by H. C. Lee et al., (1983) by measuring the uptake and discharge of two experimental probes as 9-aminoacrydine and methylamine, under various conditions, e.g., in the absence of different ionophores. The results indicate a pH increase of 0.4–0.5 units coincident with the activation of sperm motility and a further increase of 0.16 units coincident with the acrosome reaction. A reacidification follows, which is probably induced by the metabolic stimulation

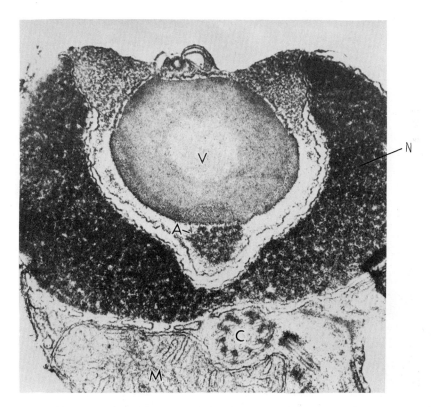

Fig. 1.7. Thin section through the apical end of a *Pisaster* sperm. The acrosomal vacuole (*V*), the nucleus (*N*), the centriole (*C*), a portion of the mitochondrion (*M*), and the actomere (*A*) are indicated. Note that the actomere of *Pisaster* consists of small dense granules arranged into rows. (× 80,000). (Tilney and Kallenbach, 1979)

caused by the Ca^{2+} influx (see also Shapiro *et al.*, 1982 for a summary of recent data on this subject).

If the acrosomal reaction is experimentally inhibited, and a jelly coat preparation is added to sperm of *Strongylocentrotus purpuratus*, the alkalinization takes place, but without subsequent re-acidification, mitochondria remain coupled and respiration and intracellular ATP levels remain high (Christen *et al.*, 1983a), thus suggesting that most of the effects of jelly on the sperm physiology are mediated through the acrosomal reaction, or are at least related to it.

An important event which occurs during the acrosomal reaction is the polymerization of actin, which forms the "skeleton" of the acrosomal tubule (Tilney *et al.*, 1973). The mechanism of actin polymerization during the acrosomal reaction has been thoroughly investigated both ultrastructurally and biochemically by Tilney and coworkers (Tilney, 1976b; 1977, 1978; Tilney *et al.*, 1973a, 1978; Tilney and Kallenbach, 1979). These authors have also described the presence in the acrosome of a new organelle, the actomere, which appears to act as a nucleation site for actin polymerization, and appears to control the polarity of the actin filaments. This polarity is demonstrated by the so-called decoration with myosin of subfragment 1, which shows an unidirectional pointing to the sperm nucleus (Figs. 1.7 and 1.8). In the opinion of Tilney et al. this is the wrong direction of one hypothesizes that it is the interaction of this actin with the egg myosin that pulls the sperm into the

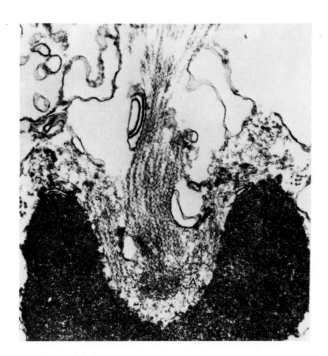

Fig. 1.8. Thin section through the apical end of a *Pisaster* sperm which had been induced to undergo the acrosomal reaction with the ionophore 537A, then glycerinated and decorated with myosin. The actin filaments which extend from the actomere (*A*) are unidirectionally polarized — they all point toward the nucleus. (× 64,000). (Tilney and Kallenbach, 1979)

egg. Tilney and S. Inoué (1982) postulate the necessity of a control mechanism which prevents a spontaneous nucleation of the diffusable actin, which at the base of the elongating process reaches concentrations over 160 mg ml^{-1}. This role might be attributed to one of the four additional proteins described by Tilney (1979) in the cup of filactin precursors. The latter author (Tilney and S. Inoué, 1982; S. Inoué and Tilney, 1982) stresses the necessity of a delicate balance between the events which appear to govern the shape of the growing acrosomal process, i.e., extending of the actin filaments, membrane addition and water influx.

A role for an acrosin-like enzyme both in the acrosomal reaction and in fertilization has been proposed by Levine and Walsh (1979, 1980). I. D. Green and Summers, following earlier experiments (1980) providing ultrastructural evidence for the presence of trypsin-like enzymes in the acrosome of *Strongylocentrotus purpuratus* sperms, have recently (1982) concluded that the effect of specific protease inhibitors suggests a role for chymotrypsin-like enzymes in the processes of acrosome reaction and sperm penetration. Also a chymotrypsin-like enzyme of probable acrosomal localization has been purified from *Hemicentrotus pulcherrimus* sperm by Yamada *et al.*, (1982) and a trypsin-like enzyme by Yamada and Aketa (1982). Finally, an arylsulfatase able to digest the jelly coat, and which may be liberated from the acrosome of *Hemicentrotus pulcherrimus* and *Strongylocentrotus intermedius* sperm, has been described by Hoshi and Moriya (1980). All these enzymes may be thought to play a role in the process of sperm penetration through the egg envelopes and membrane at fertilization (Yamada and Aketa, 1981; Yamada *et al.*, 1982).

1.3 Sperm-Egg Interaction

1.3.1 Sperm Motility

Sea urchin sperm is brought into contact with the egg surface by the active movement of the sperm flagellum, whose role, however, is only that of ensuring sperm movement; in fact sperm heads, experimentally deprived of the tails, retain the ability to undergo acrosomal reaction and to activate the egg (Gabers, 1981). The flagellum originates from a concavity at the basis of the sperm head, the so-called centriolar fossa, containing two orthogonally arranged centrioles. The position and denomination of the two centrioles in echinoderm sperms have been a controversial matter, which has been recently reviewed by Baccetti and Burrini (1983) (see Fig. 9).

A great deal of work has been carried out to clarify the mechanism of sperm movement. Afzelius (1950) has provided probably the earliest and yet very modern ultrastructural description of the sea urchin sperm flagellum. This shows in a cross section (Fig. 1.9 and 1.10) the usual 9 + 2 array of rings, representing the sections of microtubule doublets which extend along the tail and make the so-called axoneme. Each of the outer doublets is composed of one entire ring, representing the section of the microtubule A, and of a C-shaped figure attached to the ring and representing the section of the microtubule B. A appears to be composed of 13 protofilaments and B of 11 protofilaments (see Fig. 1.11).

The two central microtubules also appear to be constructed of 13 proto-

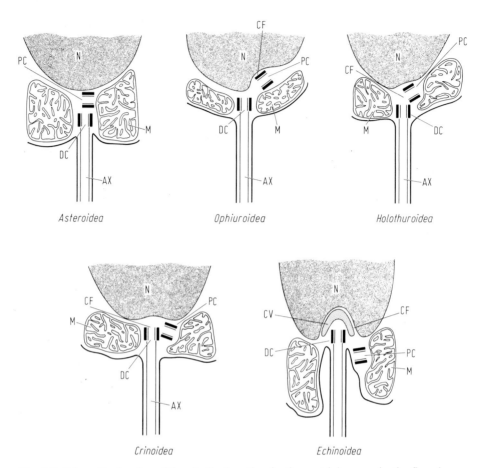

Fig. 1.9. Schematic drawing of longitudinal section in the centriolar area in the five classes of *Echinodermata* studied so far. The different position of the two centrioles in the different cases is quite evident; *AX* axoneme; *CF* centriolar fossa; *CV* centriolar vesicle; *DC* distal centriole; *M* mitochondrion; *PC* proximal centriole; *N* nucleus. (Baccetti and Burrini 1983)

filaments each (Tilney *et al.*, 1973). Two arms are seen to project from each A ring toward the next B ring in a clockwise manner, and radial projections extend from the outer doublets to the central tubules. The constituent proteins of the A and B tubules can be differentially extracted at different temperatures (Stephens, 1970a) or at different concentrations of Sarkosyl. It has therefore been possible to demonstrate that they are mainly composed of the α and β tubulin subunits (about 54,000 m.w. each) in the ratio of about 1:1 in both microtubules (Meza *et al.*, 1972). These are similar, but not identical, to the α and β-tubulin of cilia, mitotic apparati and brain cells (Bibring and Baxandall, 1971; Feit *et al.*, 1971; Yanagisawa *et al.*, 1973; Luduena and Woodward, 1973; Bibring *et al.*, 1976; Stephens, 1978; Farrel *et al.*, 1979a, b).

The tails of the sea urchin sperm have represented favored material for studies of tubulin assembly into microtubules (Borisy *et al.*, 1972; Binder *et al.*, 1975, Farrel

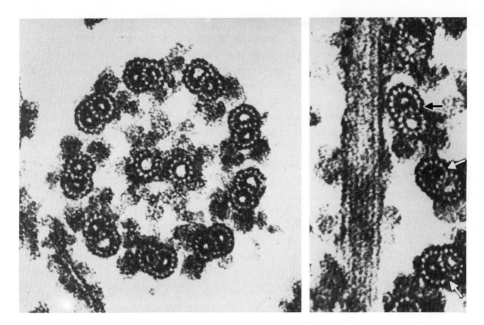

Fig. 1.10. *Left* Transverse section through the isolated flagellar axoneme of *Lytechinus*. 4%
tannic acid. (× 330,000). *Right* Section through a pellet of flagellar axonemes. The outer doublet,
cut in longitudinal section, clearly displays the protofilament structure. Two transverse sections
of portions of axonemes are seen to the right of the micrograph. The outer sides of these
doublets are toward the left; their inner sides to the right. The number of subunits making
up the A and B can be counted easily and the connection of the B to the A clearly
demonstrated. *Arrows* point toward the 11th subunit of the B. 8% tannic acid (× 360,000).
(Tilney et al., 1973a)

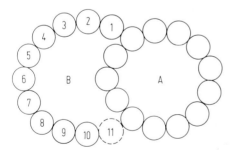

Fig. 1.11. Drawing of an outer doublet. The
A tubule (*A*) is composed of 13 subunits, the
B (*B*) of 11 subunits. The inside of the cilium
would be beneath this doublet, the outside
above it. (Tilney et al., 1973a)

and Wilson, 1977; 1978; Farrel *et al.*, 1979a, b; Binder and Rosenbaum, 1973, 1979).
The arms projecting from each outer doublet of the axoneme toward the next one are
primarily composed of the ATPase called dynein. This protein has been studied by
selective extraction and localized by means of conjugated antibodies using the transmission electron microscope (I. R. Gibbons, 1972; B. H. Gibbons and I. R. Gibbons,
1972a, 1973, 1976; I. R. Gibbons and Fronk, 1972; B. H. Gibbons *et al.*, 1975,
1976 I. R. Gibbons et al. 1976, Ogawa and I. R. Gibbons, 1976; Ogawa *et al.*,

1977a, b; Fronk *et al.*, 1975; Burns, 1975; Hayashi, 1976). At least two immunologically distinct dyneins have been identified; dynein 1 of about 500,000 molecular weight, that represents the major constituent of the arm, and dynein 2 of about 400,000.

A "soluble" dynein, i.e., not linked to the doublets, of 125,000 molecular weight, has been extracted with low salt concentrations by Y. Kobayashi *et al.*, (1979). Baccetti *et al.*, (1979) by analyzing the sperms of the gall midge, *Diplolaboncus*, and of the eel, *Anguilla*, which are naturally both devoid of central tubules and radial spokes, and of the inner arm and of the outer arm respectively, concluded that dynein 1 is localized in the outer arm of the doublets and dynein 2 in the so-called γ-links, which bind the doublets to the membrane, whereas the inner arm contains a protein corresponding to the electrophoretic band B of the five dynein bands described by I. R. Gibbons *et al.*, (1976). The localization in the outer arms, of dynein 1, was later confirmed by B. G. Gibbons and Fronk (1979) and by Hata *et al.*, (1980), who were able to extract it from demembranated axonemes, which after extraction show the disappearance of the outer arms. B. G. Gibbons and I. R. Gibbons (1979) also showed that the dynein 1 thus extracted can be added back to reactivate sperm movement, provided its ATPase function has been retained in a latent form, which therefore appears to represent the native form of dynein 1 (see also Yano and Miki-Nomura, 1981).

Part of dynein 1 may also be contained in the inner arms; this, however, shows immunological differences from that contained in the outer arms (Ogawa *et al.*, 1982).

Piperno and Fleming (1983) have now succeeded in preparing monoclonal antibodies against a 15S dynein fraction solubilized with a modified procedure from the outer arms of the axonemes of sperm of *Strongylocentrotus purpuratus*. Preliminary data obtained with this new tool indicate that some of the dynein subunits, although differing in mobility in SDS-polyacrylamide gel electrophoresis, do share antigenic determinats.

Some insight into the structure of the 21S dynein complex which can be purified from the outer arms of sea urchin axonemes (B. G. Gibbons and Fronk, 1979) has been recently provided by Sale (1983), who by means of low-angle rotatory-shadow replication has obtained high-power electron microscope images showing complexes of three or four large subunits connected by thin strands.

Finally, Fujiwara *et al.*, (1982,b) have demonstrated that a 14S dynein partially purified from spermatozoa of *Hemicentrotus pulcherrimus* is inhibited by high concentration of palmytoyl CoA, and stimulated by low concentrations of the same compound, which stimulates the authors to suggest a role for the CoA thioesters in the regulation of sperm motility.

Another protein that has been isolated from the sea urchin flagella is the so-called nexin (Stephens, 1970b); it represents about 2% of the axoneme proteins, has a molecular weight of about 165,000, and links together the adjacent A tubules. A number of minor protein components of the axoneme have been described, whose function is not yet understood (Linck, 1976).

The movement that pushes the sperm toward the egg is a particular, rhythmic, propagating bending of the flagellum, which has been described in detail in a cinematographic analysis by Rikmenspoel (1978) and by Goldstein (1979), who have shown that the bending wave starts from near the base of the tail and propagates to the tip. Current theories about the mechanism of this bending tend to compare it to the mechanism of muscle contraction, considering the flagellar tubulin analogous to muscle actin and the flagellar dynein analogous to muscle myosin. The movement would be produced by the sliding of the dynein arms along the outer doublet microtubules according to the general theory proposed by Satir (1965) and adapted to sea urchins by I. R. Gibbons (1972). Sperm motility is influenced by a variety of factors like pH, ionic concentration, and others, which have been studied either on entire sperms or on demembranated tails with the aim of dissecting the mechanism of flagellar motion and of its regulation (Morisawa and Mohri, 1972, 1974; Costello,

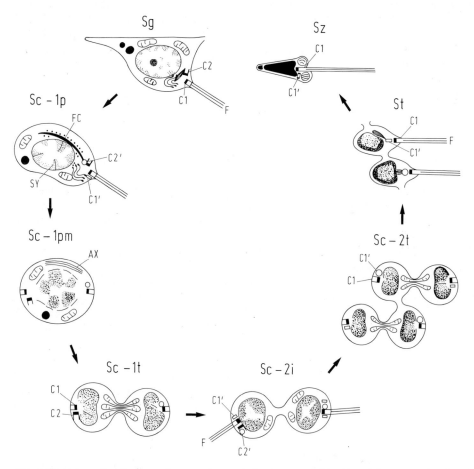

Fig. 1.12. Schematic diagram of spermatogenesis in the sea urchin *H. pulcherrimus*, showing flagellum formation and absorption, and centriolar behavior and duplication at the critical periods in meiosis. *Sg* spermatogonial cell; *Sc-lp* primary spermatocyte at prophase; *Sc-1pm* primary spermatocyte at prometaphase; *Sc-1t* primary spermatocyte at telophase; *Sc-2i* secondary spermatocytes at interphase; *Sc-2t* seconday spermatocyte at telophase; *St* spermatids; *Sz* spermatozoon; *AX* absorbed axoneme; *C1* and *C2* (parent) mature centriole, *C1'* and *C2'* (daughter) immature centriole; *F* flagellum; *FC* fibrous complex; *SY* synaptonemal complex. (Kato and Ishikawa, 1982)

1973; Brokaw, 1977, 1979; Brokaw and Simonick, 1977; Brokaw et al., 1974, B. H. Gibbons, 1980, 1982; B. H. Gibbons and I. R. Gibbons, 1977, 1980, 1981; Cosson and I. R. Gibbons, 1977; Hinkley 1979; Okuno and Brokaw, 1979; Asai and Brokaw, 1980; Yano and Miki-Nomura, 1980; Okuno, 1980; Goldstein, 1981; Pennigroth et al., 1982; Ishiguro et al., 1982; K. Takahashi and Kaminura, 1983; Cosson et al., 1983). Among the various observations reported we shall mention here only first, that variations in pH and calcium concentration act synergistically to affect sperm motility and the helical shape of the microtubules of the outer doublet (Miki-Nomura and Kamiya, 1979); and second, that, by the use of α-bungarotoxin, Nelson has concluded that acetylcholine receptors, possibly nicotinic, located at the cell surface may be involved in the regulation of sperm motility (Nelson, 1971, 1972, 1975, 1976, 1980, 1982; Nelson et al., 1980).

Little information is available on the mechanism of flagellum formation during spermiogenesis. One of the works most directly addressed to this question is that of Kato and Ishikawa (1982), whose results are schematically reported in Fig. 1.12.

It has long been known that sperm motility is lost as the sperm attaches to the surface (Lillie, 1919). Yasumasu and coworkers have investigated the hypothesis that the block of sperm motility might be due to an impairment of the energy production mechanisms. In order to circumvent the bias due to the high respiration rate of the egg to which the sperm attaches, these authors have studied the respiration of the sperm attached to fixed eggs. These studies were carried out thoroughly on a variety of Japanese species (Hino et al., 1980a, b; Yasumasu et al., 1980; Fujiwara et al., 1980, 1982a; Hiruma et al., 1982), and suggested that the mechanism of inhibition of sperm respiration following attachment to the fixed eggs, although differing slightly in the different species, results from a block in the electron transport in the mitochondrial respiratory chain; it was also suggested that the decreased motility may be cause and not effect of the decreased respiration.

Another interesting question is that of what prevents sperm movement before the sperms are diluted in sea water. They are, in fact, normally immotile in semen, thus preventing a useless and energetically wasteful movement. This old question has been recently re-investigated by C. H. Johnson et al. (1983), who found that sperm motility can be initiated in the semen by blowing a gas like N_2 over it, and suggested that the seminal fluid contains a volatile inhibitor of sperm motility. Many lines of evidence indicate that this inhibitor may be Co_2, whose release upon sperm dilution contributes to the physiological increase of the internal pH (by 0.36 units) which in turn activates sperm motility. Christen et al. (1983,b) have provided evidence that this internal pH increase acts through activating the dynein ATPase, whose activity is rate-limiting for the respiration of tightly coupled mitochondria.

In a careful and detailed analysis of the effect of the dilution in sea water on the mechanisms of sperm energy production, Mita and Yasumasu (1983) concluded that the sperm preferentially use phospholipids as energy source at 20 °C and glycogen at 2 °C. They draw attention, however, to the fact that the breeding season for the species studied (Hemicentrotus pulcherrimus) is winter and that glycogen therefore may be the preferred energy source in nature.

By means of NRM analysis Christen et al. (1983c) have confirmed the low internal pH of the nondiluted immotile sperms, which have low levels of inorganic phosphate and high levels of phosphocreatine. Upon dilution, the internal pH and the inorganic phosphate increase, while phosphocreatine decreases and respiration and motility are activated.

More information about the energy-generating processes may come in the near future from studies on isolated sperm mitochondria, which can now be started since a method of preparing purified sperm mitochondria has recently been developed (Rinaldi *et al.* 1983, b).

1.3.2 Sperm Binding

Once the sperm has reached the egg it binds to the vitelline layer in a species-specific way.

Species specificity can be experimentally overcome by a variety of means (see Giudice, 1973, for a review; Summers and Hylander, 1976, for a more recent analysis and Chap 2, this Vol.). Interspecific hybrids can be obtained, some of which are able to develop normally and even to undergo metamorphosis. Other hybrids, known as lethal hybrids, are arrested at earlier developmental stages. Species specificity of sperm binding can be retained even after fixation of eggs with glutaraldehyde (Kato and Sugiyama, 1978).

The egg surface has a large but limited number of sperm attachment sites that has been calculated by Vacquier and Payne (1973) as 1 per 28.2 μ^2 in *Lytechinus pictus*. Once the sperm is attached, its plasma membrane quickly fuses with the egg plasma membrane and the egg surface is immediately uplifted in a small cone, the so-called fertilization cone which within 3 and 5 min after fertilization becomes filled with bundles of actin filaments parallel to the cone axis (Tilney and Jaffe, 1980). This is not yet final proof that long microfilaments are naturally involved in sperm incorporation, since the experiments of Tilney and Jaffe were made in the presence of nicotine, which causes the formation of unusually large fertilization cones.

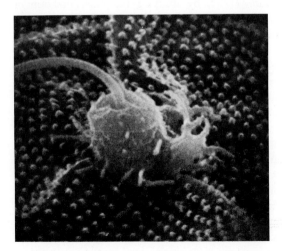

Fig. 1.13. A *Strongylocentrotus purpuratus* sperm is penetrating the egg; microvilli elongate around the sperm as the membrane appears slack and convoluted (×20,000). (G. Schatten and Mazia, 1976)

The head, the midpiece and the tail of the sperm penetrate into the egg. This process is accompanied by a wrinkling of the egg membrane with a clustering and elongation of the surrounding microvilli. This moment is accompanied by increase in the egg membrane fluidity (Campisi and Scandella, 1978). A detailed description of the process of sperm penetration can be found in the elegant scanning electron microscope figures of G. Schatten and Mazia (1976) (see figure 1.13), and in the time-lapse videomicroscopy and scanning electron microscope studies of Schatten and Schatten (1980, 1981), which were preceded by the work of Tegner (1972) and Tegner and Epel (1973), and by the ultrastructural analysis of Summers and Hylander (1974). Little information is available about the mechanism of fusion of these membranes, but it seems to be inhibited by fluorescein dyes (B. J. Carrol and Levitan, 1978a; Finkel *et al.*, 1981): it is also inhibited by quercetin, an ATPase inhibitor, in a very specific way, i.e., without affecting other steps of the fertilization process, such as sperm motility, acrosome reaction, sperm binding and cortical granule reaction. Although the mechanism of action of quercetin is still poorly understood, the specificity of its effect on gamete membrane fusion makes it a promising tool for the understanding of the biochemistry of such a process. Interestingly, a patch of sperm membrane can be detected on the surface of the fertilized egg, and can be traced as such throughout development till pluteus (Gabel *et al.*, 1979), although Longo (1982) using concanavalin A, which binds the egg and not the sperm membrane, has described a rather quick movement of the Con A reactive membrane to cover the nonreactive site occupied by the sperm membrane shortly after fertilization.

1.3.3 Sperm Internalization and Pronuclear Movement

For the forces that draw the sperm nucleus into the egg cytoplasm, a role for the actin fibers present in the cortical layer of the egg has been proposed based on the inhibitory effect of cytochalasin B (Longo, 1977, 1978, 1980; Byrd *et al.*, 1977; Byrd and Perry, 1980). Furthermore, a movement of actin fibers toward the center of the egg has been described following fertilization by P. Harris (1979), and Longo (1980); and recently Gundersen *et al.*, (1982) by studying the internalization of sperms labeled with radioactive fluorescent dyes in the presence or in the absence of cytochalasins, concluded that sperm internalization seems to be mediated by the actin-dependent cytoskeleton of the egg.

As soons as the sperm nucleus has entered the egg cytoplasm it rotates 180° (Flemming, 1881; Boveri, 1888) and starts moving toward the center of the egg while the female pronucleus moves toward it. There has been much discussion about the mechanisms regulating such a movement, but the most consistent hypothesis appears to be that of the mechanical action of the growing astral microtubules (A. Brachet, 1910; Allen, 1954). This conclusion was initially based on the inhibitory effect of high pressure (Zimmerman and Silberman, 1964), of colcemid (Zimmerman and Zimmerman, 1967), and of a wide series of microtubule inhibitors (G. Schatten and H. Schatten, 1981). Only the entrance of the male pronucleus and its rotation of 180° depend upon microfilaments, whereas the spermaster formation and pronuclear movement are microtubule-dependent (G. Schatten, 1979). The early observations have been recently confirmed by means of immunofluorescence microscopy, which showed that microtubules are responsible for pronuclear movement (Bestor and G. Schatten, 1980, 1981). This certainly occurs for the action of the spermaster, but also for the action of microtubules originating from the egg, as suggested by the fact that the egg pronucleus moves toward the center also following parthenogenetic activation, and as directly demonstrated by several authors (Mar, 1980; P. Harris *et al.*, 1980a, b; Bestor and G. Schatten, 1982). Harris *et al.*, (1980, b) also described the breakdown and disappearance of the two monasters before the formation of the mitotic apparatus. While moving toward the egg pronucleus, the sperm nucleus becomes transformed into a pronucleus. This process, together with that of nuclear fusion, has been accurately described at the level of the electron microscope by Longo and Anderson (1968, 1970a, b) reviewed by Longo and Kunkle (1978), by Longo (1980), and studied by time lapse video microscopy by Schatten (G. Schatten, 1981; G. Schatten and H. Schatten, 1981). Essentially, this process consists of the degeneration of the membrane of the male nucleus through a process of vesiculation, followed by chromatin dispersion, accompanied by the uptake of egg proteins into the sperm nucleus (Kunkle and Longo, 1975). Ultimately, the electrophoretic profile of the proteins of the male pronucleus becomes similar to that of the female one (Longo and Kunkle, 1978). Meanwhile, a new nuclear membrane is formed by pre-existing material made up by the endoplasmic reticulum of the egg (Longo, 1973); then the two pronuclei come into contact and fuse.

Little is known about the biochemical mechanisms underlying the formation of the male pronucleus from the sperm nucleus; the whole process, however, seems to depend upon the rise in pH which occurs at fertilization (Carron and Longo, 1980a, b).

It is of interest that Delgado *et al.*, (1983) have found that a glycosamineglycan-sulfate extracted from sea urchin eggs is able to cause decondensation of the nuclei of sperms of different species, although this does not necessarily imply that this mechanism it at work under natural conditions.

1.4 Some Physiological Changes that Occur at Fertilization or Following Parthenogenetic Activation

As we shall see in detail throughout this chapter, fertilization brings about a general metabolic activation of the sea urchin egg. Much work has been devoted to understanding the mechanism of such an activation, one important approach being that of investigating the temporal and possibly the causal correlations of the various metabolic events. A preliminary point to be made is that the sperm is not a necessary element to start this metabolic sequence, since parthenogenetic activation can do the same. Many parthenogenetic agents have been described (see Harvey, 1956, for a list); some of them cause development to proceed even through metamorphosis, but others, which are more interesting for this chapter, allow only some of the metabolic events following fertilization to occur, and therefore permit a dissection of the metabolic activation.

1.4.1 Ionic Movements

One of the parthenogenetic agents, ionophore A 23187, has been described by Stein-hardt and Epel (1974). This increases the rate of transportation of Ca^{2+} through the membranes and causes the elevation of the fertilization membrane, elongation of the microvilli, even if with loose internal structure, (Carron and Longo, 1982), an increase in the oxygen consumption, a change in membrane potential and initiation of DNA synthesis, (i.e., most of the events one can observe following fertilization), but causes only an abortive increase in the rate of protein synthesis and no cell division. Similar observations were reported by Chambers (1974; Chambers et al., 1974) and by Lallier (1974a). Although a dramatic increase of the influx-efflux of ^{45}Ca following fertilization has been described (Azarnia and Chambers, 1970; Chambers et al., 1970), this is not necessary for egg activation to occur (see also Paul and Johnston, 1978), provided high sperm concentrations and low temperatures are used (Schmidt et al., 1982); otherwise, the conclusion is reached that external Ca^{2+} is needed for acrosome-reacted sperms to fertilize (Sano and Kanatani, 1980). A common idea is that activation is accompanied by a liberation of the internal bound calcium (see Jaffe, 1983, for a recent review). In fact, activation can occur in the absence of external Ca^{2+}, Mg^{2+}, or Na^+. Moreover, an increase in the intracellular free Ca^{2+} has been directly demonstrated by microinjections of a Ca^{2+} photolumine-scent protein, the aequorin (Steinhardt et al., 1977; Zucker et al., 1978; Kiehardt et al., 1977); and microinjections of calcium buffers, which bring the free internal Ca^{2+} concentration above 0.2 μM cause elevation of the fertilization membrane and formation of the monaster, whereas microinjections of Ca^{2+} chelators inhibit fertili-zation (Hamaguchi and Hiramoto, 1981; Hamaguchi and Kuriyama, 1982). Naku-mura and Yasumasu (1974) have described a Ca^{2+}-binding factor within the unfertiliz-ed egg, whose dissociation constant increases by an order of magnitude following fertilization. X-ray microprobe analyses have revealed a Ca^{2+} concentration in the cortical layer of the egg (Schuel et al., 1975a; Cardasis et al., 1978). H. Inoué and Yoshioka (1982) found that the microsomal fraction isolated from fertilized eggs of *Hemicentrotus pulcherrimus* takes up Ca^{2+} five times more quickly than if isolated from unfertilized eggs, which may be related to the higher availability of free calcium after fertilization. The role of Ca^{2+} in the egg activation has been suggested also for asteroids and for vertebrates (Steinhardt et al., 1974).

1.4.2 pH Changes

Another interesting means by which a partial egg activation can be obtained is treatment with ammonia (Steinhardt and Mazia, 1973; Mazia, 1974; Epel et al., 1974). This causes an increase in the K^+ conductance, initiation of DNA synthesis and chromosome condensation (through the synthesis of some new proteins on maternal templates, according to Krystal and Poccia, 1979), RNA polyadenylation, activation of the cyclic activity of the Ca^{2+} ATPase (Petzelt, 1976) and an increased rate of protein synthesis, but no increase in the respiratory rate, no cortical granule reaction, no microvillar elongation (Carron and Longo, 1980a), nor an increase in the Na^+-dependent amino acid transport. Chromosome condensation occurs only at high amine concentrations according to Lois et al., (1983); and according to

Evans et al., (1983) activation of protein synthesis by ammonia occurs for most but not for all of the abundant proteins. These effects are due to the ammonia per se rather than to the NH_4^+ ions or to elevated pH, according to Mazia et al., (1975). By measuring the external pH, Winkler and Grainger (1978) concluded that the ammonia activation occurs following an influx of the base. Whatever the mechanism of action of ammonia, it is well known that the internal pH of the egg undergoes a transitory increase between 1 and 4 min following fertilization (J. D. Johnson et al., 1976; Lopo and Vacquier, 1977), which seems to be due to Na^+ influx and H^+ efflux, although it has been found that it occurs also at very low external Na^+ concentrations (Shen and Steinhardt, 1979). It has, in fact, been proved that Na^+ is required for the activation of DNA and protein synthesis and the initiation of cleavage (Chambers, 1975a; Nishioka and Epel, 1977), whereas the partial activation by A 23187 can occur also in the absence of external cations (Paul and Epel, 1975). Little is known about the mechanism of Na^+-H^+ exchange. Accurate experiments performed in the absence of external pH variation conclude that it does not occur through amiloride-sensitive channels (Cuthbert and Cuthbert, 1978). The extent of the pH increase at fertilization has been precisely measured by means of internal microelectrodes (Shen and Steinhardt, 1978) and found to be from 6.84 to 7.26 ± 0.06 within 5–6 min after fertilization of Lytechinus pictus eggs.

This increase in pH has been confirmed by a variety of methods; Johnson and Epel (1981), for example, measured the intra-extracellular distribution of ^{14}C-dimethyloxazolidine, which is a function of the cellular pH; Winkler et al., (1982), also confirmed the increase in pH by NMR studies of the shift of the ^{31}P peak, which depends upon pH, thus criticizing contrary results.

Lois et al., (1983) have accurately titrated the amount of acid released by Strongylocentrotus purpuratus eggs following activation with a variety of amines and found that the rate of acid release was directly proportional to the carbon number of the amine used showing an increase of 8.3-fold for methylamine and 470-fold for benzylamine. The total equivalents of acid released varied from 0.50 to 8.2×10^{-12} mol H^+ cell^{-1}, in direct proportion to the amine concentration. Na^+ influx–efflux has been measured in Paracentrotus eggs by Payan et al., (1981), who found a 30% increase of the internal Na^+ 5 min after fertilization, accompanied by H^+ efflux, and followed by a Na^+ efflux which lowers its internal concentration of 40% in 2 h. By extending their investigations, based on the measurement of the partition of a weak acid (dimethyloxazolidinedione) Payan et al., (1983) concluded that the internal pH of the unfertilized egg of Paracentrotus lividus is regulated via a permanent amiloride-sensitive Na^+/H^+ exchange.

The increase in pH which follows fertilization is maintained, in addition to the first mechanism, by an energy-dependent, acid-extruding pump, which is dependent on external Na^+ and is amiloride-sensitive. The acid production which follows fertilization is also mediated by the excretion of various substances, the nature of which has long been discussed (see Giudice, 1973). Among the most important ones are acid polysaccharides (Ishihara, 1968b). Acid production, as already mentioned, is inhibited by inhibitors of the cortical granule dehiscence (Schmidt and Epel, 1983). Preliminary experiments by Gillies et al., (1980) suggest that the external pH decrease is due to CO_2 liberation from internal Ca Co_3.

Lee and Epel (1983), by means of acridine orange as a probe for looking at acidic intracellular compartments, have observed, that following fertilization or parthenogenetic activation of eggs of *Strongylocentrotus purpuratus* or *Lytechinus pictus*, some acidic granules appear in the cortical layer, which may play a role in the intracellular pH regulation.

A special case is that of melittin, a component of bee venom, which was found to cause cortical granule breakdown and elevation of the fertilization membrane without other signs of egg activation. The correct interpretation of the authors (Shimada *et al.*, 1982) is that melittin is poisonous to the other egg functions; which represents a good warning against too simplistic interpretations of natural uncoupling of functions which can be obtained by the use of other partial activators.

1.4.3 Membrane Potential

Probably the earliest detectable physiological change that occurs at fertilization is a change in membrane potential and membrane resistance, accompanied by the development of a K^+ conductance. The first direct measurements made by the use of microelectrodes are due to Tyler *et al.*, (1956). Among the most recent measurements are those made by De Felice and Dale (1979), who have found a resting potential of -8 to -16 mV in newly shed eggs of *Paracentrotus lividus* and *Psammechinus microtuberculatus*; resting potentials of -60 to -80 mV were observed in only a few eggs and only of they had been left for several hours in the sea water before impalement. In these species, each successful sperm collision is signaled by an immediate 1–2 mV depolarization, accompanied by an increase in the noise across the egg plasma membrane (Dale *et al.*, 1978) and followed, 13 s later, by a slower depolarization to a positive value about of $+10$ mV. The membrane then repolarizes gradually to about -10 mV. These authors stress the importance of sperm concentration for the shape of this fertilization potential curve. Their results, together with those of Byrd and Collins (1975a, b), argue against a fast block to polyspermy, because many successful sperm-egg collisions are possible within the first 13 s. Chambers and De Armendi (1979) hold that the resting membrane potential is of -70 to -80 mV and that lower potentials are due to cytoplasmic leakage following electrode impalement. Accurate measurements of Dale and De Santis (1981 b), however, showed that 70–90 mV potentials are always observed in maturing oocytes. These measurements were made by means of citrate microelectrodes, which, contrary to the K^+ microelectrodes, do not damage the egg even for observations longer than 20 min. In these experiments it was found that in *Paracentrotus* and *Phaerechinus* egg depolarization occurs 2 s after sperm attachment; 11 s later the fertilization potential starts, cortical granules are discharged and the sperm becomes immobilized. The interval between these two series of events is elongated by cytochalasin B and D, thus suggesting that a microfilament-dependent process occurs in this time interval (Dale and De Santis, 1981a). All other available measurements show this depolarization wave, followed by a slower repolarization with quantitative variations in various species (Steinhardt *et al.*, 1971, 1972; Uehara and Katow, 1972; Steinhardt and Mazia, 1973; Ito and Yoshioka, 1972, 1973; Tupper, 1973; Jaffe, 1976; Jaffe and Robinson, 1978; Jaffe *et al.*, 1978; Dale *et al.*, 1978; Taglietti 1979).

What then remain controversial are two points, the first, if the resting potential of the egg is around -70 mV or around -10 mV, and the second, whether or not there is an immediate change of the membrane potential which brings about a fast block to polyspermy (see Dale and Monroy, 1981 for a discussion). Recent measurements of G. Schatten and Hüsler (1983), while confirming the values of about $-15\,mV$ for the membrane potential of the unfertilized eggs of *Lytechinus variegatus*, suggest that this low potential is the result of ionic leakage due to the microelectrode impalement, and that this, as discussed earlier in this chapter, interferes with the fast block to polyspermy. Also the data of Whitaker and Steinhardt (1983) based on the measurement of the extracellular action current favor the hypothesis that the membrane potential of the nonimpaled egg is $-50\,mV$.

The influx-efflux of K^+ at fertilization has been studied by means of ^{42}K or by the use of K microelectrodes. All these studies agree that fertilization is immediately followed by an increase in K^+ permeability (Tyler and Monroy, 1956, 1959; Tyler, 1958; Steinhardt *et al.*, 1971, 1972; Tupper, 1973, 1974; Chambers, 1975b; Robinson, 1976; Jaffe and Robinson, 1978), which appears to depend upon the pH increase, since it is inhibited by weak acids and stimulated by alkalis (Shen and Steinhardt, 1980). This increase does not seem to be responsible for the activation of amino acid transport and protein synthesis, since these events can be independently inhibited (Tupper, 1974). In spite of the increased K^+ exchange, the balance between the external and internal K^+ remains essentially unchanged following fertilization (Steinhardt *et al.*, 1971). The idea of a compartimentalization of the K^+ before fertilization does not seem to withstand the criticisms raised by Robinson (1976).

1.4.4 Other Permeability Changes

An increase in permeability following fertilization has also been observed for other ions, notably PO_4^{3-}. This was first demonstrated by Brooks (1943) by means of ^{32}P. This early observation has been repeatedly confirmed. Among the most careful studies are those of Litchfield and A. H. Whiteley (1959), A. H. Whiteley and Chambers (1960), Chambers and A. H. Whiteley (1966) and A. H. Whiteley and Chambers (1966), who concluded that at fertilization a phosphate transport system is synthesized by means of an energy-requiring process. The phosphate transport system consists of a surface-located carrier, which is not energy-dependent.

Another means to study the ionic requirements for egg activation has been developed by Yoshioka and H. Inoué (1981), by means of a cholesterol reagent, amphoptericin B, which causes the formation in the plasma membrane of 80 nm pores through which ions and neutral molecules can pass. It was confirmed by this method that no external Ca^{2+} is required for the activation of protein synthesis, whereas external Na^+ is needed in order to initiate egg activation in *Hemicentrotus pulcherrimus* (at least under the conditions tested by the authors) and for pronuclear development (Carron and Longo, 1980b). No information, of course, can be obtained about internal ion movements.

In addition to the permeability to different ions, the permeability to amino acids and nucleosides is greatly increased following fertilization or parthenogenetic activation. This has been observed by many authors (Hultin, 1952, 1953d; Giudice *et al.*, 1962; Nemer, 1962). Later works have tried to shed some light on the mechanism of uptake of amino acids and nucleosides (Piatigorsky and A. H. Whiteley, 1965; Mitchison and Cummins, 1966; Piatigorsky and Tyler, 1968; Epel, 1973).

Criticisms to earlier conclusions have been made by Doree and Guerrier (1974), who hold that erroneous results are obtained when the influx of radioactive leucine is measured by collecting the eggs on Millipore filters, which cause amino acid leakage; it should be observed, however, that this leakage does not necessarily occur if filtration is done carefully and without removal of the fertilization membrane. From the effects of the concentration of the external substrate, and from the lack of effect of adenine analogs and DNP, Doree and Guerrier conclude that uptake of adenine in *Sphaerechinus granularis* eggs occurs through a process of simple physical diffusion, whereas leucine uptake occurs through a carrier-mediated transport. The increased uptake which follows fertilization has to be looked for in an increase of the general membrane permeability, which does not involve de novo synthesis (lack of effect puromycin) or recruitment of previously inactive carriers. Tyler *et al.*, (1966) found that amino acids belonging to the same group (i.e., acidic, alkaline and neutral) compete among themselves for uptake, thus suggesting the existence of partially specific amino acid carriers. In conclusion, whereas agreement exists about nucleoside transport, the question is still open for the transport of amino acids.

Experiments with radioactive ^{14}c-leucine in competition with exogenous ^{12}c-leucine have demonstrated that exogenous amino acids, once taken up, pass through an endogenous pool, which can be expanded severalfold (Fry and Gross, 1970a, b; Berg, 1970; Berg and Mertes, 1970). The passage through the endogenous pool does not occur if the exogenously supplied amino acid is ^{35}s-methionine, according to Ilan and Ilan (1981).

Manahan *et al.* (1983) have recently measured the rate of amino acid uptake with a different approach, i.e., by measuring its disappearance from the medium due to the uptake by larvae derived from gametes collected in a sterile way and cultured in a sterile medium without antibiotics. It is unfortunate that these measurements have been made only on late plutei.

The sperm have also been shown to have a natural amino acid carrier (Gache and Vacquier, 1983).

Other important physiological activations which occur at fertilization include a respiratory increase, an increase in the rate of protein synthesis, and the starting of DNA synthesis. These will be described in detail later in specific chapters.

In an attempt to find a causal relationship between all these metabolic changes, we will follow a temporal scheme indicating the course of the main physiological changes which follow fertilization (see Epel, 1978 for a detailed discussion, and Shapiro *et al.*, 1981 for a review). Probably the first change is that of membrane potential (about 3 s); the second is Ca^{2+} release and cortical granule exocytosis (about 25 s), immediately followed by acid release and beginning of the increase in intracellular pH; activation of NAD kinase and O_2 consumption start at about 30 s; it is only after 300 s that the so-called late changes occur, i.e., increase in permeability to ions, amino acids, and nucleosides, activation of protein synthesis; the latest event is activation of DNA synthesis concomitant with pronuclear fusion (about 1200 s). These values are valid for *Strongylocentrotus purpuratus*.

Chapter 2

Embryonic Morphogenesis

2.1 General Description

Once fertilized, the sea urchin egg starts to cleave at a very high frequency. Figures 2.7 and 2.8 describe in detail the pattern of cleavage and the morphogenetic movements which bring about the formation of the larva in the Mediterranean species *Paracentrotus lividus*, when cultured in sea water at 20 °C with gentle stirring. Very similar developmental patterns are observed in other sea urchins: see Mortensen (1921), for the earliest descriptions, E. B. Harvey (1956) for the development of *Arbacia lixula*; Stephens (1972a, b) for that of *Strongylocentrotus purpuratus*; Gustafson and Wolpert (1967) for a critical description of the development of *Psammechinus miliaris*, and Amemiya and coworkers for the early development of *Anthodaris crassispina, Hemicentrotus pulcherrimus* and *Pseudocentrotus depressus* (Amemiya *et al.*, 1982a, b; Akasaka *et al.*, 1980). Schroeder (1981a), on the other hand, has described the different developmental pattern of a primitive sea urchin (*Eucidaris tribuloides*), which shows peculiarities in the micromeres and hyaline layer, absence of the primary mesenchyme and reduction of the skeleton (Fig. 2.1).

As we shall see later, the egg has an animal and a vegetal pole, which are already determined before fertilization. The first cleavage occurs within about 60 min along a plane going from the animal to the vegetal pole; two cells are thus produced. The second cleavage plane is again meridional, and perpendicular to the first one, the third is equatorial and brings the embryo to eight cells; the fourth division cuts the upper (animal) quartet meridionally, while a horizontal plane, close to the vegetal pole, divides the embryo into four smaller cells, the micromeres, and four larger ones, the macromeres, the cells of the upper quarters being termed mesomeres (see Prothero and Tamarin, 1977, for a computer analysis of the first four cleavages of the sand dollar *Dendraster excentricus*) (Fig. 2.2). Cell divisions occur frequently (about every 30 min) at this stage, which is called the cleavage stage, bringing the embryo through the morula stage to that of early blastula. At this stage, the cells, ordered in one layer around the blastocoel cavity, acquire the shape of a cylindrical epithelium, and quickly grow cilia, by means of which the embryo starts to rotate inside the fertilization membrane. Wolpert and Gustafson (1961a, b) explain the formation of the blastocoel by the attachment of the cells to the hyaline layer, together with a radial pattern of cell division, whereas earlier authors add to this mechanism an osmotic factor (Monné and Harde, 1950; K. Dan, 1952; 1960). An enzymatic complex is secreted at this point (about 10 h after fertilization in *Paracentrotus*), by which the blastula hatches and starts to swim freely. The hatched blastula has a tuft of long stereocilia restricted to the animal pole leading the way, whereas smaller cilia are distributed all around the embryo. We shall discuss the mechanism of synthesis of the ciliary proteins, and of the skeleton matrix, when talking of the synthesis of stage-specific proteins. We shall only recall here that the skeleton is made of calcium carbonate with some magnesium carbonate (Okazaki and S. Inoué, 1976), is surrounded by a membrane (Gibbins *et al.*, 1969), and contains an organic matrix (Okazaki, 1960; Millonig, 1970; Benson *et al.*, 1983). The vegetal pole at this point flattens, and some cells derived from the micromeres, the primary mesenchyme cells, glide inside the blastocoel (Fig. 2.3). The vegetal pole starts invaginating to form the archenteron,

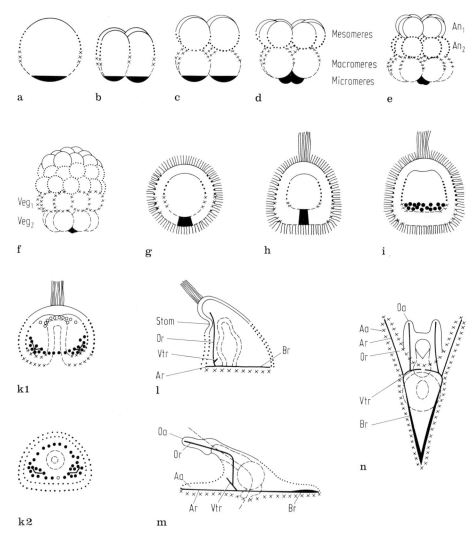

Fig. 2.1a–n. Diagram of the normal development of *Paracentrotus lividus*. Indication of the layers: an_1 continuous lines; an_2 dotted; veg_1 crosses; veg_2 broken lines; micromeres, *black*; **a** uncleaved egg; **b** 4-cell stage; **c** 8-cell stage; **d** 16-cell stage; **e** 32-cell stage; **f** 64-cell stage; **g** young blastula; **h** later blastula, with apical organ, before the formation of the primary mesenchyme; **i** blastula after the formation of the primary mesenchyme; **k1** gastrula: secondary mesenchyme and the two tiradiate spicules formed; **k2** transverse optical section of the same gastrula: bilateral symmetry established; **l** the so-called prism stage; stomodaeum invaginating; **m** pluteus larva from the left side; *broken line* indicates the position of the egg axis; **n** pluteus from the anal side. *aa* anal arm; *ar* anal rod; *br* body rod; *oa* oral arm; *or* oral rod; *stom* stomodaeum; *vtr* ventral transverse rod. (Hörstadius 1939)

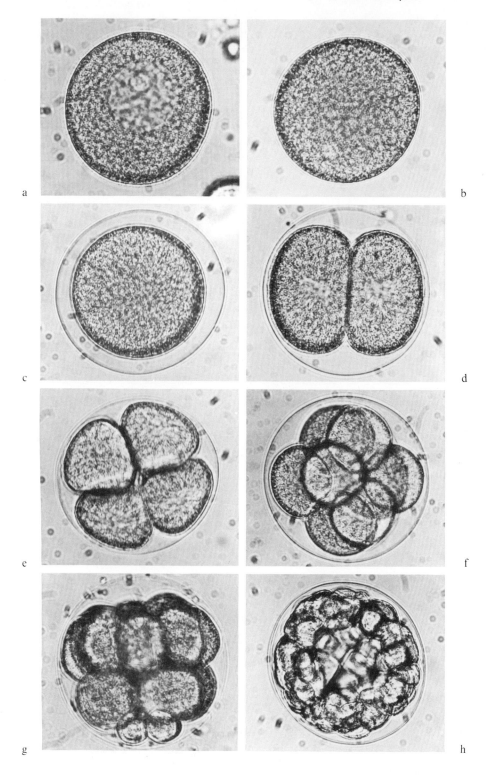

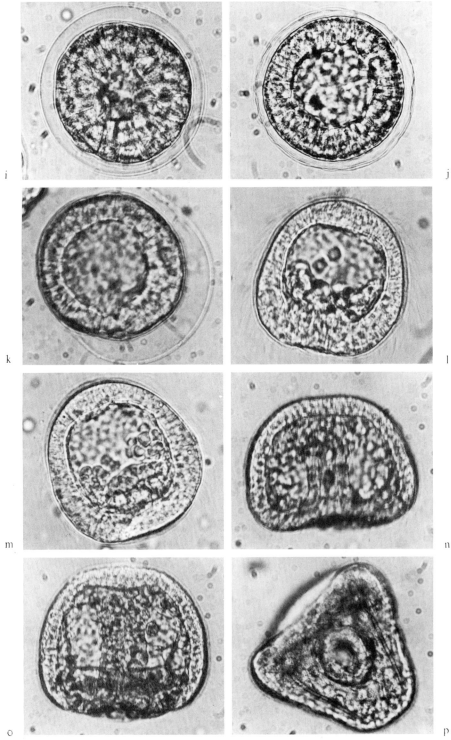

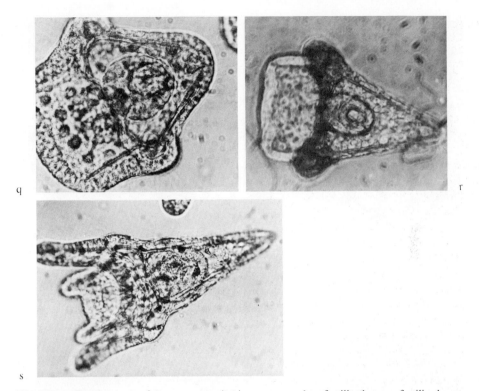

q

r

s

Fig. 2.2a–s. Development of *Paracentrotus lividus*: **a** oocyte; **b** unfertilized egg; **c** fertilized egg; **d** the 2-cell stage; **e** the 4-cell stage; **f** the 8-cell stage; **g** the 16-cell stage; **h** the morula stage; **i** the early blastula stage; **j** the blastula membrane starts to be digested; **k** the hatching blastula stage; **l** the mesenchyme blastula stage; **m** the onset of gastrulation; **n** the mid-gastrula stage; **o** the late-gastrula stage; **p** a prism viewed from the vegetal face; **q** a more advanced prism viewed from the same face but slightly rotated toward the oral face; **r** early pluteus viewed from the same side as in *q*; **s** late pluteus, viewed from the same face. From *A* through *Q* × 600; R = × 400; S = × 180

and the primary mesenchyme cells fuse with each other and form a ring around the basis of the archenteron. From this, two cellular columns elevate at one side of the embryo, which has started to flatten and will become its ventral side. Triradiated spicules originate within the mesenchyme cells, while the top of the archenteron has reached the animal pole, partially by the actions of the secondary semenchyme cells, which originate at its tip. Because of the shape of the embryo, this stage is called the prism stage. The archenteron is bent toward the ventral side and opens into a stomadaeum, while the invagination site represents the anus. Between these two, the intestine differentiates into an esophagus, a stomach, and intestine; with the secondary mesenchyme forming a sac at both sides of the esophagus. In the meanwhile, the two spicules have grown to form two oral arms, two anal arms, two transverse rods and two body rods. This is the pluteus stage. The pluteus swims and captures food by means of a band of cilia along the arms and the circumference of the main body region (see Strathmann, 1975, for a recent work on pluteus feeding). The mechanics of pluteus swimming have been recently reviewed by Emlet (1983). The ciliary move-

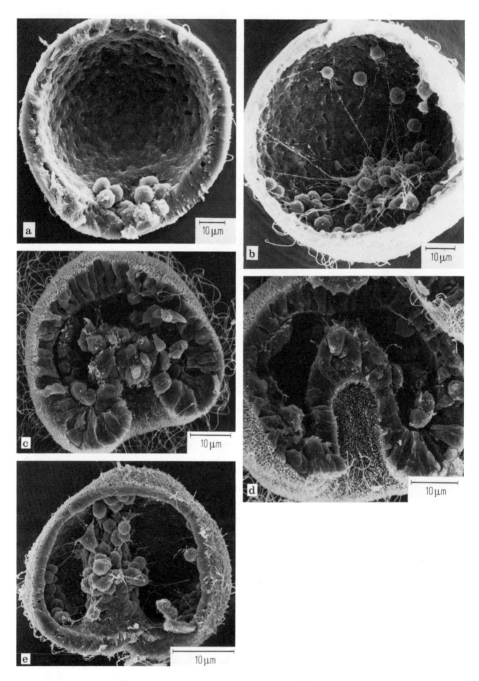

Fig. 2.3a–e. The formation of the primary mesenchyme and the process of gastrulation as seen at the scanning electron microscope: **a** and **b** mesenchyme blastulae of *Pseudocentrotus depressus*. (Akasaka et al., 1980); **c**, **d**, **e** gastrulae of *Authocidaris crassispina* (*C* and *D* from Amemiya et al., 1982a); *E* courtesy of Dr. Amemiya)

ment appears to be coordinated by some kind of nervous system, as repeatedly suggested in the past due to the existence of biogenic amines and acetylcholinesterase in the pluteus (Buznikov *et al.*, 1964, 1968, 1972; Sakharova *et al.*, 1972; Gustafson and Toneby, 1970, 1971; Gustafson *et al.*, 1972a, b; Ryberg, 1974; Ozaki, 1974, 1976). The existence of some kind of nervous system in the pluteus appears quite clear after the descriptions of Ryberg (1977) (Figs. 2.4 and 2.5).

The echinopluteus also contains some heavily pigmented cells which already start to differentiate at mid gastrula stage within the secondary mesenchyme cells (Ryberg and Lundgren, 1979). The pigment is composed of carotenoids and naphthoquinones (see Fox and Hopkins, 1966, for a review, and Hallenstret *et al.*, 1978, for more recent work). Ryberg and Lundgren (1979) suggest that the pigment is involved in some photo-activated processes which are relevant to larval migration and/or morphogenesis.

The pluteus at this stage is able to feed on plankton. The intestine shows coordinated peristaltic movements and contains secretory cells as shown by Ryberg and Lundgren (1979). Another special feature of the intestinal cells, probably related to their digestive function, is the high content of alkaline phosphatase (Evola-Maltese, 1957), an enzyme which has been thoroughly studied by Pfohl (1965, 1975).

Plutei have been reared through metamorphosis since the last century, but it was only in 1969 that Hinegardner developed a method by which large and uniform populations of metamorphosized animals can be obtained under laboratory conditions. This might for the first time offer the possibility of studying the sea urchin offspring and therefore of genetic studies. The most recent descriptions of sea urchin metamorphosis are also due to Hinegardner, and coworkers (Cameron and Hinegardner, 1974, 1978), who have also focused their attention on the early stages of germ cell differentiation (Houk and Hinegardner, 1980, 1981).

What is important to bear in mind for the understanding of the sea urchin metabolism is that the embryo does not "grow" till at least the prism stage; rather, cells only divide into smaller cells, transforming the raw material stored during oogenesis into other molecules.

2.2 Hybrids

We have summarized here the development of the normal larvae. The morphogenesis of the hybrids obtained by crossing different species or different genera has also been thoroughly investigated for several reasons: one, the possibility of obtaining information about the inheritance of the parental morphologic characters, of which the shape of the skeleton is one of the most important; and two, the fact that some of these hybrids develop till the onset of gastrulation and then stop. This has for a long time represented the main, perhaps the only, indication of the fact that genetic activity in the sea urchin is not needed for the embryo to develop until the gastrula stage.

The extensive older literature on the subject has been reviewed by E. B Harvey (1956), by A. H. Whitely and Baltzer (1958) and by Giudice (1973). We shall only report here a summary of the main possible interspecific crossings and recall some more recent papers, like those of Osanai (1972, 1974) and of Ozaki (1975), which describe crossings of Japanese species.

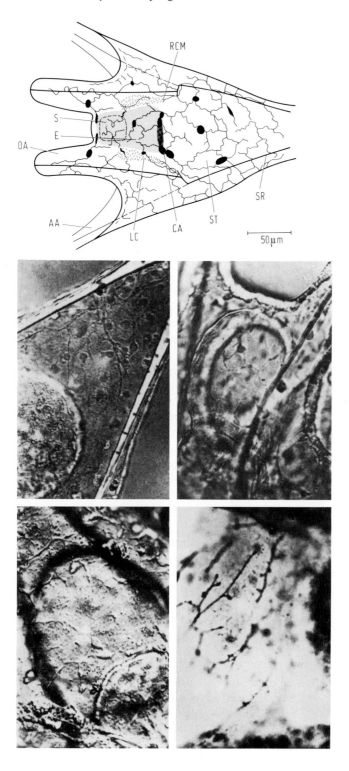

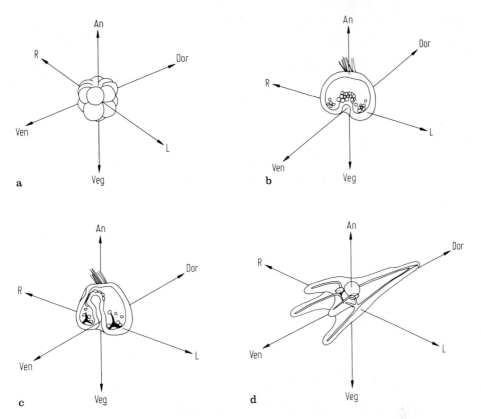

Fig. 2.5. Diagram showing the position of the embryonic axes in **a** 16-cell embryo; **b** an early gastrula; **c** an early prism; **d** a pluteus. *An* animal; *Veg* vegetal; *Ven* ventral; *Dor* dorsal; *R* right; *L* left. (Giudice, 1973)

2.3 Embryo Polarity

A sea urchin larva has two main axes, one running from the animal to the vegetal pole and the other, the ventrodorsal axis, going from the stomodaeum to the opposite side.

2.3.1 Animal-Vegetal Axis

The animal vegetal polarity is already established in the egg before fertilization (see Hörstadius, 1939, 1973, for reviews of the enormous amount of literature on this

◀ **Fig. 2.4. a** Dorsal view of a day 3-old pluteus based on camera lucida drawings. *RCM* right coelom muscle; *S* stomadeum; *E* esophagus; *OA* oral arm; *AA* anal arm; *LC* left coelom; *CA* cardia; *ST* stomach; *SR* skeletal rod. **b** Nerve net between stomach and skeleton in a 6-day-old larva. Nomarski interference contrast. (×590). **c** Dorsal wall of stomodaeum with nerve net in a pluteus, $3^{1}/_{2}$ days old. (×590). **d** Dorsal outer wall of stomach with nerve net. Nomarski interference contrast. Larva 63 h of age (×590). **e** Dorsal wall of stomodaeum. Nerve net stained according to Cajal's silver staining method. 10 μ paraffin section of a 47-h-old pluteus. (×1460). (Ryberg, 1977)

Table 2.1. Hybrids Found in Nature or Produced Experimentally

---

Arbacia lixula ♀ × *Paracentrotus lividus* ♂ = pluteus (1)[a]

Arbacia punctulata ♀ × *Echinaracnius parma* ♂ = *pluteus* (2,3); × *Lytechinus variegatus* ♂ = pluteus (4); × *Mellita pentafora* ♂ = pluteus (5); × *Moira atropos* ♂ = pluteus (5); × *Strongylocentrotus dröbachiensis* ♂ = pluteus (2)

Asterias forbesii ♀ × *Arbacia punctulata* ♂ = a few gastrulae (6)

Cidaris tribuloides ♀ × *Lytechinus variegatus* ♂ = gastrula (7); × *Tripneustes esculentus* ♂ = gastrula (7)

Echinaracnius parma ♀ × *Arbacia punctulata* ♂ = pluteus (2,3); × *Strongylocentrotus dröbachiensis* ♂ = pluteus (2)

Echinometra mathaei ♀ × *Comatula pectinata* ♂ = blastula (24); × *Comatula purpurea* ♂ = blastula (24)

Echinus acutus ♀ × *Echinus esculentus* ♂ = pluteus (8); × *Psammechinus miliaris* ♂ = pluteus (8)

Echinus esculentus ♀ × *Echinus acutus* ♂ = pluteus (8); × *Psammechinus miliaris* ♂ = pluteus (8)

Lytechinus (Toxopneustes) variegatus ♀ × *Arbacia punctulata* ♂ = pluteus (4); × *Mellita pentafora* ♂ = pluteus (5); *Moira atropos* ♂ = pluteus (5); × *Tripneustes (Hipponoe) esculentus* ♂ = pluteus (5)

Mellita pentafora ♀ × *Moira atropos* ♂ = pluteus (5)

Moira atropos ♀ × *Tripneustes esculentus* ♂ = pluteus (5)

Paracentrotus lividus ♀ × *Arbacia lixula* ♂ = gastrula (9–12, 25, 26); × *Psammechinus microtuberculatus* ♂ = pluteus (1); × *Sphaerechinus granularis* ♂ = pluteus (9, 13)

Psammechinus microtuberculatus ♀ × *Arbacia lixula* ♂ = gastrula (9); × *Paracentrotus lividus* ♂ = pluteus (25)

Psammechinus miliaris ♀ *Arbacia lixula* ♂ = gastrula (12); × *Echinus acutus* ♂ = pluteus (8); × *Echinus esculentus* ♂ = pluteus (8)

Sphaerechinus granularis ♀ × *Paracentrotus lividus* ♂ = pluteus (13, 14)

Strongylocentrotus franciscanus ♀ × *Asterias ochracea* ♂ = pluteus (15); × *Strongylocentrotus purpuratus* ♂ = pluteus (15–19)

Strongylocentrotus pallidus ♀ × *Strongylocentrotus dröbachiensis* ♂ (or vice versa) = pluteus (20)

Strongylocentrotus purpuratus ♀ × *Asterias ochracea* ♂ = pluteus (15); × *Dendraster excentricus* ♂ = pluteus (21, 22); × *Strongylocentrotus franciscanus* ♂ = pluteus (2, 16–19, 23)

Tripneustes (Hipponoë) esculentus ♀ × *Lytechinus variegatus* ♂ = pluteus (5)

Hemicentrotus pulcherrimus ♀ × *Glyptocidaris crenularis* ♂ and viceversa (possible after protease treatment of the eggs) = pluteus (27)

Strongylocentrotus intermedius ♀ × *Strongylocentrotus nudus* ♂ (possible after protease treatment of the eggs) = pluteus (28)

Strongylocentrotus purpuratus ♀ × *Strongylocentrotus* droebachiensis ♂ = pluteus (28)
Strongylocentrotus purpuratus ♀ (or droebachiensis ♀) × *Dendraster excentricus* ♂ = gastrulae (28)

---

subject and later on in this Chapter for some modern data). As shown in Fig. 2.1, the animal half of the egg gives rise to two-third of the ectoderm and to the apical tuft of cilia. Therefore, the main criterion for evaluating the development of the animal potentialities of the egg is that of measuring the degree of development of a ciliated ectoderm, whereas that for evaluating the development of the vegetal potentialities is that of measuring the development of the intestine, which is the structure originating from the vegetal half. In extreme cases of "vegetalization" the intestine protrudes to form the so-called exogastrula (Fig. 2.6, last row).

Although the skeleton is produced by micromere-derived cells, the development of this structure apparently requires some interaction of the mesenchyme and ectoderm.

Hörstadius (1928) gave the strongest proof that the animal-vegetal polarity of the egg is irreversibly laid down already during oogenesis, mainly by showing that isolated animal halves of microsurgically cut eggs, once fertilized, give rise to animalized halves, although the vegetal halves are usually able to give rise to small but otherwise normal embryos. In the vegetal part of the egg, the vegetal potentialities become irreversibly fixed quite soon, as shown by the vegetal effect of micromeres in microtransplantation experiments (see Fig. 2.6), or by the fact that isolated micromeres are able to develop into mesenchyme cells (Pucci-Minafra *et al.*, 1968; Harkey and A. H. Whiteley, 1983) able to generate triradiated spicules (Okazaki, 1975a, b).

The sum of the many experiments of experimental embryology fits in a general scheme proposed by Runnström (1928a, b) on the existence of two opposite gradients, animal-vegetal and vice versa, running from one pole of the egg to the other.

What brings about the formation and maintenance of animal-vegetal gradients? The experiments aimed at answering these questions fall into two categories. Those in the first one tend to modify the gradients by the addition of foreign, or more properly, endogenous (Hörstadius and Josefsson, 1977; Fujiwara and Yasumasu, 1974) chemicals. Unfortunately, a tremendous list of substances able to shift the animal-vegetal equilibrium in favor of either side has accumulated (see for example Tamini, 1943; Lallier, 1964, 1978; Gustafson and Toneby, 1970; Hoshi, 1979). One conclusion which can be drawn is that there is a very unstable equilibrium between these two potentialities. Almost any stimulus can perturb it and shift development toward animalization or vegetalization by acting through very generic properties such as very high or very low hydration radius of lithium and thiocyanate, respectively (Tamini, 1943). Indeed, sea water containing 0.066 M LiCl is frequently used to cause vegetalization (Lindahl, 1933), and zinc ions (Lallier, 1955), Na

[a] Key: (1) Baltzer *et al.* (1961); (2) E. B. Harvey (1942); (3) Matsui (1924); (4) Tennent (1912b, c); (5) Tennent (1910); (6) Morgan (1893: questioned by Mathews, 1901); (7) Tennent (1922); (8) Shearer *et al.* (1913); (9) A. H. Whiteley and Baltzer (1958); Baltzer *et al.* 1954, 1959); Baltzer and Bernhard (1955); Baltzer and Chen (1960); Chen and Baltzer (1962); (10) Ficq and Brachet (1963); (11) Hagström (1959); (12) S. Denis (1968); H. Denis and Brachet (1969a, b); (13) Vernon (1900); (14) Steinbruck (1902); (15) Hagedorn (1909); (16) Swan (1953); (17) J. Loeb *et al.* (1910); (18) Barrett and Angelo (1969); (19) Chafee and Mazia (1963); (20) Vasseur (1952c); (21) Flickinger (1957); (22) Moore (1957); (23) Moore (1943); (24) Tennent (1929); (25) Harding and Harding (1952a, b); Harding *et al.* (1954, 1955); (26) Geuskens (1968a); (27) Osanai (1972); (28) Osanai (1974); (29) Ozaki (1975).

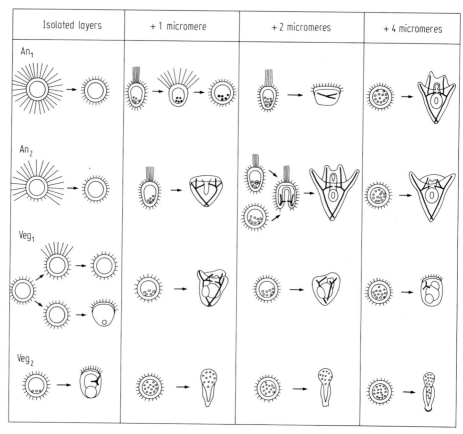

Fig. 2.6. Diagram of the development of the layers an_1, an_2, veg_1, and veg_2, isolated (*left column*) and with 1, 2 and 4 implanted micromeres. (Hörstadius 1935)

thiocyanate (Herbst, 1895; Runnström, 1967), and trypsin (Hörstadius, 1949; Riederer-Henderson and Rosenbaum, 1979) are often used to cause animalization.

The last observation has been revived by Ortolani (1982) who, by exposing the unfertilized eggs to low doses of trypsin (25 µg ml^{-1}), has obtained dwarf but otherwise normal plutei from animal halves microsurgically isolated at the 16-cell stage, thus suggesting that the trypsin effect varies with the dosage; high doses being animalizing and low doses being vegetalizing.

The second type of experiment is aimed at finding ultrastructural or metabolic differences between the different embryonic territories. There are, however, no obvious ultrastructural differences between the animal and vegetal halves of the egg (Berg et al., 1962; Berg and Long, 1964). Only high speed (40,000 g) centrifugation of the entire egg is able to change its polarity, always moving the vegetal pole to the centripetal side (Motomura, 1949). Lower speed centrifugation, while enough, if operated under proper conditions, to split the egg into a centripetal fragment which contains the nucleus, and a centrifugal one which contains most of the mitochondria

(E. B. Harvey, 1956), also causes, however, a disturbance in the centripetal (i.e., nucleated) fragment. When this is fertilized, in fact, some giant cells containing giant nuclei arise from its centripetal side in Japanese sea urchins (Osanai, 1970). Rinaldi, however, (personal communication) found an apparently normal development of the centripetal halves of *Paracentrotus* following fertilization; but development was arrested at the blastula stage in *Paracentrotus*, although small plutei were originated in *Arbacia*.

Among the most important metabolic differences between the animal and vegetal regions of the embryo is a difference in the mitotic rhythm, which has been described by Parisi *et al.* (1978, 1979). They described a higher mitotic frequency of the micromers with respect to the other cells, which decreases toward the animal pole. A partial inhibition of DNA synthesis, obtained by hydroxyurea, counteracts the vegetalizing effect of cAMP phosphodiesterase inhibitors (Yoshimi and Yasumasu, 1978, 1979).

No great differences in the overall rate of RNA synthesis per cell have been found between the various embryonic territories of the 16-cell stage embryo (Hynes and Gross, 1970; Hynes *et al.*, 1972b; M. Spiegel and Rubinstein, 1972; Czihak, 1977). No substantial differences in the rate of protein synthesis between micromeres and other cells, when calculated per unit of volume, have been found by M. Spiegel and Tyler (1966), nor by Hynes and Gross (1970).

The decrease of the protein synthesis rate which is known to occur after the gastrula stage in lithium-vegetalized embryos has recently been confirmed by Wolcott (1981, 1982), who has found that it can be reversed by potassium, and that it does not seem to depend upon inhibition of RNA synthesis, which appears to remain normal if one correlates the measurements for permeability to the external precursors and for variations in the internal pool.

More, however, than the overall rate of synthesis in different embryonic territories, the synthesis of some specific class of RNA or proteins should be looked into. Few experiments of the latter kind are available, due to the difficulty of obtaining enough material from the different embryonic territories. These experiments have, however, become possible since clones of simple genes have become available. Showman *et al.* (1983) in fact have been able to prove that the maternal histone messenger RNA is specifically located in the cytoplasm that surrounds the nucleus, thus showing, for the first time to our knowledge, a specific territorial localization of a messenger RNA. An electrophoretic analysis by Senger and Gross (1978) has revealed only quantitative differences between the major proteins synthesized by the micromeres and the other blastomeres, the former being more active in histone synthesis, which is in line with their higher mitotic activity (Parisi *et al.*, 1978, 1979). Chamberlain (1977), in a two-dimensional electrophoretic analysis of the proteins synthesized by isolated micromeres, found only one polypeptide different from those synthesized by other blastomeres. No qualitative differences between the proteins synthesized by the different blastomeres of the 16-cell stage have been reported by Tufaro and Brandhorst (1979) following a highly resolutive electrophoresis. Many specific differences, however, will arise later during micromere differentiation (Harkey and A. H. Whiteley, 1983). Looking at a specific protein, Hynes *et al.* (1972a) found no differences in the rate of tubulin synthesis between the micromeres and the other blastomeres.

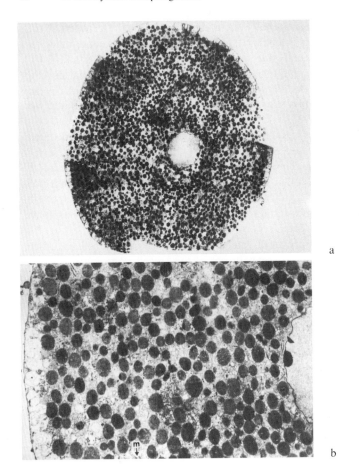

a

b

Using hybridization competition experiments, Mizuno *et al.* (1974) found few differences between the RNA's present in the micromers and in the rest of the embryo which are complementary to repetitive sequences; if, on the other hand, one looks at the RNA's made on single-copy DNA sequences, an inhomogeneous distribution among the blastomeres is found at the 16-cell stage, also if the embryos are developed in the presence of actinomycin D (Rodgers and Gross, 1978). The lower complexity of the RNA of the micromeres with respect to the other blastomeres was confirmed by Ernst *et al.* (1980), especially for the nuclear RNA and also indicated by in situ hybridization experiments with ^{3}H poly(U), showing a lower amount of poly(A$^+$)-RNA in micromeres (Angerer and Angerer, 1981). All this is in line with the observations of Cognetti and Shaw (1981) of a lower sensitivity to the micrococcal DNAase of the micromere chromatin, although these authors have recently found (personal communication) that this difference disappears if the nuclei are prepared by another method. How relevant, however, are these differences in RNA synthesis for the animal vegetal gradient to be established? Not very, if one looks at the experiments on the inhibition of RNA synthesis done by Giudice and Hörstadius (1965), which suggested that during the period between fertilization and the 16-cell

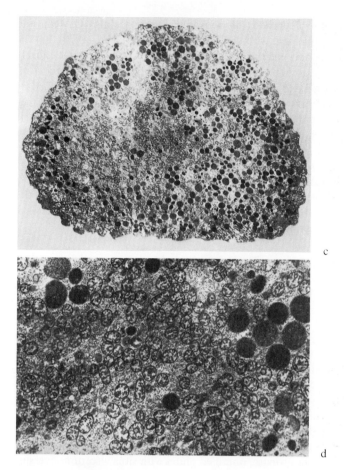

c

d

Fig. 2.7. Electron micrographs of a nucleated (**a**, **b**) and non-nucleated (**c**, **d**) half of *Paracentrotus lividus* egg; **a** and **c** = × 1000; **b** and **d** = × 5000. (Rinaldi et al., 1979 b)

stage, no RNA synthesis is required for the animal and vegetal potentialities to be fixed in the embryonic territories. On the other hand, treatment with actinomycin D between fertilization and the 16-blastomere stage prevents the formation of the mitotic gradient (Parisi *et al.*, (1979), and Senger and Gross (1978) found that the higher rate of histone synthesis does not arise in the micromeres formed in the presence of actinomycin D. Eckberg and Ozaki (1975) reported that the increased expression of unique DNA sequences, which normally takes place from blastula to prism, does not occur in animalized embryos. This is not surprising in view of the fact that practically no morphogenesis occurs in the animalized embryos. More interestingly, the results of A. G. Carrol *et al.* (1975) show differences between the electrophoretic patterns of the total proteins synthesized by normal and animalized embryos, and those of Galau *et al.* (1976) show differences between the mRNA's of normal embryos and exogastrulae.

The only results concerning a specific RNA class have been for a long time those regarding ribosomal RNA: Pirrone *et al.* (1970) found that animalization causes a quite specific inhibition of the synthesis of ribosomal RNA (see also O'Melia and Villee, 1972a), which is in agreement with the finding of Roccheri *et al.* (1979) that the synthesis of ribosomal RNA between the blastula and pluteus stages is almost entirely localized in the cells of the archenteron. More recently, however, Shepherd *et al.* (1983) have found that at least two individual polysomal mRNA species, identified by hybridization to a cDNA library of *Strongylocentrotus purpuratus*, increase ten times in abundance during the 5 h preceding the mesenchyme blastula stage, whereas their synthesis is severely repressed in experimentally animalized embryos. In the latter, on the other hand, at least one mRNA species dramatically increases in abundance.

Lack of the increase in some enzymatic activities, which normally accompanies gastrulation, has been detected in vegetalized embryos by Gustafson and Hasselberg (1951). Here, however, one wonders whether gastrulation does not occur because of the defect in these enzymatic activities or if, conversely, these activities do not increase because of the defective morphogenesis induced by the vegetalizing agent.

2.3.2 Ventrodorsal Axis

Less attention has been devoted to the determination of the ventrodorsal axis; whose position does not seem to become fixed probably until the early gastrula stage. This was first demonstrated by Driesch (1891, 1892, 1900, 1903), who showed that if the first two blastomeres are separated, each one will give rise to a dwarf pluteus (see also Mazia, 1958), and, more recently, by experiments of exchanges of longitudinal halves of 16-cell embryos cut by microsurgery (Hörstadius, 1957). For a complete review of the subject the reader is referred to Hörstadius (1973). Studies aimed at finding metabolic differences between the ventral and the dorsal side of the embryo have, to date, failed to give meaningful answers (see Guidice, 1973, for a review).

Interestingly again, modifications of the cell surface can alter the embryonic symmetry, as shown by Marcus (1981), who, by treating the embryos of *Arbacia punctulata* before cleavage with pronase, obtained plutei with situs viscerum inversus, or with visceral duplication.

2.3.3 The Orientation of the Mitotic Furrow

Related to the problem of the determination of the embryonic axes is that of the orientation of the first mitotic furrows. A long series of papers devoted to the solution of such a problem lead to the conclusion that it is the aster that, by metaphase, induces into the egg cortex the ability to contract along a specific circular line. Once the tips of the aster rays have conveyed this information to the cortex, this is able to contract by itself. These conclusions are mainly based on the results of microsurgery experiments in which the mitotic apparatus is removed at various time intervals after fertilization (Hiramoto, 1956, 1965, 1971), or the distance between the aster and the cortical layer of the egg has been altered by flattening or perforating the egg (Rappaport, 1961, 1964, 1965, 1967, 1968, 1969a, b, 1975, 1982; Rappaport and Ebstein, 1965; Rappaport and Ratner, 1967; Timourian and Watchmaker, 1971). The question of what brings about the position of the first cleavage furrow therefore has to be rephrased by asking what brings about the position of the first astral rays, although the theory that the astral rays sum up their action at the equator needs to be partially modified, because they do not overlap at the equator, as more recently shown by Asnes and Schroeder (1979).

Schroeder (1980a, b) has recently revived the old problem of whether it is possible to recognize the animal vegetal axis already before fertilization. The answer is again positive, but more than the Boveri band, i.e., a subequatorial band consisting of 7000 granules of an orange yellow pigment, which is present only in some batches of *Paracentrotus* eggs, Schroeder found it useful to identify by means of china ink treatment the position within the jelly coat of large oocytes or eggs of the "jelly canal", which is located in the animal pole, and is independent of the point of sperm penetration. G. Schatten (1981), on the other hand, has reported that the first cleavage usually occurs at $8°$ from the direction of pronuclear movement, which seems again to re-evaluate the importance of the point of sperm entrance. A possible way of reconciling the latter result with the previous ones is that the animal vegetal axis, already present in the unfertilized egg, specifies a large number of radii around the egg poles through which the first cleavage plane may occur. The site of sperm incorporation then selects which of these polar diameters is actually used for cytokinesis. The point of sperm entrance, in other words, brings about a finer control on a coarsely established animal-vegetal axis.

Special attention within this problem has been merited by the case of the unequal cleavage leading to the formation of the micromeres. K. Dan (1978) suggested that the resting nuclei move toward the vegetal pole of the 8-blastomere embryo, and that this determines the unequal cleavage. What, then, causes this movement of the nuclei toward the vegetal pole? This is not known, but according to K. Dan and Ikeda (1971) it seems to depend on a time-schedule connected to rhythmic fluctuation of the SH group (Sakai, 1968). Czihak (1974) hypothesizes that once the astral rays are in contact with the vegetal cortex they dissolve, thus becoming shorter and causing the movement of the nucleus toward the vegetal pole. Differences between the cortices of the animal and vegetal blastomeres have been described by K. Dan *et al.* (1983) in *Hemicentrotus pulcherrimus* and in *Clypeaster japonicus*: the former, in fact, are limited by a continuous row of vesicles, whereas the latter show some gaps between the same vesicles. It is to these gaps that the spindles point in the vegetal blastomeres. Schroeder (1982), although acknowledging that micromeres show peculiar cortical and surface characteristics such as, e.g., lack of pigment granules and microvilli, warns against too easy conclusions regarding their relevance to the formation and function of micromeres, because the intermesomeric division also occurs between areas devoid of microvilli; e.g., the area which is cut by the cell division which divides micromeres from macromeres is also devoid of microvilli.

As already mentioned, once it has received the information, the egg cortex is able to cleave by itself by an active construction along a contractile ring, which contains actin (Schroeder, 1972, 1973; Kane, 1974, 1975a, b, 1976; Bryan and Kane, 1977, 1978; Begg et al., 1977; A. Spudich and J. A. Spudich, 1979). Additionally, the cortical layer of the egg contains an ATPase (Miki, 1964; Mabuchi, 1973), later recognized as dynein (Y. Kobayashi et al., 1978), which may be involved in the mechanism of furrow contraction. It is interesting in this respect that a rise of pH to values comparable to those achieved after fertilization causes polymerization of actin in isolated cortices (Begg and Rebhun, 1979).

An active expansion of the polar cortex has also been suggested as a mechanism relevant to cleavage (Mitchison and Swann, 1955; Swann and Mitchison, 1958). Polar relaxation could be due to local plasma "solation" through the concentration of cellular granules containing a heparin-like substance according to Kinoshita (Kinoshita, 1969; Kinoshita and Yazaki, 1967). These granules migrate to the poles during anaphase; furthermore, if the eggs are centrifuged and the contents are stratified, furrowing occurs at the level of the lowest content in "relaxing granules" (Kinoshita, 1968). This theory contrasts with the experiments of Hiramoto (1974), which show that the stiffness of the cortical plasma, as measured by means of magnetic particles microinjected into the eggs, is the same at the equator and the poles, but has been revived by Schroeder (1981b), who, by measuring the size variation following experimental compression of the egg, concluded that the eggs undergo a strong cortical contraction till the anaphase, especially in the zone of furrow formation, then a general contraction is observed followed by a "polar relaxation" induced by the mitotic stimulus. Usui and Yoneda (1982) have investigated by electron microscopy the physical basis of the equatorial contraction in egg cortices of *Hemicentrotus pulcherrimus* and *Toxopneustes pileolus*. They found that already during anaphase, when the cortical tension increases, but the furrow is not yet formed, a meshwork of fine filaments of 70–90 Å in diameter appears beneath the cell membrane. This meshwork becomes denser in the zone where the furrow will be formed, and disappears when the furrow becomes visible. These filaments are cytochalasin B-sensitive, and the authors suggest that they consist of actin microfilaments which are responsible for the increase in cortical tension at anaphase and represent the precursors of the contractile ring of telophase.

Also Begg *et al.* (1983), by the combined use of high hydrostatic pressure, which disrupts the cortical actin organization, and electron microscopy, concluded that a network of actin stabilizes the egg cortex and contributes to the formation of the furrow.

2.3.4 The Egg Cortex

We have already spoken of the egg "cortex" as if the peripheral layer of the egg constituted a special identifiable entity. This has indeed been the idea of several authors in the past (see Runnström, 1966 and Giudice, 1973 for reviews), who held either that the egg cortex has a gelatinous consistency, or rather that it jelifies upon homogenization in 0.002 M CaCl (Kane and Stephens, 1969), at least in some egg species. The idea of the cortex as a special physical entity has resisted time (see Vacquier, 1981 for a recent review), and the cortex has been found to contain 7 major proteins, one of which amounting to about 12–27 %, seems to be actin (Vacquier and Moy, 1980). Also Kane (1980, 1982) has found actin in a part of the egg cytoplasm which gelates in the presence of ATP. Bryan (1982) proposes that this gelation is due also to the presence of some Ca-deficient G. actin-binding proteins. One of these is fascin, a 58,000 molecular weight protein, which is known to cross-link adjacent actin filaments (Bryan and Kane, 1978) and which concentrates in the egg cortex following fertilization, contributing to microvillar elongation.

The mechanism which causes the polymerization of actin responsible for the transient increase in the rigidity of the cortex, which occurs upon fertilization, has been thoroughly investigated by Spudich and coworkers (A. Spudich *et al.*, 1982).

These authors found a value of actin concentration of 1.5×10^{-9} μg μ^{-2} of apparent surface area in the *Strongylocentrotus purpuratus* unfertilized eggs, which increases to 4×10^{-9} μg μ^{-2} in the fertilized egg. This increase is in keeping with the descriptions of other authors (see, e.g., Vacquier and Moy, 1980), and seems due to the gathering into the cortex of molecules from a soluble pool of the egg. Begg and Rebhun (1979) proposed that actin polymerization occurring in the cortex of the fertilized egg is due to the increase in pH. A. Spudich *et al.* (1982) have partially purified a protein which plays a role in the regulation of such a polymerization. Morton and Nishioka (1983), by treating *Strongylocentrotus purpuratus* eggs with cytochalasin B, obtained indirect evidence that cortical granules are held in place by a microfilamentous network which at the same time excludes other intracellular inclusions from the cortex. Other authors, on the other hand, have described by immunofluorescence and electron microscopy the presence of microtubules in the egg cortex which depart radially toward the egg center, and then, when the pronuclei fuse, become spirally orientated, to disappear at the so-called streak stage, i.e., shortly before the prophase.

2.3.5 The Mitotic Apparatus and the Mechanism of Cleavage

The problem of the composition of the mitotic apparatus, as well as that of its functioning, have received much attention in sea urchins since Mazia and Dan (1952) described a method for purifying it in bulk. More recent data (see Petzelt, 1979, for a review) have confirmed that the mitotic spindle is made of tubulin (Inoué and Sato, 1967; Sato, 1983), and probably also of actin and myosin, as parts of the contracting machinery. The tubulins in the mitotic spindle have been thoroughly studied by Rebhun *et al.* (1982), and found to be represented in the proportions of 2 or 3 α-tubulins and 2 β-tubulins (Suprenant and Rebhun, 1983). A dynein-like ATPase has also been described by Pratt *et al.* (1980), in mitotic apparati isolated from *Strongylocentrotus droebachiensis* eggs. Hisanaga and Sakai (1983) have purified the dynein near to homogeneity from the cytoplasm of *Hemicentrotus pulcherrimus* eggs. This dynein showed several similarities to that of the axonemes, i.e., high specificity for ATP, inhibition by low concentrations of vanadate, but was activated not only by Ca^{2+} but also by K^+. It bound to calmoduline-Sepharose 4 B columns in the presence of Ca^{2+}, and was eluted by EGTA. Cytoplasmic dynein can be distinguished from the dynein of sperm flagella by means of monoclonal antibodies (Asai and Wilson, 1983). Preliminary evidence for the existence of a rich assortment of microtubule-associated proteins within the mitotic apparatus of *Lytechinus variegatus* and *Strongylocentrotus purpuratus* eggs has been presented by Bloom *et al.* (1983) by a method involving the use of taxol and monoclonal antibodies. These proteins appear to be stored already in the unfertilized egg. A full report of these studies has been more recently published (Vallee and Bloom, 1983), and four monoclonal antibodies reacting specifically with microtubule-associated proteins (MAP's) of 37,000; 150,000; 205,000, and 235,000 molecular weight have been described, which all strongly and specifically stain the mitotic apparatus of dividing eggs of *Lytechinus variegatus*. Hiramoto *et al.* (1981 a, b), by means of a special microphotometric birefringence system, have described, in the sand dollar *Clypeaster japonicus*, changes in the amount of the polymerized tubulin which suggest both assembly–disassembly of microtubules as well as their dislocation during the mitotic cycle. The results of the micromanipulation experiments carried out by Hiramoto and Shôji (1982) suggest that the forces which pull the chromosomes during anaphase are generated in the region of the spindle close to the chromatids, while the intermediate system contributes to spindle elongation. Detrich and Wilson (1983), with the aim of understanding the mechanism of regulation of tubulin polymerization, have described the mode of assembly of tubulin purified from *Strongylocentrotus purpuratus* eggs and found that it polymerizes into normal microtubules when warmed at 37 °C at the concentration of 0.12–0.15 mg ml^{-1} in the presence of guanosine triphosphate. An in vivo approach to the same problem has recently been described by Wadsworth and Sloboda (1983), who observed repeated rounds of assembly and disassembly of tubulin in association with mitotic spindles, following microinjection of fluorescent tubulin into *Lytechinus variegatus* eggs.

Coffe *et al.* (1983) have also reported cycles of tubulin polymerization in eggs of *Paracentrotus lividus* with peaks of about 1 and 2 h after activation with parallel variations of cytoplasmic cohesiveness.

The role of Ca^{2+} ions in the retraction of the spindle fibers has been confirmed by Salmon and Segall (1980), who were able to isolate from *Lytechinus variegatus* egg mitotic spindles (complete with central and astral fibers, centrosomes and chromosomes, but without interfibrillar membrane) which can be stored for weeks and retain their Ca^{2+} sensitivity (see also Salmon, 1982). These authors suggest that the spindle-associated membranes are responsible for binding and releasing the Ca^{2+} in vivo. This is an idea which has recently been confirmed by G. Schatten *et al.*, (1982), who were able to demonstrate the presence of Ca^{2+} in the mitotic apparatus membrane, by means of the chlorotetracycline fluorescence. It has, moreover, been shown (P. Harris, 1983) that caffeine, which is known to cause Ca^{2+} release from the sarcoplasmic reticulum, induces monaster cycling in fertilized *Strongylocentrotus* eggs. It has to be kept in mind that Ca^{2+} ions play a delicate and complex role in the process of cell division because while Ca^{2+} release is required in order to activate the egg, some Ca^{2+} sequestration must occur in order to initiate mitosis, according to Wagenaar (1983b), as indicated by the fact that an experimentally induced increase in free Ca^{2+} in eggs already fertilized or parthenogentically activated delays the onset of mitosis. The observation of Keller (T. C. S. Keller and Rebhun, 1982; T. C. S. Keller *et al.*, 1982) has to be kept in mind when considering the role of Ca^{2+} and of the microtubule-associated proteins in tubulin polymerization, that this is evident only at the temperature that is physiological for the considered species. These authors (Suprenant and Rebhun, 1983) have recently been able to purify the α- and β-tubulins of *Strongylocentrotus purpuratus* eggs and to obtain microtubule assembly in a cell-free system. Their results have contributed some hints as to the problem of what prevents tubulin from spontaneous polymerization in view of its high concentration within the egg. One probable answer is the absence or the compartmentalization of the assembly-promoting factors; in fact, they obtained spontaneous assembly of microtubules at a tubulin concentration of 0.79 mg ml^{-1} at 18 °C, but this concentration drops down to 4–20 μg ml^{-1} if heterologous microtubule-associated proteins (from pig's brain) are added.

Among the various means used in order to understand the mechanism of microtubule assembly and disassembly during mitosis, D$_2$O is worth mentioning. The results have recently been reviewed by Sato *et al.* (1982) (see also Giudice, 1973, for

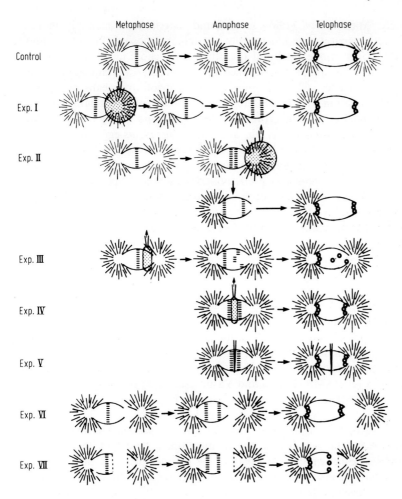

Fig. 2.8. Microoperation of the mitotic apparatus in eggs of *Clypeaster japonicus*. *Exp. I* removal of one of the asters during metaphase. *Exp. II* removal of one of the asters during anaphase. *Exp. III* removal of the region between the chromosomes and the pole of the spindle during metaphase. *Exp. IV* removal of the interzonal region of the spindle during anaphase. *Exp. V* insertion of a thin glass plate into the interzonal region of the spindle during anaphase. *Exp. VI* separation of one of the asters from the mitotic apparatus. *Exp. VII* separation of one of the asters and the polar part of the spindle from the mitotic apparatus. (Hiramoto and Shôji, 1982)

the older literature). Among the most interesting points of the new results is the observation that the known delay of the mitotic process brought about by D_2O mainly occurs in the prophase to metaphase, and not through a disturbance of chromosome movement in anaphase. This implies that D_2O in vivo may act by elevating tubulin concentration, but still not affect tubulin dissociation. T. C. Takahashi and Sato (1982) have also carried out a thermodynamic analysis of the effect of D_2O on the mitotic spindle of a variety of sea urchin eggs.

Interestingly, Mazia *et al.* (1981) have found that in experimentally induced monopolar mitotic apparati one kinetochore of each chromosome faces the spindle fibers (which are only in one side) and the other kinetochore faces the other side, although this contains no spindle fibers. Since it is possible in sea urchin eggs to cause the formation of monasters which do not contain centrioli, by means of a partial parthenogenetic activation, this has offered the possibility of investigating the role of centrioli in the nucleation of microtubules. Kuriyama and Borisy (1983), in fact, by treating *Strongylocentrotus purpuratus* and *Lytechinus pictus* eggs with a variety of agents, such as acidic or basic sea water, or procaine or thymol, obtained the formation of monasters which appeared to lack centrioles and, by means of a second treatment like D_2O-containing, ethanol-containing, or hypertonic sea water, the formation of multiple cytasters, each containing one or more centrioli. It was found that microtubules radiate from the center of isolated asters, whether they contained centrioli or not, thus indicating that centrioli are not necessary for microtubule organization.

Centrioli can be formed de novo in the egg since they also appear following parthenogenetical activation of non-nucleated egg halves (Kato and Sugiyama, 1971).

What triggers mitosis in the developing sea urchin? As already discussed in the chapter on sperm-egg interaction, the rise in pH seems to be crucial for the activation of mitosis. Although the activation of the synthesis of DNA is not the only requirement for mitosis to occur, it is certainly a necessary step, whose activation will be described later, in the chapter on DNA. Suffice it here to recall the role proposed for cyclic AMP (Fujino and Yasumasu, 1975; Ishida and Yasumasu, 1981, 1982), and cAMP-dependent protein kinase (Fujino and Yasumasu, 1978, 1981, 1982), for polyamines and ornithine decaboxylase (Pirrone *et al.*, 1983, Kusunoki and Yasumasu, 1976, 1978, a, b, 1980), and for tyrosine protein kinase (Dasgupta and Garbers, 1983). The fact is also worth mentioning here that the inhibition of protein synthesis blocks mitosis if operated by the metaphase of each cell cycle (Wagenaar, 1983, a), probably by inhibiting the synthesis of some division protein, such as, e.g., the "cyclin" described by T. Evans *et al.* (1983), which is a protein synthesized at each cell cycle and disrupted thereafter, which might be involved in the breakdown of the nuclear membrane at each cell cycle. We shall also recall here that Renaud *et al.* (1983) have purified serotonin and 5-methoxy-triptamine from *Paracentrotus lividus* and *Sphaerechimus granularis* eggs and proposed a role for these compounds in the stimulation of cell cleavage, based on the observation that serotonin antagonists such as gramine and metergoline cause a cleavage delay and stimulate a Ca^{2+} efflux from the egg with a strong decrease in cAMP concentration.

2.4 Cell Interactions

2.4.1 The Process of Gastrulation

One way of approaching the problem of morphogenesis is to study the mechanism of interaction of the cells belonging to different embryonic districts. This was first done by Gustafson and Wolpert by means of time lapse cinematographic analysis of larval development (Gustafson and Wolpert, 1961a, b, c, 1962, 1963a, b; 1967; Wolpert and Gustafson, 1961a, b). These authors have recognized the differing adhesiveness of the cells among themselves or to the basement membrane as the key mechanism for the morphogenetic movements. It is of interest in this respect that Fink and McClay (1980) have experimentally proved that the mesenchyme cells lose adhesivity to the hyaline layer when they start their migration inside the blastocoel. Time-lapse cinematography, coupled to an ultrastructural analysis, has recently been used by Katow and Solursh (1981, 1982), who have stressed the importance of sulfate (see also Lallier, 1980) for the adhesiveness of the primary mesenchyme cells to the basal lamina. Six types of cellular processes have been described by these authors, through which the mesenchyme cells can migrate and adhere to the basal lamina of the ectoderm, which shows in the points of adhesion granulations 30 nm wide. The importance of sulfate ions for the formation of pseudopodia and filopodia in the secondary mesenchyme which are responsible for the second part of gastrulation has also been stressed by Akasaka et al. (1980). These authors have also described the adverse effect on gastrulation of *Pseudocentrotus* embryos of aryl-β-xyloside and tunicamycin, which is in agreement with ultrastructural observations of Kawabe et al. (1981) of the presence of filaments, fibers and granules to line the internal blastocoelic wall during gastrulation, with a layer which appears to be made of sulfated glycosaminoglycans and seems to guide archenteron invagination. The relevance of glycoproteins for the process of gastrulation has been stressed by several authors; Kinoshita and Yoshi (1979), for example, described the inhibitory effect of exogenous proteoglycans; Carson and Lennarz (1981) found that the inhibition of the synthesis of dolichol, a saccharide carrier in glycoprotein synthesis, also inhibits gastrulation, and proposed a role for the dolichol phosphate as a regulator of glycoprotein synthesis and of gastrulation. It is in fact only at gastrulation that N-linked glycoproteins start to be synthesized (Heifetz and Lennarz, 1979, Carson and Lennarz, 1981; Lau and Lennarz, 1983). A role for sulfated mucopolysaccharides in mediating the adhesion of primary mesenchyme cells is suggested by the biochemical and scanning electron microscope observations of Karp and Solursh (1974) and of Kinoshita (1974a, b). Preliminary experiments of Rosenberg and Wallace (1973) suggest that at least *Arbacia punctulata* embryos show no disturbance of gastrulation in the absence of external sulfate, which only causes disturbance of the attachment of the archenteron to the ventral wall. According to Gezelius (1974a, b; 1976) the lack of sulfate ions does not disturb gastrulation in *Paracentrotus lividus* or *Psammechinus miliaris*, but interferes with development through a disturbance of the transport of the RNA from the nucleus to the cytoplasm. Heifetz and Lennarz, on the other hand, have shown that specific inhibition of the synthesis of sulfated (but also unsulfated) N-linked glycoproteins and of oligosaccharide-linked lipids, obtained with low doses of tunicamycin, inhibits gastrulation. Akasaka and Terayama (1983), by means of light microscope observations of Alcian blue-stained gastrulae of *Hemicentrotus pulcherrimus* and by scanning electron microscopy, have observed the presence of highly acidic glycans in the internal surface of the archenteron and in the secondary mesenchyme cells. An EDTA extract of the embryos contains, among other glycans, a component called F, which is absent when the embryos are cultured in sulfate-free sea water, which causes disturbance of gastrulation and weakening of the

Alcian blue stainability. The F component is made of sulfated fucan and acid muco-polysaccharide chain linked to a protein core. This factor is able to stimulate the reaggregation of cells dissociated from mesenchyme blastulae (Akasaka and Terayama, 1984). Only preliminary experiments on the kinds of glycoprotein synthesized by the different blastomeres are currently available (Brown and Rosman, 1978). The sum of the above data points, however, to a role for glycoproteins and collagen in the process of gastrulation.

Membrane glycotransferases have been described and analyzed during early development (Schneider and Lennarz, 1976; I. M. Evans and Bosman, 1977), but not yet in connection with the mechanism of intercellular adhesion.

Since it has long been known (Moore and Burst (1939) that isolated vegetal plates can undergo invagination, at least the onset of gastrulation cannot depend upon the interactions of the invaginating cells with the internal blastocoel wall, but rather, as hypothesized by Gustafson and Wolpert (1963, b), may be explained by changes of the adhesiveness of the vegetal cells between themselves or to the supporting membrane, which is represented by the so-called apical lamina, which is immediately underlying the hyaline layer, and whose composition has been recently described by Hall and Vacquier (1982). Again the presence of glycoproteins, which contain sulfate but not sialic acid, and which are different from collagen and free of glycosaminoglycans, was stressed. It is interesting to mention the appearance of an extracellular arylsulfatase in coincidence with the morphogenetic cell movements which cause gastrulation (Rapraeger and Epel, 1980, 1981). Solursh and Katow (1982) have described a method for isolating the blastocoelic matrix of mesenchyme blastulae. The analysis of such a matrix shows dermatan sulfate and chondroitin sulfate in *Strongylocentrotus*, whereas little or no chondroitin sulfate is found in *Lytechinus*. E. Spiegel *et al.* (1980, 1983) have identified fibronectin and laminin in the extracellular matrix, in the basement membrane, on the cell surface of the outer epithelial layer of the secondary mesenchyme cells of *Sphaerechinus granularis* and *Arbacia punctulata* blastulae or prisms. They propose a role for these glycoproteins in the morphogenetic movements through the formation of a continuous matrix surrounding the cells and connecting the hyaline layer to the basement membrane. The role of fibronectin in cell adhesion is also supported by the findings of Vittorelli (personal communication) that its addition to cells dissociated from stages later than early blastula stimulates reaggregation.

Mizoguchi and Yasumasu (1982a, b; 1983; Mizoguchi *et al.* 1983), furthermore, have also proposed a role for collagen in gastrulation, since the inhibitory effect of the absence of sulfate ions in sea water can be reversed by ascorbate and α-keto-glutarate, which are activators of protocollagen prolyne hydroxylase. Moreover, the incorporation of ^{14}C hydroxyproline into collagen is reduced and gastrulation is inhibited in the presence of inhibitors of prolyl hydroxylase such as α, α dipyridyl or Zn^+. The effect of α,α-dipyridyl can be cancelled by the addition of Fe^{2+} and that of Zn^+ by the addition of ascorbate or α-ketoglutarate. Nakano and Iwata (1982) have described the synthesis of some collagen-binding proteins during early development, and suggested that they may be involved in the processes of cell interactions. E. Spiegel and M. Spiegel (1979), who hold that the hyaline layer plays a central role in cell interactions and reaggregation, found that the hyaline contains collagen. Vittorelli (personal communication) also finds stimulation of intercellular adhesion by collagen addition to sea urchin cells dissociated from stages later than early blastula.

A very ingenious method for studying cell adhesion in gastrulation has been developed by M. Spiegel and Burger (1982). These authors have microinjected a variety of compounds within the blastocoelic cavity of gastrulating embryos and observed their effect on cell adhesion and gastrulation. They found by such a method that proteases and lectins cause specific filopodical detachment in the secondary mesenchyme cells and archenteron regression. Interestingly, fluorescein-conjugated lectins such as, e.g., concanavalin A and wheat germ agglutinin show a very specific binding to specific regions of the blastocoel wall or of the invaginating archenteron or mesenchyme cells, thus offering the possibility of probing the relevance of specific changes in the surface composition to the morphogenetic movements.

2.4.2 Cell Dissociation-Reaggregation Studies

A different approach to the problem of cell interactions was followed by Giudice (1961 a), who developed a method, based on the mechanical treatment of embryos in a sucrose solution, that allows the complete dissociation of embryos into single cells. These cells are able to reaggregate and to differentiate into structures closely resembling normal larvae (see Figs. 2.9 and 2.10). This method applies to embryos from the blastula till the pluteus stage. When cells are dissociated from early blastulae till late mesenchyme blastulae, the pattern of reaggregation is always the same: formation of solid clumps that then become hollow spheres by means of a process of internal cavitation, appearance of cilia, with the formation of structures closely resembling "swimming blastulae". Groups of cells inside the blastocoel form tubular structures which then attach to the blastula wall and, by opening outside, form one intestine-like structure. Triradiated spicules appear at this point, which elongate and bend in the characteristic way of the pluteus, while pigmented cells differentiate. The shape and size of the reformed larvae vary with the aggregation conditions, i.e., concentration of the cells, rate of stirring or lack of stirring, of the sea water, and so on. The efficiency of formation of new larvae also varies, but can be close to 100% under optimal conditions (see Giudice and Mutolo, 1970, and Giudice, 1973, for reviews). If the embryos are dissociated at the gastrula stage and the cells reaggregated, the reaggregation pattern is the same as that described for blastulae, but the skeleton usually does not elongate to form rods. If cells are dissociated from the young plutei, they again reassociate according to the blastula cell pattern, but the skeleton is never formed.

Embryos can also be dissociated into cells during stages earlier than blastula, i.e., before the cells have differentiated an epithelial shape and are still in the form of rounded blastomeres. In this case treatment with Ca-free sea water followed by repeated pipeting is enough to achieve dissociation, as already described by Herbst (1900), because the main element which keeps the blastomeres together throughout the "cleavage" period is the hyaline layer (Turner 1980; Watanabe et al., 1980). Septate junctions are already present at the 4-cell stage (Chang and Afzelius, 1973), but they form a continuous layer only much later in development (Gilula, 1973). The development of cell junctions in sea urchin embryogenesis has recently been reinvestigated by E. Spiegel and Howard (1983) by means of more modern methods like freeze fracture and lanthanum impregnation. These authors describe six types of cell junction: three types of desmosome, i.e., belt- spot- and hemi-desmosomes; two types of septate junction, double septum septate, straight and unbranched, the first type, and pleated anastomosing single septum septate, the second type, and finally the tricellular junctions, which join the bicellular junctions of three adjoining cells.

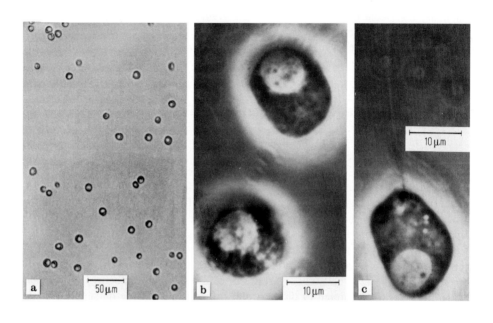

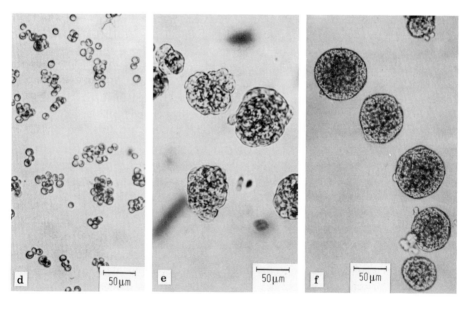

The belt desmosomes appear at the 4-cell stage, and surround each cell at the blastula stage, when the hemi-desmosomes also appear, joining the cells to the basal membrane. The spot desmosomes are already frequent at the 4-cell stage, and appear to decrease later on. The unbranched biseptated junctions are present from the 4-cell stage to pluteus, and are replaced by the pleated junctions in those cells which migrate

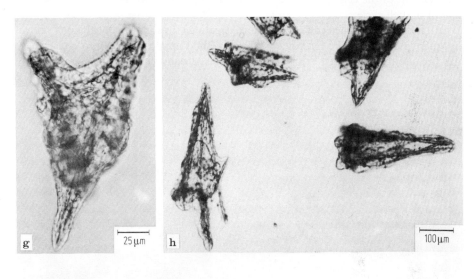

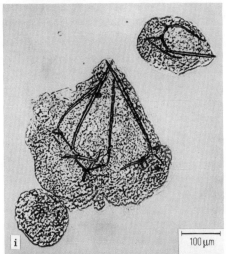

Fig. 2.9a–i. Rotation-mediated aggregation of cells dissociated from mesenchyme blastulae of *Paracentrotus lividus*. **a–c** freshly dissociated cells; **d** after 1 h of aggregation; **e** after 6 h of aggregation; **f** after 9 h of aggregation; **g** after 3 days of aggregation; **h** after 5 days of aggregation; **i** pluteus-like and a small aggregate flattened to show details of skeleton. **a–h** Giudice and Mutolo, 1970; *I* Giudice, 1962a)

inside the blastocoel during the process of gastrulation. Since these pleated junctions have also been described in the intestine of adult sea urchins (C. R. Green, 1981; C. R. Green *et al.*, 1979), it can be supposed that they play a role in the digestive processes.

Intercellular communications between the blastomeres at the 16-cell stage have also been described by Andreuccetti *et al.* (1982). An electrical coupling of the blastomeres in the first cell divisions has been reported by Dale *et al.* (1982) during the first half of the cell cycle. This coupling disappears, however, during the second half of the cell cycle.

All this explains the need for mechanical treatment in addition to calcium removal in order to dissociate the blastomeres. This method applies less well to later stages, also because the prolonged exposure to Ca-free sea water causes damage and loss of

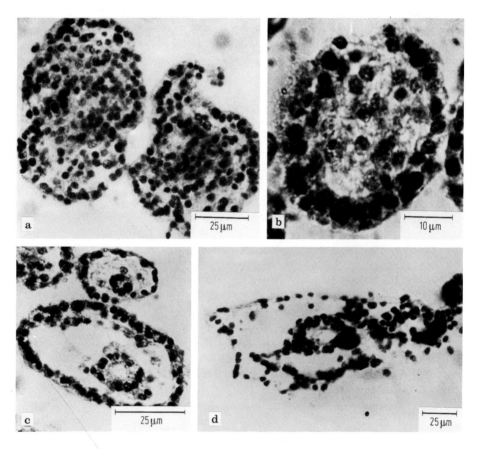

Fig. 2.10. Histological section of reaggregates deriving from cells dissociated from mesenchyme blastulae of *Paracentrotus lividus* (**a** and **d**) and *Arbacia lixula* (**b** and **c**). **a** after 12 h of aggregation; **b** after 20 h of aggregation; **c** after 30 h of aggregation; **d** after 3 days of aggregation. (Giudice 1962a)

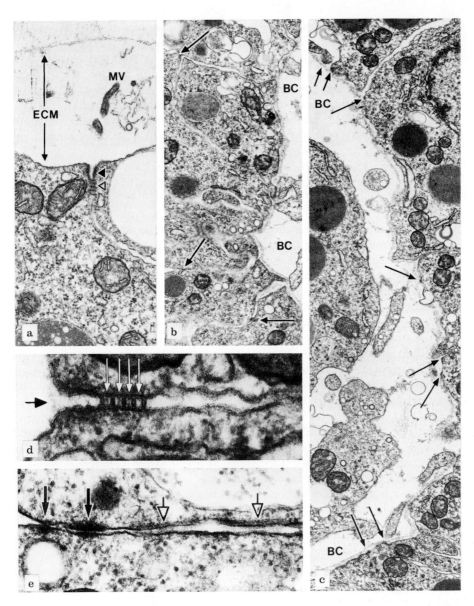

Fig. 2.11. a to **d** late blastula stage embryos. **a** Outer cell surface of two adjoining cells showing belt desmosome (*black arrowhead*) and septate junction (*open arrowhead*). Extracellular matrix (*ECM*) surrounds cells and several microvilli (*MV*) are seen within the *ECM*. (×26,300). **b** Blastocoel cavity (*BC*) in interior of embryo is lined with basement membrane (*arrowheads*); spot desmosomes are indicated by *arrows*. (×13,650). **c** Blastocoel cavity (*BC*) in interior of embryo is lined with basement membrane (*arrowheads*); hemi-desmosomes are indicated by *arrows*. (×13,650). **d** High magnification of junctional are showing desmosome and septate junction with extracellular dense material between desmosome (*black arrows*) and between septa (*white arrows*). (×130,000). **e** Four-cell stage with spot desmosome (*black arrows*) and intercellular dense areas (*open arrows*), where additional spot desmosomes appear to be forming, with electronlucent areas in between. (×77,000). (E. Spiegel and Howard, 1983)

ability to reaggregate. Reaggregates obtained by cells dissociated at the 16-blastomere stage have been reared through metamorphosis by Hinegardner (1975).

Other methods to dissociate cells from embryos at stages later than blastula have been described (see, e.g., Kondo and Sakai, 1971; Kane, 1973; Sano, 1977). These are essentially based on treatment of the embryos with Ca^{2+}-free sea water followed by a mechanical treatment or on treatment with 1 M glycin plus 2 mM EDTA. These methods may be applied with differing success to different species, but an accurate analysis of the ability of reaggregated cells to differentiate after dissociation using these methods has not always been made.

Morphological analyses of the reaggregation process have been made by time lapse cinematography (Timourian and Watchmaker, 1975) and by electron microscopy (Millonig and Giudice, 1967; E, Spiegel and M. Spiegel, 1977,a; M. Spiegel and E. Spiegel, 1975; 1978a, b). The latter authors stress the importance of microvilli and of the hyaline layer in the process of reaggregation, as was observed with the scanning electron microscope (E. Spiegel and M. Spiegel, 1977b).

Several questions relevant to the mechanism of morphogenesis have been posed using the system of dissociated and reaggregated cells. We will list the most important ones here: First, is the reaggregation stage-specific? That is, do ·the cells acquire new surface recognition properties during early development? Giudice et al. (1969), by reaggregating radioisotopically labeled blastula cells with nonlabeled prism cells, found no strict stage specificity in the reaggregation process. More recently, however, M. Spiegel and E. Spiegel (1978b) were able to prove that vitally stained micromeres dissociated from the 16-blastomere stage, if reaggregated with macro- and mesomeres, sort out and group separately from the latter blastomeres within the context of the same aggregate. Related to this cell-type specificity observed during reaggregation, McClay and Chambers (1978) have been able to identify new surface antigens appearing at gastrulation, and to show that different surface antigens can be detected in different germ layers, at least at the pluteus stage (McClay and Marchase, 1979; McClay, 1979; 1982). When, in fact, viable hybrids between *Lytechinus variegatus* and *Tripneustes esculentus* are dissociated into cells, these aggregate only to cells of the maternal genotype if the dissociation stage was before gastrulation, but they aggregate to cells of both genotypes if the dissociation stage was after gastrulation. Changes in cell surface properties during the course of development have been described by several other authors. Krach et al. (1973, 1974) describe a decrease in cell agglutinability with Con A and *Ricinus communis* agglutinins during the second day of development. Roberson and Oppenheimer (1975) and Neri et al. (1975) found that micromeres are more agglutinable with Con A than the other blastomeres and that, unlike the other blastomeres, they appear "capped" upon Con A treatment (Roberson et al., 1975). It has been only recently, however, that an analysis of the in situ distribution of the Con A-binding sites in mesenchyme blastulae and early gastrulae has been reported (Katow and Solursh, 1982), which showed a correlation between the loss of Con A-binding sites on the ectoderm and sites of migration of the ectodermal cells. Further developmental changes in the properties of cell membranes have been reported by Sano (1977), who described an increase in the electrophoretic mobility of the entire cell as development proceeds. Timourian et al., (1973) distinguish four cell types already at the blastula stage with respect to the ability to reaggregate after dissociation; Turner et al. (1977) report an increase in the ability of cells to perform the first phases of reaggregation from the hatching to the gastrula stage, and finally Sasaki and Aketa (1981), following fluorescence microscope observations, reported the appearance on the egg surface after fertilization of a lectin which is located within the cytoplasm in the unfertilized egg (see also Tonegawa, 1982, for the presence of lectins in sea urchin embryos).

The second question solved by reaggregation is that of species specificity of cell interaction. Unlike vertebrate cells, sea urchin embryonic cells reaggregate in a species-specific way. This is demonstrated by the reaggregation of cells of differently pigmented species (Giudice, 1962a) or by the lack of formation of microvilli between cells of different species (M. Spiegel and E. Spiegel, 1975, 1978a, b). Cells derived from interspecific hybrid embryos, as already mentioned, adhere preferentially to cells of the maternal type until the gastrulation stage and to cells of the paternal type as well, after this stage. This shows that paternal genome activation contributes to the membrane specificity after the gastrula stage (McClay and Hausman, 1975; McClay et al., 1977).

The third area of investigation involving cell reaggregation has been directed at establishing the kinds of molecule which are involved in intercellular adhesion. Experiments performed with the aid of inhibitors of reaggregation seem to indicate the importance of the galactopyranosyl groups of the cell surface because of the inhibitory effect of specific carbohydrates (Asao and Oppenheimer, 1979). The effect of Con A points to the importance of glycopyranosyl or mannopyranosyl groups (Lallier, 1972), while the general role of proteoglycans in cell adhesion is indicated by the experiments of Kinoshita and Saiga (1979), showing that gastrulation does not occur when the synthesis of proteoglycans is inhibited (see also the experiments discussed earlier, in Sect. 2.4). The role of sterol-like substances is also suggested by experiments based on the use of mycostatin (Oppenheimer et al., 1973). Finally, Tonegawa (1982) has extracted some lectin-type molecules from embryos.

The most promising approach, however, is that of a direct isolation or reaggregation-promoting substances. Toward this goal, methods have been developed for isolating the sea urchin embryo plasma membranes (Cestelli et al., 1975; McCarthy and Brown, 1978). Noll et al. (1979) have been able to extract with butanol a factor from the membranes, the absence of which prevents the cells from reaggregating. If, however, this "butanol factor" is added back to the cells, they reaggregate and differentiate into pluteus-like larvae. Experiments in progress by Vittorelli et al. (1980) show that the "butanol factor" is a glycoproteic complex, one fraction of which seems to retain all the activity. Univalent antibodies against the "butanol factor" dissociate the embryos into cells in a species-specific way.

A factor which accelerates the adhesion between sea urchin embryonic cells, as measured by a electronic particle counter, has been isolated by Oppenheimer and Meyer (1982a, b) by treatment of the embryos with Ca^{2+} and Mg^{2+} free sea water. This factor is trypsin-sensitive and acts also on gluteraldehyde-fixed cells, therefore suggesting that it is involved in the first steps of cell adhesion, although cells aggregated in its presence are able to develop into swimming embroids. This factor is also species-specific and, less clearly in our opinion, stage-specific.

Another aggregation-promoting factor has been extracted from sea urchin cells and partially characterized by Kondo, who termed it ovacquenin (Kondo and Sakai, 1971; Kondo, 1973; 1974). This was first identified with the hyaline layer, but this was later discounted. Its study is rendered more difficult by its being closely associated with a reaggregation inhibitor. Also Tonegawa (1973; 1982) has extracted a factor from *Hemicentrotus pulcherrimus* embryos which stimulates cell aggregation.

This factor is a large glycoprotein, containing sialic acid, uronic acid and sulfate. Caution has to be used for this as for other aggregation factors before accepting their in vivo role, because many nonspecific macromolecules can cause cell agglutination.

It has finally to be recalled that, as already mentioned, Akasaka and Terayama (1984) have found that the 10^6 nulecular weight sulfated glycoprotein extracted with EDTA from the *Hemicentrotus* gastrule, the so-called F factor, is able to stimulate the reaggregation of dissociated blastula cells.

Although data on phospholipid biosynthesis are available (Pasternak, 1973; Schmell and Lennarz, 1974; Byrd, 1975b), they do not yet allow any conclusion about the biogenesis of specific cell surfaces.

2.4.3 Metabolism and Cell Interactions

A fourth pertinent question is that of the relevance of cell interactions for the development of morphological as well as metabolic patterns characteristic of the specific embryonic stages. One way of reformulating the same question is to ask if each embryonic territory is endowed with the properties that will lead it to develop these patterns or if it needs interaction with other embryonic parts. As we have seen, an animal-vegetal axis is already present in the unfertilized egg, so the first question that has been asked is if the isolated micromeres are able to develop into vegetal structures in the absence of interactions with the other blastomeres. The answer is that they reaggregate to form solid spheres which do not form ectoderm, which is a sign of vegetalization (Pucci-Minafra *et al.*, 1968), they do not form skeleton, unless serum is added to the culture (Okazaki, 1975a, b). The active principle of serum seems to be a peptide of 10,000 molecular weight, according to preliminary experiments of Blaukenship and Benson (1980). The fact that a nonspecific stimulus, like serum addition, elicits the formation of spicules, recalls the situation of the primary induction in the amphibian embryo, where the neural ectoderm is already con petent to react to a variety of stimuli by forming neural structures, although in nature the stimulus is the interaction with the chordomesoderm, and which for sea urchins might be represented by the basal lamina, since mesenchyme cells isolated together with the basal lamina do form spicules in culture (Harkey and A. H. Whiteley, 1980). This reaction of micromeres to serum seems to involve gene activity, since it is blocked by actinomycin D (Rosenspire *et al.*, 1977). Once the induction has taken place, the isolated mesenchyme cells are able to differentiate the spicules even in the absence of serum, although this still exerts a stimulatory effect on the process (Mintz *et al.*, 1981). McCarthy and M. Spiegel (1983) have investigated in detail at which time the cells derived from isolated micromeres need serum addition in order to differentiate the spicules, and found that the optimum time for serum addition is 36 h after fertilization, at which time a 1-h pulse of serum is enough to elicit a full response. These authors found that in the absence of serum the micromere-derived cells fail to form the pseudopodial network which precedes skeleton formation.

Since micromeres respond to the induction from other blastomeres, and since the animal-vegetal potentialities are already laid down in the egg, then significant territorial differences exist in the embryo. How, then, do cells dissociated from blastulae and randomly reaggregated find out the correct way of generating embryos which show a balance between animal and vegetal structures? Two theories can be proposed: one, the cells sort out and find their correct position in the aggregate, grouping among themselves according to cell types; two, the cells dedifferentiate upon dissociation and redifferentiate again upon reaggregation, according to the position occupied in the new aggregate. Two experiments strongly favor the first hypothesis. First, that of M. Spiegel and E. Spiegel (1978b), who followed the movements of cells derived from labeled micromeres during cell reaggregation; Second, the observation by Giudice (1963) that cells of chemically animalized or vegetalized embryos do not lose their chemically acquired potentialities during the dissociation reaggregation process (see also I. Kobayashi and Kimura, 1976). Amemiya *et al.* (1979) have demonstrated that cells of isolated pluteus guts, which have their cilium facing the intestinal lumen, shortly after isolation show a cilium which no longer faces the intestinal sac. The authors suggest a reversal of cell polarity, but the interpretation that cells turn themselves upside down, due to the new stimulus of the sea water from outside the intestine, is still possible. Relevant to this question are also the experiments of Timourian and Watchmaker (1975), suggesting that if the embryos at the blastula stage are dissociated into cells and then reaggregated by letting them sit on the bottom of a Petri dish, those cells which derive from the animal part of the blastula form cilia and become epithelial, while those deriving from the vegetal part of the blastula flatten on the bottom and become polyfilamentous. What is certain from these experiments is that epithelial-type and mesenchyme-type cells are formed before reaggregation is completed. What is only interpretation, although the most probable, is the derivation of these cells. Timourian and Watchmaker (1975) also observed that calcium and hyaline are required for the formation of the ciliated cells and their blastulation, whereas they are not for differentiation and interaction of the mesenchyme type cells.

If one looks at metabolic changes occuring after the blastula stage and investigates their dependence upon correct cell interactions, the most clear-cut answer comes, perhaps, from the experiments of Pfohl and Giudice (1967), who looked at the sharp increase in alkaline phosphatase activity occurring at the late pluteus stage in connection with intestine development (Evola-Maltese, 1957; Pfohl, 1965). They found that if cells are dissociated at the early pluteus stage, they will undergo a sharp increase in alkaline phosphatase at the same time as the entire embryos, although still in the form of amorphous aggregates. If, on the other hand, cells are dissociated from the blastula stage and then reaggregated, they show no increase in alkaline phosphatase activity when intact embryos become late plutei; they show such an increase only much later, when the aggregates form an intestine, and never (Pfohl and Giudice, unpublished) if they are prevented from reaggregation by overdilution. These experiments indicate that at some time between the blastula and the early pluteus stage, cells become committed to increase their alkaline phosphatase activity. Once committed, they do not need correct cell interactions to undergo the increase in enzyme activity. They also indicate that the commitment requires the correct cell interactions which lead to intestine formation.

Experiments performed with the same rationale indicate that the cells at the hatching blastula stage are already committed for the increase in the rate of labelled precursor incorporation into mature ribosomal RNA observed after the mesenchyme blastula stage (Sconzo et al., 1970b). If, however, embryos at the 4–8-cell stage are dissociated into cells, they undergo an immediate activation of the synthesis of ribosomal RNA, almost silent at that stage (Arezzo and Giudice, 1983), thus indicating a role for cell interactions in the regulation of rRNA synthesis.

The rate of protein synthesis of cells dissociated after the mesenchyme blastula stage has been followed only for the first 20 min following dissociation, and found

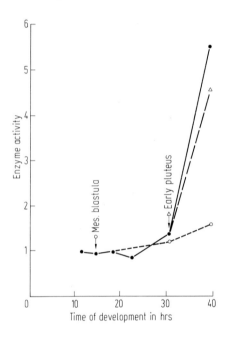

Fig. 2.12. Alkaline phosphatase activity of normal embryos and reaggregating cells. The values have been anormalized assuming as 1.0 the value for the swimming blastulae. Control embryos (*filled circles*); cells dissociated from mesenchyme blastulae (*open circles*); cells dissociated from early plutei (*open triangles*). *Arrows* point to the moment of cell dissociation. (Pfohl and Giudice, 1967)

to remain characteristic of the developmental stage of dissociation (Giudice, 1962b). Also, oxygen consumption remains so for the first 4 following the dissociation at the mesenchyme blastula stage (Giudice, 1965). The same kind of question has been asked with respect to the synthesis of specific proteins and the answer was that cells dissociated from embryos at 16 cells (Arcèci and Gross, 1980c) or at 64–128 cells (Brookbank, 1980) undergo the same type of change in the synthesis of histones as the entire embryos, even in the absence of reaggregation. Again the same answer was obtained by Sconzo et al., (1983) about the ability to respond to a rise in temperature with the synthesis of the so-called "heat shock proteins", an ability which the embryos develop only the swimming blastula stage. If they are dissociated into cells, these develop this ability at the same time as the entire embryos, even if prevented from reaggregating. It seems to us that a plausible hypothesis may be that these metabolic changes may arise in the embryonic cells after a certain number of cell divisions. A most detailed analysis of possible variations of the pattern of protein synthesis in isolated micromeres has been more recently presented by Harkey and A. H. Whiteley (1982b, 1983), who found that isolated micromeres, if allowed to reaggregate in the presence of horse serum, underwent the same development changes of the pattern of protein synthesis, judged by two-dimensional electrophoresis, as their in situ counterparts in the undissociated embryos.

The case of DNA synthesis in dissociated cells deserves a separate description. If embryos are dissociated into cells at the blastula stage and the cells reaggregated, DNA synthesis proceeds normally. If, on the other hand, the cells are prevented from reaggregating by overdilution, then the synthesis of DNA is halted (Giudice and Mutolo, 1970; Sconzo *et al.* 12970b). As expected, the same inhibitory effect on DNA synthesis is obtained if the blastula embryos are extracted with butanol or treated with univalent antibodies against the butanol extract (Vittorelli *et al.*, 1980). The hypothesis was then put forward that a signal leaves the membrane to tell the cells to stop DNA synthesis because of the lack of cell contact. In order to test this hypothesis Vittorelli *et al.* (1973) treated the dissociated and nonreaggregated cells with trypsin, with the aim of interfering with the production of the signal from the membrane. Such a treatment was actually effective in restoring DNA synthesis in the dissociated and nonreaggregated blastula cells. Matranga *et al.* (1978) have shown that trypsin removes a high molecular weight glycoprotein from the plasma membrane, which might be involved in the regulation of DNA synthesis in a fashion similar

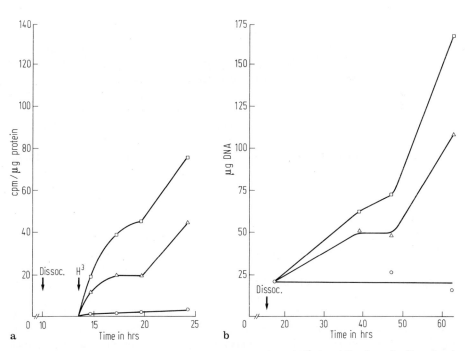

Fig. 2.13a, b. Effect of trypsin treatment. **a** Incorporation of H³-thymidine by cells dissociated from *Paracentrotus lividus* swimming blastulae. **b** Total DNA content of the same cell culture.. O———O dissociated nontreated cells; △———△ dissociated cells treated with 0.15 mg ml⁻¹ of trypsin for 5 s at room temperature; □———□ nondissociated embryos. The dissociated cells were prevented from reaggregating by overdilution. *Abscissa* time after fertilization. (Vittorelli et al., 1973)

to what the LETS proteins do in other systems, which is in agreement with the demonstration of fibronectin, by immunofluorescence, on the cell surfaces and between the cells of *Sphaerechinus granularis* embryos (E. Spiegel *et al.* 1980). However, Di Liegro *et al.* (1978) found that no histone synthesis is elicited by trypsin treatment under the above conditions, which makes it possible to dissect the regulatory effect of membrane proteins on DNA synthesis from that on other molecules which are usually synthesized before mitosis.

The dependence of DNA synthesis upon cell contact varies with development, being low at 32 blastomeres, very high at early blastula, and decreasing again at gastrula (De Petrocellis and Vittorelli, 1975): therefore, the highest sensitivity to cell dissociation is in the periods in which the embryo synthesizes DNA at a very low rate, thus indicating that the lack of further intercellular contacts, such as those brought about by the gastrulation process, prevents the cells from entering a new phase of DNA synthesis. In agreement with this interpretation of the data are the results of Scarano *et al.* (1964), and of De Petrocellis and Vittorelli (1975). They found that deoxycytidylate aminohydrolase activity remains frozen at the level of the stage of dissociation, if cells do not reaggregate.

Much more complex and difficult to explain are the results of the same authors of the effect of cell dissociation on other enzymatic activities, such as thymidine kinase, thymidilate kinase, DNAase and DNApolymerase, all related to DNA synthesis.

Another metabolic and morphogenetic phenomenon that is regulated by cell interactions is ciliogenesis. Amemiya (1971) has found that if swimming blastulae are chemically deciliated and then dissociated into cells, ciliogenesis will occur again if the cells are reaggregated, but not if prevented from reaggregating.

Finally, the reverse question can be asked: How relevant is metabolism for cell interactions? Experiments of Giudice (1965) show that reaggregation does not occur at low temperatures, or in the presence of general metabolic inhibitors such as sodium azide and dinitrophenol. Nor does reaggregation occur in the absence of external K^+, Ca^{2+} or Mg^{2+}, which are metabolically important ions. Inhibition of protein synthesis permits the reaggregation of sea urchin cells until it causes an inhibition of oxygen consumption, at which point the aggregating cells disaggregate again. It is relevant under this respect to mention that Watanabe *et al.* (1982) distinguish two types of cell adhesion in sea urchins, one from cell to cell, which leads to the formation of functional aggregates and which is inhibited by inhibitors of the protein synthesis, and the other from the cells to the substrata, which is independent of the protein synthesis, although sensitive to protease treatment; the extraction with butanol prevents only the first type of adhesion.

We cannot close the chapter on cell interactions without mentioning two extremely interesting original approaches. One is that used by Bennett and Mazia (1981 a, b), who succeeded in fusing eggs of different species and unfertilized with fertilized eggs, thus studying the effects of either cytoplasm on the state of the surface of the other egg. It is hoped that this promising approach can be extended to other developmental stages.

The other interesting, although preliminary, approach is that of Kew *et al.* (1979), who were able to study the phagosome membranes of embryos at different developmental stages by isolating them by the use of latex microbeads.

Chapter 3

Energy Metabolism

3.1 Oxygen Uptake

Warburg (1908) was the first to describe a burst of oxygen consumption following fertilization of *Arbacia* eggs. This observation was followed by a number of investigations (see Giudice, 1973, and Yanagisawa, 1975a, b for reviews), which confirmed this discovery. Recent studies have often employed the oxygen electrode (Nakazawa *et al.*, 1970), which avoids errors due to CO_2 displacement from sea water. The burst of oxygen consumption occurs within 1 min after fertilization, reaching, in two steps, values about 15 times higher than those of the unfertilized eggs. Oxygen consumption then decreases and reaches a plateau at a variable value generally a few times higher than that in the unfertilized egg (Fig. 3.1).

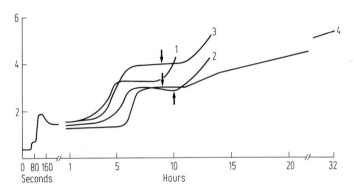

Fig. 3.1. Rate of oxygen uptake by developing sea urchin embryos. The curves are redrawn from the data of several authors: *1* from Lindahl, 1939; *2* from Giudice et al., 1968; *3* from Borei, 1948; *4* from Immers and Runnström, 1960. The curve relative to the first 200 s, following fertilization is redrawn from Epel (1964b). *Arrows* point to the moment of hatching. (Giudice, 1973)

What causes such a burst? It is concomitant but not related to membrane elevation, because the latter can be inhibited and the respiratory burst still occurs (Nakazawa *et al.*, 1970). Nor is it necessary for the activation of DNA or protein synthesis, because these also occur when eggs are activated by ammonia, which does not cause the respiratory burst. It temporally follows the increased concentration of intracellular free Ca^{2+} and it is at least in part related to the increase in carbohydrate metabolism. A great deal of attention has therefore been dedicated in the past to try to find out what causes the increase in carbohydrate metabolism following fertilization, and therefore we shall come back to this problem under Section 3.2. However, 30 % of the egg respiration is cyanide-insensitive and this increases following fertilization to 50 %. This part of nonmitochondrial activation of respiration might represent the consequence of a calcium-stimulated activation of a lipoxygenase, as suggested by the activation of oxydation of molecules like arachidonic acid (Perry and Epel, 1977; Epel, 1978). The picture is complicated in pigmented eggs like those of *Arbacia* by the oxydation of echinochrome which is stimulated by Ca^{2+}, at least in vitro (Perry and Epel, 1975; Perry, 1977).

Following the respiratory burst of fertilization, there are other changes in oxygen consumption during development, as depicted in Fig. 3.1. As observed by Monroy and Maggio (1963), the oxygen consumption roughly parallels the increase in cell number per embryo, which might account for the higher energy demand. When development proceeds with the translation of primarily maternal messages, these variations in oxygen consumption are actinomycin D-insensitive; but after hatching, when translation utilizes newly synthesized messages, the changes in oxygen consumption are blocked by actinomycin D (Giudice et al., 1968). Earlier observations of a parallel increase in the number of mitochondria should be abandoned, because recent electron microscope measurements demonstrate that it does not change from the blastula to the prism stage (Bresch, 1978). In agreement with the ultrastructural observations, Matsumoto et al., (1974), as well Rinaldi et al., (1979b), did not find any synthesis of mitochondrial DNA from fertilization until the pluteus stage.

The existence of cyclic variations in respiratory activity during early development has been a debated question (see, e.g., Zeuthen, 1960; and Scholander et al., 1958). Løvtrup and Iverson, (1969), who reinvestigated the question by an automatic diver balance, confirmed the cyclic variations but failed to correlate them with the mitotic cycle.

Finally, the kind of metabolite preferred by the developing embryo for oxidation is still controversial. The overall picture emerging from the literature on the respiratory quotient speaks for a preferential utilization of carbohydrates in the unfertilized eggs, that starts to change progressively in favor of lipids after fertilization with an inversion of this proportion at hatching (see Isono, 1963a). Løvtrup-Rein and Løvtrup (1980) observe that the total lipid content of the egg is such-about 2 mg mg^{-1} N according to Isono and Isono (1975), with a decrease of 10–20 % during 24 h of development-as to account for all the energetic needs of the embryo; whereas the decrease in carbohydrate content would not indicate these as the main energetic source, also because some of them are utilized for synthetic purposes.

3.2 Carbohydrate Metabolism

Carbohydrates, irrespective of their utilization relative to other oxydizable metabolites, certainly represent an important energetic resource for the sea urchin embryos. Their metabolism has been studied in such detail that it deserves a separate discussion. The egg contains a storage of glycogen, the utilization of which is accelerated following fertilization according to many early reports (Örström and Lindberg, 1940; Blanchard, 1935; Hutchens et al., 1942; Lindberg, 1945; Cleland

and Rotschild, 1952a). The increased utilization is in agreement with the release of a glycogen phosphorylase from a particulate fraction (Bergami et al., 1968) and with the activation of this enzyme described by Yasumasu et al., (1973, 1975) and by Hino et al. (1978). This activation might be brought about by the increase in Ca^{2+} ions, and at least for some of the enzyme, by cyclic AMP (Shoger et al., 1973). The activation of phosphorylase can therefore be considered to be the first stage in the post-fertilization activation of glycolysis. The finding of Goudsmith (1972) that glycogen content is constant until the gastrula stage are, however, incompatible with the above results, unless one postulate a glycogen turnover as suggested by ^{14}C incorporation. In view of these conflicting results, Hino and Yasumasu (1979) have more recently revived the question and found that different sea urchin species behave differently with respect to glycogen utilization: for example, it was found that the level of glycogen remains constant from fertilization to swimming blastula and then decreases sharply at early gastrula in *Pseudocentrotus depressus* and *Anthocidaris crassispina*, whereas it decreases slowly from fertilization to mesenchyme blastula and more rapidly during gastrulation in *Clypeaster japonicus*, in *Hemicentrotus pulcherrimus* and *Mespilia globulus*. The amount of glycogen utilized accounts for 30, 62 and 67% of the oxygen consumption up to morula in the three last species respectively, and only for 0–4% in the first two. These authors, however, also acknowledge that these data should be corrected for the active glycogen synthesis that demonstratedly occurs in the early sea urchin development.

It is interesting that glycogenolysis is stimulated not only by increase in cyclic AMP, which occurs after fertilization, but also by many glycolytic intermediates, thus suggesting a possible "feed back" regulation, (Fujino and Yasumasu, 1975).

Glucose-6-phosphate is generated through two different pathways according to Hino et al., (1978): glucosidase and hexokinase in all the examined Japanese species and also by the combined action of phosphorylase and phosphoglucomutase in all these species except *Anthocidaris crassispina* and *Pseudocentrotus depressus*, probably because these two species present a low level of acid-soluble glycogen; most of their glycogen being protein-bound and therefore not utilized by these enzymes.

That carbohydrate can be oxydated via the glycolytic pathway in sea urchin eggs was first demonstrated by Cleland and Rotschild (1952a,) and by Krahl et al., (1954). It was, however, only in 1964 that Aketa et al., tried to explain the mechanism of glycolysis activation that follows fertilization. They measured the levels of carbohydrate phosphate esters, in a search for a rate-limiting reaction before fertilization. The results of their study supported an activation of glycogenphosphorylase at fertilization. A more detailed analysis of the amount of glycolytic intermediates and adenine nucleotides in eggs and embryos (see Tables 3.1 and 3.2), followed by calculations of mass action ratios and free energy changes for each enzymatic spep of glycolysis, was undertaken by Yasumasu et al., (1973). They concluded that, besides glycogenphosphorylase, pyruvate kinase was activated at fertilization, and 20 min later, phosphofructokinase was also activated. In a further investigation, Yasumasu et al., (1975) found that, upon fertilization, glucose-6-phosphate deydrogenase and some aldolase activity are transferred from a particulate fraction to the soluble cytoplasm. A decrease in the activity of fructose 1,6 biphosphatase has also been described during the first 30 min after fertilization, as a consequence probably of two facts, one the increase of the intracellular free calcium concentration, and two,

88pon't88

88888

88pon't

Table 3.1

Expt No....	Unfertilized	Time after fertilization				
		10 min	20 min	30 min	40 min	60 min
	6	5	5	5	4	5
G6P	11.6 (0.5)	45.2 (1.0)	40.6 (2.4)	32.4 (3.1)	30.6 (5.0)	21.3 (3.5)
F6P	2.1 (0.2)	8.6 (1.0)	7.4 (1.0)	6.2 (1.8)	6.8 (1.1)	4.0 (0.9)
FDP	3.6 (0.2)	3.9 (0.7)	8.9 (2.1)	9.5 (1.3)	9.0 (2.0)	7.8 (1.1)
DHAP	4.4 (0.5)	4.3 (0.5)	9.8 (2.6)	12.3 (2.8)	14.9 (1.0)	9.4 (1.9)
GA3P	2.1 (0.3)	2.2 (0.3)	3.4 (0.4)	3.6 (0.3)	3.9 (0.3)	2.7 (1.2)
3PG	2.0 (0.6)	2.5 (0.3)	6.6 (1.5)	8.9 (2.3)	7.8 (1.6)	5.0 (0.8)
2PG	1.0 (0.8)	0.7 (0.3)	1.2 (0.1)	1.3 (0.4)	1.2 (0.4)	1.0 (0.3)
PEP	27.3 (4.6)	23.1 (4.0)	24.2 (6.6)	23.9 (6.2)	21.7 (5.1)	12.9 (0.4)
Pyr	9.1 (1.7)	29.4 (3.7)	60.3 (10	50 (13)	50.2 (7.6)	79.1 (10)
Lac	289.3 (51)	280.4 (46)	245.6 (33)	206.2 (22)	189.2 (21)	128.3 (14)
αGP	0.6		2.4			
6PG^a	0.7		2.3			
Pi*	43 (16)	39 (11)	37 (10)	44 (16)	43 (14)	47 (12)
ATP*	1.72 (0.21)	1.74 (0.24)	1.79 (0.15)	1.53 (0.21)	1.47 (0.12)	1.06 (0.13)
ADP	123.5 (23)	186.2 (34)	255.3 (32)	266.4 (30)	252.0 (20)	236.2 (14)
AMP	71.0 (15)	98.3 (14)	99.8 (14)	98.6 (11)	95.4 (12)	77.5 (21)

(from Yasumaso et al., 1973)

Table 3.2

	Time after fertilization			
	3 h	5 h	8 h	20 h
Expt No. ...	3	3	3	3
G6P.	22.1 (5.1)	26.3 (6.3)	34.2 (7.0)	32.9 (8.1)
F6P	2.6 (0.9)	2.7 (1.1)	3.2 (1.3)	3.4 (1.5)
FDP	7.5 (2.5)	6.9 (2.8)	6.6 (2.1)	20.7 (6.0)
DHAP	10.2 (2.2)	9.8 (3.0)	12.1 (3.4)	25.6 (7.1)
GA3P	2.9 (0.3)	3.3 (0.4)	4.6 (0.7)	8.2 (0.3)
3PG	50.9 (1.0)	4.8 (0.6)	2.7 (6.4)	10.9 (2.5)
2PG	1.2 (0.2)	1.1 (0.6)	2.6 (0.6)	2.9 (0.4)
PEP	12.2 (0.3)	11.6 (3.5)	23.2 (5.1)	28.4 (8.6)
Pyr	81.1 (9.8)	36.4 (3.7)	24.3 (4.2)	44.6 (7.3)
Lac	109.4 (16)	65.8 (11)	34.9 (7.0)	41.2 (10)
ATP*	1.01 (0.20)	1.05 (0.19)	1.10 (0.28)	0.93 (0.20)
ADP	228.4 (50.4)	221.6 (39.7)	204.3 (49.4)	296.2 (50)
AMP	79.3 (10.6)	97.6 (7.6)	104.4 (20.4)	195.6 (22.6)

(From Yasumasu et al., 1973)

the increase in the phosphofruktokinase activity, which generates an excess of substrate for the 1,6-biphosphatase, thus inhibiting its activity.

Linkage of glycolysis to the tricarboxylic acid cycle is blocked before fertilization in *Pseudocentrotus depressus* because of the low activity of pyruvate dehydrogenase, which is enhanced 5 min after fertilization, apparently as a consequence of a mitochondrial protein dephosphorylation which, in turn, is activated by the increase in the intracellular calcium concentration (Yasumasu, 1976).

Intermediate steps of the tricarboxylic acid cycle in sea urchin eggs have been described by many authors (Lindberg, 1943; Crane and Keltch 1949; Keltch et al., 1950; Cleland and Rothschild, 1952b; Krahl et al., 1942; Ycas, 1950, 1954; Goldinger and Barron, 1946). Aketa and Tomita (1958) and Aiello and Maggio (1961) provided the first direct evidence for oxidative phosphorylation by isolated egg mitochondria.

The pentose phosphate cycle is the predominant pathway of carbohydrate breakdown in at least some embryological stages of sea urchins. This was first suggested by Örström and Lindberg (1940) by Lindberg (1943, 1945) and later demonstrated by Krahl et al. (1954, 1955; Krahl, 1956). Bäckström et al., (1960) and Isono and Yasumasu (1968) subsequently investigated the relevance of this pathway throughout development and found that its activity undergoes an increase from fertilization up to gastrulation (or to hatching according to Bäckström) and then declines in favor of glycolysis and of the tricarboxylic acid cycle. Also Løvtrup-Rein and Løvtrup (1980) found a similar behavior of the shunt index in *Paracentrotus lividus*, although describing a continuous increase of the $^{14}CO_2$ produced from ^{14}C-l-D-glucose. The activation of the pentose cycle might be brought about by the release into the cytoplasm of glucose-6-phosphate dehydrogenase (Isono, 1963b; Isono et al., 1963; Yasumasu et al., 1975), whose activity can be regulated by the interplay between palmitol coA, and polyamines, which reverse such an inhibition. (Mita and Yasumasu, 1979). A sign for the shift to the pentose pathway at fertilization is the immediate generation of NADPH, within 1 min of sperm addition, which represents the very first detectable change in respiratory activity. This was shown by Epel (1964a, b), who confirmed and extended earlier reports by Krane and Crane (1960). The increase of NADPH is apparently generated through NAD phosphorylation accompanied by NADP reduction (which is in agreement with the activation of macromolecular biosyntheses occurring at fertilization) and is accompanied by an increase in NAD kinase activity, which seems to be indirectly activated by the increase in Ca^{2+}, but not by the increase in pH according to Wiley et al., (1977). In agreement with this hypothesis, Epel et al., (1981) found that those parthenogenetic agents which act through calcium mobilization do activate the NAD kinase; the others, such as ammonia, do not. These authors also provided evidence that NAD kinase is in vivo linked to calmodulin, which confers upon it sensitivity to calcium.

The cytochrome system has been demonstrated both in unfertilized and fertilized sea urchin eggs (see Rothschild, 1949 for a review of early results, and Borei and Björklund, 1953; Maggio and Ghiretti-Magaldi 1958; Ghiretti et al., 1958). Its involvement in the respiratory activation at fertilization has also been suggested: Maggio (1959) has found a 25% increase in cytochrome oxidase activity at fertilization, and Maggio and Monroy (1959) have found an inhibitor of cytochrome oxidase in the soluble cytoplasm of unfertilized eggs, which disappears following fertilization. A partial purification of this inhibitor (Maggio et al., 1960) has unfortunately never been followed up by further studies investigating its chemical nature and possible role in vivo.

Much work has been devoted to establishing the concentration of the adenine nucleotides before and after fertilization, with conflicting results. Among the most detailed analyses are those of Yanagisawa and Isono (1966) and Yasumasu et al., (1973), both of which demonstrate an increase in the ADP/ATP ratio at fertilization.

Conversely Innis et al., (1976) find a decrease of this ratio, which is consistent with the ultrastructural modifications of mitochondria found by the authors.

Additionally, a doubling of the arginine phosphate concentration following fertilization was reported by Yanagisawa (1968), who also found a decrease in the concentration of this energy storage compound from the 16-cell stage up to the gastrula stage.

Many inhibitors of carbohydrate metabolism have been studied in an attempt to understand the key steps of energy metabolism in sea urchins. For example, the inhibitor DNP has been used to reveal the maximal electron transport capacity at different embryonic stages. Immers and Runnström (1960) were among the latest investigators to use this tool. They found that the increase in oxygen consumption following DNP administration is high in fertilized eggs, and the increase becomes progressively lower by the mesenchyme blastula. Furthermore, this increase is higher in animal halves than in vegetal halves (De Vincentiis et al., 1966).

We have already spoken about the respiratory metabolism of the sperm in the first chapter. We shall only add here that W. A. Anderson and Perotti (1975) have demonstrated that sperm mitochondria remain active within the fertilized eggs until at least the 8-cell stage; and shall recall that the respiratory activity of the sperm is decreased following the acrosomal reaction (Kinsey et al., 1979).

Part II

Nucleic Acids and Proteins

This subject has been divided for the sake of order into three different sections: DNA (Chap. 4), RNA (Chap 5) and proteins (Chap 5). Since the function of DNA is that of providing the information for making proteins, these subjects are necessarily intermingled. The reader should, therefore, go through all of them before he can receive complete information on each one.

A fourth Section on nucleo-mitochondrial interactions (Chap. 7) has been added to this part because the kind of interaction treated here refers to the synthesis of nucleic acids and proteins.

Chapter 4

Deoxyribonucleic Acid

4.1 Organization of the Genome

Because of the ease of its extraction, DNA from sea urchin sperm has been used for many important studies. Indeed, Chargaff, used DNA from sea urchin sperms to establish his rules on the equivalence of A + G and T + C (see Chargaff and Davidson, 1955). It was also with sea urchin sperm DNA that Britten and Kone (1968) established some of their evidence for the existence of repetitive sequences. A series of subsequent papers have illuminated the general organization of the sea urchin genome. Weinblum et al., (1973) analyzed the kinetics of DNA reannealing and concluded that up to 19% of *Sphaerechinus granularis* DNA contains sequences repeated 800 times, with a kinetic complexity of 1.8×10^5 nucleotide pairs (b.p.); 31% of DNA contains sequences repeated 100 times, with a kinetic complexity of 1.7×10^6 b.p.; 20% of the DNA contains sequences repeated 3 times, with a kinetic complexity of 5.2×10^8 b.p.; and 30% of the DNA contains unique sequences, with a kinetic complexity of 2.3×10^8 b.p. Further analyses (Graham et al., 1974) have established that about 50% of genome of *Strongylocentrotus purpuratus* contains unique sequences $1.2–1.5 \times 10^3$ b.p. long separated by repeated sequences 300–400 b.p. long; 20% is organized as above, but is made of longer sequences, the repeats of which have little homology with the shorter repeats (Eden *et al.*, 1977); 22% is made of unique sequences, with little or no repeats interspersed; and 6% is made of longer repeated sequences (Fig. 4.1). Most or all of the polyribosomal messenger RNA of the gastrula stage transcribed from unique sequences appears to be synthesized by those arranged in the form A of Fig. 4.1, which represents about 30% of the unique sequences of the genome (Davidson *et al.*, 1975).

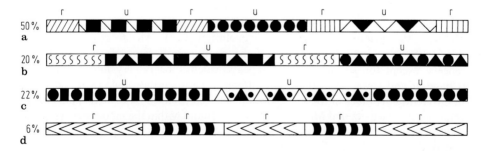

Fig. 4.1. Schematic representation of the genomic organization in *Strongylocentrotus purpuratus*, as essentially deduced from the data of Graham et al. (1974); *r* repetitive sequences; *u* unique sequences

These figures refer to the order of *Echinoida*; Yanagisawa and Amemiya (1982) have also reported an analysis of the genome complexity of the order of *Echinothurioida*, i.e., the most primitive among the subclass *Euechinoidea*. Values similar to those of *Strongylocentrotus* were obtained, but with about 30% of single copy sequences and with a fraction of fold back sequences (in *Aerosoma*) and of repetitive sequences (in *Asthenosoma*), larger than in *Strongylocentrotus*.

The possibility of cloning fragments of DNA (Scheller *et al.*, 1977) has permitted the confirmation of the above genomic structure (A. S. Lee *et al.*, 1977a). Furthermore, electron microscopic analysis of the base pairing within one single clone has demonstrated that the repeated sequences within one Eco RI fragment (of 7×10^3 b.p.) do not belong to the same family (Lee *et al.*, 1977b). These short interspersed repeats can occur from a few times to thousands of times and show little heterogeneity within each family (Klein *et al.*, 1978). D. M. Anderson *et al.*, (1981) have analyzed the distribution in the genome of *Strongylocentrotus purpuratus* of three specific repetitive families; They are found far apart in the genome, one is interspersed to longer repeats, a second and shorter one (200–300 b.p.) is generally flanked by single copy sequences, and the third one is within repetitive sequences in regions of the genome which are distant from each other.

The long repeats (about 6000 b.p.) are represented only 4–7 times in the haploid DNA and may contain sequences similar to those of the short repeats, according to Craig *et al.*, (1979). The analysis of the reannealing of long repeats experimentally fragmented to various lengths has revealed that they are composed of units of 150–300 b.p., which are similar in the different repeats and arranged in the same order (Chaudari and Craig, 1979b).

No single prominent family, of the type described in mammals, has been found among the dispersed repeats in sea urchins (Scheller *et al.*, 1981; see also Jelinek and Schmid, 1982, for a review).

What is the function of the interspersed repeats? If they have an important biological role one might expect their sequences to have undergone little change during evolution. It has actually been found that they are highly conserved in evolution: 0.1% divergence in 10^6 years according to Harpold and Craig (1977), especially those moderately repeated, which represents a low degree of divergence compared to the average of the nonrepetitive sequences, which deverge at a rate of 0.22% per 10^6 years except for a small fraction of these latter (14%), which is highly conserved (Harpold and Craig, 1978). Even in closely related species, the degree of repeatedness has been found to change, in a way that indicates that repeated sequences are being added to the genome in the course of evolution (G. P. Moore *et al.*, 1978). No differences have been found between the evolutionary rates of short and long repeats, except for 1% of the long repeats which appears very highly conserved (Chaudari and Craig, 1979a). A 4% difference between the single copy sequences of different individuals of the same species has been detected by Britten *et al.*, (1978). Interestingly, Pearson and Morrow (1981) have found within the moderately repeated sequences one class, whose constant length and high fidelity of sequence suggest that it may duplicate like a transposable element. Does it play a role in the evolution? It is too early draw conclusions on this point.

The possible role of the repeated sequences will be discussed again in Chapters 5 and 6, to which the reader is referred.

4.2 Some Specific Loci

As has happened with other eukaryotic cells, the sea urchin genes coding for ribosomal RNA have been the first to be isolated. The first attempts, using simple CsCl gradients, date back to 1969 (Stafford and Guild). Later, differential denaturation followed by differential partition in polyethylene glycol-dextran were tried (Patterson and Stafford, 1970), as were CsCl gradients (Patterson and Stafford, 1971; Mishra, 1978), after a treatment with Sl nuclease of the partially denatured DNA (Joseph and Stafford, 1976). Later, Griffith and Humphreys (1979) cloned a fragment of ribosomal DNA in *E. coli*. The conclusion of such studies is that rDNA is repeated 250-fold per haploid genome in *Lytechinus variegatus* and 50-fold in *Tripneustes gratilla*. The average GC content is 61 %. An electron microscope analysis reported by Wilson *et al.*, (1976, 1977) shows a total length of 11.33 (± 0.64) $\times 10^3$ b.p., 5.36 (± 0.53) of which belong to the external spacer and 1.03 (± 0.24) to the spacer between the DNA sequences coding for the 26S and 18S ribosomal RNA's. It contains three sites for the Eco RI nuclease. A finer map of an rDNA fragment by 1.9 kb was provided by the use of four restriction enzymes by Bieber *et al.*, (1981), who identified a 270 b.p. fragment which contains the 5' end of the sequence of the rRNA precursor. Finally Passananti *et al.*, (1983) have analyzed two ribosomal DNA clones from *Paracentrotus lividus* sperms (see Fig. 4.2), which together cover the entire ribosomal repeating unit, and found that there is a marked individual variability in the sequence of the spacers (up to 200 base pairs).

Based on the assumption that the eye forms represent the replication origins, Botchan and Dayton (1982) suggest that these are located within the nontranscribed spacer in *Lytechinus*. Indirect evidence by in situ hybridization experiments suggests some rDNA amplification during oogenesis (Sconzo *et al.*, 1978).

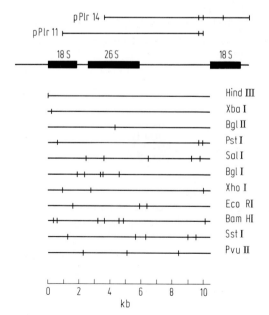

Fig. 4.2. Map of the repeating unit of the *Paracentrotus lividus* rDNA clones pPlr11 and pPlr14 with restriction endonucleases analysis. Southern hybridization with 18S and 26S rRNA and S1 protection mapping. The drawings, indicated as pPlr11 and pPlr14, show the map position and the PstI recognition sites of the inserts contained in the recombinant plasmids. (Passananti et al., 1983)

Region   A
  TTTCTATGCAGCCTGCCCCTCCTTCTAATCCAATTCCATTTGTTGGTATCCGTTCTTCTTCTCGTT
  AAAGATACCTCGGACGGGGAGGAAGATTAGGTTAAGGTAAACAACCATAGGCAAGAAGAAGAGCAA

  CGATGTCCGCCAGTAAAGGCAATGTTTGCGAGCTCAAGGTCTCGGGACGAGG
  GCTACAGGCGGTCATTTCCGTTACAAACGCTCGAGTTCCAGAGCCCTGCTCC
Region   B   Alu I
  TCAGAAAAATGACGTAATGGTGTTAAAAGGCAAAACTGCAGGGCTAGATAACATATAGAAATGCCA
  AGTCTTTTTACTGCATTACCACAATTTTCCGTTTTGACGTCCCGATCTATTGTATATCTTTACGGT

  ACGAATACATCTTTCGACACCATGCCGTCCCGACGCCAAATGAGAATGCTGCCATCTTGCGGCATA
  TGCTTATGTAGAAAGCTGTGGTACGGCAGGGCTGCGGTTTACTCTTACGACGGTAGAACGCCGTAT

  CGTTTTXXGGCCATTTGACTCAAAAAAAAAAATCTGAACTGTGAGGTCCCACTTCTCACCTGCGTCT
  GCAAAAXXCCGGTAAACTGAGTTTTTTTTTTAGACTTGACACTCCAGGCTGAAGAGTGGACGCAGA
          Hae III Hinf I
  GTGATATGTTACAGAAATGAAGCCCAGAAAAAAATGCGCCCGAAAATCGGCAACCAGTGCT
  CACTATACAATGTCTTTACTTCGGGTCTTTTTTTACGCGGGCTTDTAGCCGTTGGTCACGA
Region   C   Hha I
  Hinf I   -250   -240   -230   -220   -210   -200
  GAXTCAAAAGCATTTCTTTTGACTTTTGGATTTCTTATGGTTTTAGAGTTTTTATATAGTATGAT
  CTXAGTTTTCGTAAAGAAAACTGAAAACCTAAAGAATACCAAAATCTCAAAAATATATCATACTA
      -190   -180   -170   -160   -150   -140   -130
  AGAAAAAAATCCATGTACATATTTCATCTCTCCTCGTTCTTTGGTTTGCTTTAAAAGAAAATTAC
  TCTTTTTTTAGGTACATGTATAAAGTAGAGAGGAGCAAGAAACCAAACGAAATTTTCTTTTAATG
      -120   -110   -100   -90
  ATATGTCTATGACGCTAACTTATGAAATGACTGAAGACATCAATC
  TATACAGATACTGCGATTGAATACTTTACTGACTTCTGTAGTTAG
Region   D   Mbo II
  -80   -70   -60   -50   -40   -30 Alu I -20
  AACGAATAACTTCCAGGGATTTATAAGCCGATGACGTCATAACATCCCTGACCCTTTAAATAGCT
  TTGCTTATTGAAGGTCCCTAAATATTCGGCTACTGCAGTATTGTAGGGACTGGGAAATTTATCGA
      -10   1   10   20   30   40
  TAACTTTCATCAAGCAAGAGCCTACGACCATACCATGCTGAATATACCGGTTCTCGTCCGATCAC
  ATTGAAAGTAGTTCGTTCTCGGATGCTGGTATGGTACGACTTATATGGCCAAGAGCAGGCTAGTG
      50   60   70   80   90 Hpa II 100   110
  CGAAGTCAAGCAGCATAGGGCTCGGTTAGTACTTGGATGGGAGACCGCCTGGGAATACCGGGTGT
  GCTTCAGTTCGTCGTATCCCCAGCCAATCATGAACCTACCCTCTGGCGGACCCTTATGGCCCACA
      120   +10   +20   +30   +40   Hpa II
  TGTAGGCTTTTTTTTCTCCCCCCCCCCCCTCTTTGCTTCATGAAATGCCTCTT
  ACATCCGAAAAAAAAGAGGGGGGGGGGGGAGAAACGAAGTACTTTACGGAGAA
Region   E
  +50 Mbo II +60   +70   +80   +90   +100   +
  TGTTCATTTCTTCGTTTAAAGTATGCTTTCCTAATAGTATCACTTTTGCCTCTATTCTTTTCTCG
  ACAAGTAAAGAACGAAATTTCATACGAAAGGATTATCATAGTGAAAACGGAGATAAGAAAAGAGC
  110   +120   +130   +140   +150   +160   +170
  TCTCCTTCTTGATAGAGCAGGGGAAAAGGGAAAAGAAAAGAAAAAGAGAGGAAACTATCGTAACA
  AGAGGAAGAACTATCTCGTCCCCTTTTCCCTTTTCTTTTCTTTTCTCTCCTTTGATAGCATTGT
  +180   +190   +200   +210   +220   +230
  AATAAAGGGTTTTCTTGTTTACCCGTTCCTTTTTTGTTGTCTGCATGATGGAAAATGAAXAGCT
  TTATTTCCCAAAAGAACAAATGGGCAAGGAAAAAAACAACAGACGTACTACCTTTTACTTXTCGA
Region   F   Alu I
  GTCTCTTTTTTGGTCTGTTTCTTTGGTCGCTTGTTCATTCTATCGTTAATTTGGACGTCCACTCA
  CAGAGAAAAAACCAGACAAAGAAACCAGCCAACAAGTAAGATAGCAATTAAACCTGCAGGTGAGT
       Eco RI*Hpa II   Hae III
  ATACTCAATAAAATTCCGGATGAGCATTCATCAGGCGGGCAAGAATGTGAATAAAGGCCGGATAA
  TATGAGTTATTTTAAGGCCTACTCGTAAGTAGTCCGCCCGTTCTTACACTTATTTCCGGCCTATT
  5S rDNA ←————————|————→ pACYC 184

Fig. 4.3. DNA sequence of an *L. variegatus* 5S rRNA gene. The sequence is divided into six regions (*A–F*). Both strands of region *D*, which contains the coding sequence for 5S rRNA, were sequenced. The coding region for 5S rRNA is numbered *1–120*, and the *plus* and *minus signs* before the numbers represent the number of nucleotides before and beyond the coding region, respectively. *Xs* denotes unknown nucleotides. The Eco R1 site at the right junction between the 5S rDNA insert and pACYC 184 is shown in region *F*. (Lu and Stafford, 1982)

The best information about the 5S rDNA comes from the work of Lu *et al.*, (1980), who were able to clone and to sequence an Eco RI fragment, then (Lu *et al.*, 1981; Lu and Stafford, 1982) to analyse by means of four restriction enzymes four clones of 5S DNA, and finally to obtain the sequence of an entire 5S rDNA gene plus its spacer regions (Fig. 4.3). It is interesting to note that the sea urchin 5S rDNA shows a high degree of homology with that of other eukaryotes, for its coding region, but not for its flanking sequences. Here, too, however, as in other eukaryotes, a cluster of Ts, preceded by a relatively short GC-rich region is present in the 3'-noncoding region. These sequence studies were indicative of the existence of two families of 5S rDNA adjacent to each other but not interspersed. Are they both expressed at the same time, or at different moments of development, as in amphibians? This question is at present unanswered.

More information in this respect is available for the structural genes of actin. The analysis of these genes was initially facilitated by the possibility of cross-hybridizing them with a probe available from *Drosophila* (Durica *et al.*, 1979, 1980) and by the extraordinary abundance of the actin messenger RNA in the adult sea urchin tube feet, about 80% of the total mRNA, which made it easy to prepare a homologous cDNA (Kabat-Zinn and Singer, 1981). With these probes it was possible to isolate from a DNA library of *Strongylocentrotus purpuratus* four different actin genes (Overbeek *et al.*, 1981); they appear clustered at an electron microscope analysis of the heteroduplexes with the same orientation, 8 kb spaced and with two introns. Three of these genes have different restriction maps and the fourth one seems to be an allele; the hybridization of the clone with HpaII fragments reveals the existence of 5 different actin genes; most conspicuous differences, however, are found between the regions flanking the actin genes. Similar results have been reported by Durica *et al.*, (1980), who evaluated the number of actin genes to 5–20 per haploid genome, and spoke of an introm of about 200 nucleotides revealed by the partial sequencing of an actin clone, as also described by Schuler and E. B. Keller (1981) and By Schuler *et al.*, (1983), who sequenced two linked actin genes of sea urchins. Schuler and E. B. Keller (1981) suggested that actin genes might have evolved from primitive genes made up of 6 exons and 5 introns. The developmental control of the synthesis of the different forms of actin, however, seems to lie more at a translational than at a transcriptional level (Infante and Heilmann, 1981), as will be described later. It is interesting that the location of the intervening sequences of the actin genes in sea urchin is the same as in mammals and birds, and different from in *Drosophila*, i.e., in a protostomial invertebrate (Davidson *et al.*, 1982).

Ruderman and Alexandraki (1983) have also studied the genes for tubulin both on clones selected from a genomic library cloned in phage λ and on cDNA clones. The main conclusions of their studies are: first, that the gene copy number, estimated by "dot-blot" hybridization, using a cDNA probe, is of about 10–15 copies of α-tubulin sequences and 10–15 copies of β-tubulin sequences per genome; second, that the sequences for α- and β-tubulins are, for the most part, not organized as tightly clustered pairs, although some clustering of genes within each family has already been observed (Alexandraki and Ruderman, 1981); third that different genes appear to be expressed at different developmental stages.

The availability of genomic libraries now allows the isolation and the study of several other genes; however, the most detailed information available is still for the genes coding for histone proteins (see for review Kedes, 1976, 1979; Hentschel and Birnstiel, 1981), which were the first structural genes to be cloned in bacteria. The study of sea urchin histone genes initially stemmed from the knowledge that the most abundant class of messenger RNA synthesized during cleavage in histone mRNA (Kedes and P. R. Gross, 1969b). This synthesis accompanies the intense nuclear DNA synthesis of this development stage. As a result of its abundance, it was relatively easy to isolate a 9S RNA, highly enriched in histone mRNA, from polyribosomes.

It was then possible to show that this RNA specifically hybridized to a DNA with the G + C content and the molecular weight that one would expect of histone genes (Kedes and Birnstiel, 1971). The kinetics of hybridization indicated that histone genes are repeated about 1000 times each, although more recent figures give a repetition of 460 times in sperms and plutei of *Lytechinus pictus* (Marco *et al.*, 1977). Further electrophoretic subfractionation of the 9S RNA permitted more refined RNA–DNA hybridization experiments, which showed a degree of homogeneity between the different sets of histone genes higher than what would be required to maintain the observed histone protein conservation. This suggests the presence of a correction mechanism (Weinberg *et al.*, 1972; Farquhar and McCarthy, 1973). The technique for purifying histone genes made a new advancement with the use of actinomycin, which enhanced the buoyant density difference between histone DNA and bulk DNA. This made possible the discovery that between the single genes for the different histones, all of which have a high G + C content, there is a spacer of low G + C (54% and 37% G + C respectively, according to Birnstiel et al., 1974). The next important step came from the use of restriction enzymes, which made it possible to start an analysis of the histone gene positions within the respective units coding for all five histones (Weinberg *et al.*, 1975). Kedes (Kedes *et al.*, 1975a) first succeeded in cloning the sea urchin DNA in *E. coli*, and isolating the purified sea urchin histone genes from the bacteria. This permitted more detailed analyses of the organization of the histone genes using a variety of techniques. For example, new analyses with restriction enzymes were immediately started on cloned histone genes (Kedes *et al.*, 1975b; Cohn *et al.*, 1975). Also, improvement of the electrophoretic analysis made it possible to purify the m-RNA for each histone protein (K. Gross *et al.*, 1976a), which permitted clarification of the arrangement of three of the five cloned histone genes and their spacers (Schaffner *et al.*, 1976) as well as to establish their polarity (K. Gross *et al.*, 1976b). A denaturation map viewed in the electron microscope confirmed the alternating of structural genes and spacers (Portman *et al.*, 1976; Wu *et al.*, 1976) and the R-loop technique showed the position of the restriction enzyme sites in cloned histone genes (Holmes *et al.*, 1977). The combined use of restriction enzymes and exonucleases also showed the reciprocal positions of all five histone genes and that they all lie on the same DNA strand (Cohn *et al.*, 1976). The sum of these analyses generated the picture given in Fig. 4.4. Finally, the development of new sequencing techniques gave the start to the sequence analysis of histone gene clones. At first only about 100 base pairs were sequenced (Sures *et al.*, 1976), then 2/3 of a gene (Birnstiel *et al.*, 1977), about 500 nucleotides at the 5′ terminal (Grunstein and Grunstein, 1977), and then 3 entire genes and spacers (Sures *et al.*, 1978) and about 3000 base pairs (Schaffner *et al.*, 1978), and finally the complete sequence of an entire histone gene repeat, as mentioned by Bryan *et al.* (1983). These analyses allowed

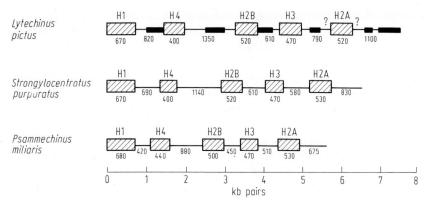

Fig. 4.4. Molecular maps of histone gene repeat units of three sea urchin species. The histone gene organizational maps are calibrated in base pairs. The mRNA coding regions, including their leader and trailer sequences, are depicted by the *hatched boxes*. Spacer DNA sequences are represented by the *lines connecting the boxes*. Regions of nonhomology, drawn to scale, are depicted by the *thick lines* in spacer regions. The "?" symbols around the *L. pictus* H2A gene signify unexamined regions. (Kedes, 1979)

some important conclusions: (1) there are no intervening sequences, or histone precursor proteins; (2) there is a heterogeneity in spacers and also in length and restriction sites of the different gene repeat units (Overton and Weinberg, 1978); (3) the sequence (CT/GA) × 25 is present in the spacer between H2A and H1, where, according to Hentschel (1982), a homocopolymer made by 5′ (GA) × 16 is contained, which is single-stranded and might represent a *focus* of recombinational events, important for the evolution of the histone genes; (4) there is an extensive evolutionary conservation especially of some of the sequences immediately preceding and following the coding regions; (5) there is more homology between the histone H4 sequences expressed in the oocyte and during cleavage than between the latter and those expressed after the blastula; these differences lie both in the transcribed and untranscribed parts of the genes (Grunstein *et al.*, 1981); (6) a sequence is presente close to the 3′ end of H1, H3 and H4 gene, and another close to the 3′ and of H2A and H2B, which, if transcribed, would generate a hairpin loop in the corresponding mRNA's (Busslinger *et al.*, 1979). This, in Busslinger's opinion, might serve as a recognition signal for a regulatory protein. As will be described in greater detail in Chapter 5, the terminal RNA stem-loop structure and 80 base pairs of the spacer DNA are necessary for the correct termination of the mRNA for the H2A histone, and a sequence at −165 to −111 nucleotides before the H2A histone mRNA cap site is especially important for gene expression (Grosscheld *et al.*, 1983).

Interestingly enough, Kedes (1979) has reported the existence in *Lytechinus pictus* individuals of two major classes of nonallelic histone gene repeat units. A map of homologies in coding and spacer DNA, as well as the polarity and gene organization of such nonallelic genes were later described (Cohn and Kedes, 1979 a, b). Busslinger *et al.*, (1982) have now reported a peculiar evolutionary behavior stemming from the analysis of histone gene sequences: A nontranscribed spacer of the clone h19 of *Psammechinus miliaris* DNA evolves 100–200 times mare slowly than the other spacers and the structural genes for H3 and H4 histones, which evolve in their silent sites (generally the third base of the codon) at the same rate as the single copy protein, i.e., 0.5–0.7% bases/million years. This discovery has also elicited excitement as one possible indication of horizontal gene transfer (see Lewin, 1982b, for a stimulating comment).

The picture that emerges becomes more interesting every day, and recently Childs et al. (1982) have discovered that those genes coding for H3 and H4 which are expressed late in development, although clustered, are not tandemly linked, and show a higher variability of the nontranscribed spacer. Maxon et al. (1983), moreover, have found that the late H2B histone genes are neither clustered nor tandemly arrayed, and present in the amount of 5–12 copies per genome in Strongylocentrotus purpuratus. The clustered arrangement of the histone genes expressed early in development is therefore not shared by those histone genes that are expressed later. Whether this is of some significance for the regulation of gene activity is not known at present. We can only note the fact that the early histone genes are expressed when the mitotic activity is very high, while the late ones are expressed when this is much lower.

In yet another instance, the sea urchin has presented the biological material for an important discovery in molecular biology, that of the "orphons": Childs et al., (1981), by cutting the Lythechinus pictus DNA with the restriction enzyme Bam H1, which does not cut inside the histone gene family, discovered that more than 50 single histone genes, which were named orphons, are not clustered with the anothers, but lie isolated and scattered in the genome. This observation, together with that of the homo copolymer sequences of the spacers (Hentschel, 1982), and of the transposon-like sequences of Pearson and Morrow (1981), promises to offer a clue to the problem of gene evolution in eukaryotes.

It is of interest in this respect that Vitelli and Weinberg (1983) reported that the breakpoints between two histone genes found in inverted position in Strongylocentrotus purpuratus, contain short direct repeats, and lie several bases proximal to the cap sites of the two different histone genes. Moreover, Liebermann et al. (1983) have found within a histone H2B pseudogene of the same sea urchin species a 3-kb DNA segment, called TU1, with the characteristics of the transposable elements described in Drosophila. TU1 is divided in to an inner domain made of nonrepeated sequences with two long terminal inverted repeats of about 840 b.p. and an outer domain made of 15 b.p. tandem repeats. The sequence CCATGGTC of the H2B pseudogenes is found at both sides of TU1.

An analysis of the distribution of TU1 within the genome suggests that about 200–400 copies of TU1 are present in a sea urchin genome differently located in

Fig. 4.5a–d. Nucleotide sequences of selected regions of TU1. **a** The sequences of the outer domain (OD) of the left inverted repeat segment of the two IVR's are identical for at least the first 99 bases. These 5-base segments allow us to define and identify a left and right IVR as IVR_1 containing the sequence TCCAT, and IVR_r containing the sequence AACCC. **b** The canonical sequence of the internally repetitive outer domain of the IVR was derived by scoring the base assignment at every position shown. **c** The secondary structure formed by one strand of TU1 after it is allowed to snap back. The 8-bp repeat of the H2B-coding sequences is boxed and the potential base-paired and 5-base keyhole loop-out structures are shown. **d** Analogous sequences of the inner domains of IVR_1 and IVR_r were aligned by the IntelliGenetics implementation of the Smith-Waterman modification of the Needleman-Wunsch algorithm. (Liebermann et al., 1983)

IVR$_L$		1	2	3	4	5	6	7	8	9	10	11	12	13	14	15
		G	A	C	A	A/T	T	T	G	C	T	C	C	G	C	C
GGTtccat																
GACACATGCTCCGCC	G	28			1	2			27					28	1	
GACGATTGCTCCGCC	A		26		26	12	2		1		1				2	2
GACATTTACTCCGCAAAACGC	T		1			8	20	26			27				1	
GACAATTGCTCCGC	C			28		3	1	1		28		28	28		24	23
GACATTTGCTCCGC	N		1		1	3	.5	1								3
GACATTTGCTCCGAC							**b**									
GTCAAATGCTCCGCC																
GACAGTTGCTCCGCC																
GACAGTTGCTCCGCC									C	G						
GACAATTGCTCCGCC									C	G						
GACAATTGCTCCGCCTGAAGC									T	A						
GACAATTGCTCCGCC									C	G						
GACATTTGCTCCGCC									G	C						
GACAATTGCTCCGCC									T	A						
GACATTTGCTCCGCC GAAGC									A	T						
G CACCTGCTCCGCC									C	G						
GACATTTGCTCCGCC									A	T						
GACAA TGCTCCGCC									C	G						
GACAA CGCTCCGCCTGAAGC									A	T						
GACAA TGCTCCGCC									G	C						
GAC TGCTCCGCCTGAAGC									T	G						
GACAA TGCTCCGCC								A		G						
GACA TTGCTCCGCC								C		G						
GACAATTGCTCCGAC								C		T						
GACA TNGCTCCGCATGAAGC								T	T							
GACACTTGCACCGGNGT								T	A							
GACATTTGCTCCGTC								G	C							
GACATTTGCTCCGCCC								G	C							
AAAAAGTGTCCCATTAAATACAAATGCT...					...CCATGGTC			CCATGGTC ...								

a   **c**

```
                1129       1139       1147
    Left   IVR  AATATTTTTT TTTCACGGGT TGTT--TTTG
                 **     *  *** *** *  * **   *
           Right IVR  -ATATTTCTT ACGCCTANTT TTTTCGTTTA
  1156       1166       1176       1188       1196
TTTCTTTCT TTTTTTCTTC ATTTTTTGAA TGCATTTCAT GACATTATGT   L
 *  *   **   **      *          *  *   **
TTTATTT-T TTACTTCTTC ATTTTCTGAA TGCATTTCTT GAAATAATGT   R
  1206       1216       1226       1236       1246
AAATGAAATG AAATGAATGA ATGTTGATAA TGTGGGAGGA AAAGCAGTTT   L
 *                    *  *     *       *
AGNTGAAATG AAATGAATGA ATGTTGATGA AGTGGGTGGA AAAGGAGTTT   R
  1256       1266       1276       1286       1296
GCCTAAAAGT CATTTCGAAT ATACTGAAAT AACTTTCCAT CTATCACTGT   L
* *      *
ACCT-AAAGT TATTTCGAAT ATACTGAAAT AACTTTNCAT CTATCACTGT   R
  1306       1315       1325       1335       1345
ACGTCCCACA TAAT-AAACA CTAAACACTC ATTACATGTC ATGGGAAGTT   L
  *         ** **                        *         *
ACGTCCCATA TAATGTNGTA CTAAACACTC ATTACATGTC GTGGGAAG-T   R
  1355       1364       1372       1382
TTTTTTTTTA AAGTT-AGAG -CAGCTAAG CCACATCCCC CGYTG        L
 *****   *  **    * * *               *       *
TTTTAAAAAA ATGTTGTGNG TCTGC-NAG CCACATTCCC CGGTG        R
```
d

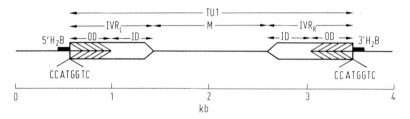

Fig. 4.6. Schematic summary diagram of the structure of TU1. *Large open arrows* represent the terminal long IVRs with their outer (*OD*) and inner (*ID*) domains demarcated. The middle segment (*M*) lies between the IVRs. The internally repetitive nature of the OD is represented by the *arrowheads*. The 8-bp direct repeat of the "host" DNA is indicated. The scale is in kb. (Liebermann et al., 1983)

different individuals, located in Eco RI DNA fragments of various size, and with a highly heterogeneous structure.

The hypothesis is appealing, although still only a conjecture, that transposable elements may also play a role in embryonic differentiation.

On the basis of the thermal stability of DNA-RNA hybrids obtained between a cRNA synthesized on a *Strongylocentrotus purpuratus* clone and the DNA from *Arbacia lixula* and *Paracentrotus lividus,* Deretic and Glisin (1982) found that the differences between the early histone genes of the closer species are larger than those of the more distant ones.

4.3 Chromatin

Whereas the sea urchin embryo has provided leading material for research on isolated genes, the research on sea urchin chromatin has followed the new technical break-throughs developed in other experimental materials. Quite a number of papers are, however, available now on this subject, the majority of which deal with the analysis of histones, which has stemmed from a double source of interest, first because of the possibility of isolating the specific mRNA's and genes, and second because of the early proposal of Hnilica and coworkers (Orengo and Hnilica, 1970; Hnilica and A. W. Johnson, 1970; A. W. Johnson and Hnilica, 1970, 1971) that true histones appear in the sea urchin chromatin only after the blastula stage. We believe that this concept needs to be modified now by saying that histones which are found in the nuclei till the blastula stage are different from those of later stages. Actually Ruderman and P. R. Gross (1974) demonstrated synthesis and incorporation into chromatin of all five major histones already at the morula stage, and Cohen et al., (1975) by the use of much more resolutive electrophoreses (i.e., including Triton × 100, beside acetic acid and urea) were able to show that only a specific variant, the so-called subtype of H2A and H2B is synthesized and incorporated into chromatin up to blastula, whereas variants β, γ and δ are synthesized thereafter. A closer analysis of the histones synthesized in the early cleavage (Newrock et al., 1978a, b) demonstrated that three different subtypes of H2A and H2B called Cs, are synthesized before the α-subtypes, during the second and the third S phases, and that H21 is the main subtype of H2, as revealed by the amino acid composition (Newrock et al., 1982). Herlands et al., (1982), however, find that the α-subtype is synthesized already before the second S phase, but incorporated into the nuclei only later. R. S. Wu et al., (1982) have described a further variant of H2A, named H2Z, which is continuously synthesized in all the stages examined by the authors, from morula to gastrula. That the Cs proteins are to be considered true histones is also strongly suggested by the work of Shaw et al., (1981), who were able to show their presence within the nucleosomal structure.

Also L. Levy and Moav (1982), by means of bidimensional electrophoreses, showed a shift of the synthesis of H1, H2A and H2B at the morula stage from a cleavage type to a new type.

Analyses of the histone electrophoretic pattern of the unfertilized egg nuclei (L. B. Evans and Ozaki, 1973; A. G. Carroll and Ozaki, 1979) show that they contain typical H3 and H4, but no bands with the electrophoretic mobility of gastrular H1, H2A and H2B; three bands with different electrophoretic mobility are found in their place, which is in agreement with the high conservation of H3 and H4 and the very low conservation of H1 during evolution.

If one looks at the histones of the zygote nucleus, one finds that they are of egg type and not of sperm type, thus indicating that the sperm does not contribute histone to the embryo (see also Imschenetzky et al., 1980). Sperm histones indeed show an electrophoretic pattern that is easily distinguished from that of egg and embryos (Easton and Chalkey, 1972; Ruiz-Carrillo and Palace, 1973; Cohen et al., 1975; A. G. Carrol and Ozaki, 1979; Geraci et al., 1979; L. Levy and Moav, 1982; Azorîn et al., 1983), the major differences from the embryonic pattern consisting in a slower migration of H1 and of H2B, which in sperm divides into two bands. The transitions of the histones of the male pronucleus to the egg type actually occurs before its fusion with the female pronucleus, as studied in detail by Poccia et al., (1981), who isolated the male pronuclei from Strongylocentrotus purpuratus eggs made polyspermic through NH_4Cl washing, and analyzed their histones by bidimensional electrophoresis, thus showing that already before sperm chromatin decondensation, H1 is replaced by CS1 and then the sperm type H2B by two histone spots named O and P; during the S phase CS2A, CS2B and a subspecies of H3 start to accumulate, so that by the beginning of chromosome condensation the histone array is CS2l, CS2A (+Sp2A), CS2B, O, P, 3 H3s and 4 H4s. All these substitutions occur even in the absence of protein synthesis, which suggests a maternal histone storage sufficient for 30–150 haploid nuclei, a figure well in agreement with that given by Cognetti et al., (1974, 1977a), who directly measured the rate of histone synthesis in the growing oocytes. Also Kuhn and Wilt (1980, 1981), by labeling the chromatin proteins in vivo with ^{14}C and in vitro with ^3H, by reductive methylation of the lysine NH_2 groups, concluded that about one half of the chromatin proteins derive from a maternal store.

Many authors have studied the developmental changes of nuclear histones after the blastula stage (Thaler et al., 1970; Benttinen and Comb, 1971; Voroboyev, 1969, Voroboyev et al., 1969a, b; Subirana, 1970; Subirana et al., 1970). The main changes concern H1 (Hill et al., 1971; Seale and Aronson, 1973b; Ruderman and P. R. Gross, 1974; Poccia and Hinegardner, 1975; Brookbank, 1978; Treigyte and Gineitis, 1979), so that two types of H1 have been distinguished, a maternal one, H1m, and a gastrular one H1g. To be precise, H1m is maternal in that it is synthesized on maternal messenger, albeit at the morula stage. Only quantitative variations can be detected for what concerns the other four main histones (Sevaljevic, 1974a), unless the electrophoresis is run in the presence of urea and Triton X 100 (Cohen *et al.*, 1975; Newrock *et al.*, 1978a; Geraci *et al.*, 1979; Treigyte and Gineitis, 1979; L. Levy and Moav, 1982), which permits the resolution of minor histone variants. It has also been shown by the analysis of the cell-free translational products of mRNA's extracted from embryos at different stages between early blastula and gastrula, that at least some of the variants of H1, H2A and H2B are due to differences in the primary structure (Weinberg *et al.*, 1978). This is true for the α-subtypes of H1, H2A and H2B, which are synthesized up to blastula and β-, γ- and δ-subtypes, synthesized thereafter. Also analyses by electrophoresis and by hybridization to selected probes of histone messenger RNA extracted from mesenchyme blastulae demonstrate the transcription of a new set of histone genes at this stage (Spinelli *et al.*, 1979; Childs *et al.*, 1979b). Since, however, it has been demonstrated that phosphorylation (Platz and Hnilica, 1973a; Garling and Hunt, 1977) and acetylation (Yukawa and Koshihara, 1973; Taylor and Burdick, 1975; Burdick and Taylor, 1976) of histones occur during development, it follows that some of the electrophoretic histone variants must arise from post-translational modifications.

More information and the regulation of the synthesis of histones will be discussed in Chapters 5 and 6.

It has been shown (Garling and Hunt, 1977) that the cytoplasm of the unfertilized egg is able to specifically catalyze phosphorylation of H1 and H2B sperm histones, which suggests a mechanism for removal of sperm histones in favor of egg histones from the zygote nucleus (see A.G. Carroll and Ozaki, 1979; Imschenetzky *et al.*, 1980; Poccia *et al.*, 1981).

Sperms have represented a suitable source of biological material for the study of chromatin proteins for a double reason: first, the ease with which uncontaminated nuclei can be obtained in good amount; second, the peculiar state of the chromatin of the sperm, inactive and in a state of compaction, from which it is awoken by fertilization, with mechanisms still under investigation and that must involve chromatin proteins. Several authors have therefore studies sperm histones: Puigdomenech *et al.*, (1980) have isolated from the tryptic digest of H1 from *Arbacia punctulata* sperm two peptides, named Gφ1 and Lφ1, which maintain the same structure to the circular dicroism and to NMR analysis, as observed in situ, thus suggesting that they represent the compact core of H1, with G1 in the innermost part, and suggesting a resemblance of H1 with the erythrocyte H5. This similarity is also stressed by Strickland *et al.* (1980a, b), who determined the primary structure of H1 from *Parechinus angolosus* sperms, and found a highly hydrophobic central part, with a COOH terminus very rich in lysine and alanine; the central part, but not the peripherical one, is highly conserved when compared to that of *Psammechinus miliaris*. Strickland *et al.* (1980c) have also determined the complete sequence of H2A from *Parechinus angolorus* sperms and found that the N-terminus is acetylated.

Interestingly De Petrocellis *et al.* (1980) found, beside a species specificity among different sea urchin species, and individual variability of H2B subclasses from sperm.

The other chromatin proteic component, i.e., the acidic proteins has also received some attention, because they can be expected to change during development if they play a role in the regulation of transcription. Some quantitative (Sevaljevic 1974a; Sevaljevic and Koviljka, 1975) but also qualitative (Cognetti *et al.*, 1972; Seale and Aronson, 1973a; Gineitis *et al.*, 1976a, b) changes have indeed been detected during early development. They might theoretically arise from synthesis of different proteins or from post-translational modification, since both synthesis (McClure and Hnilica, 1972; Sevaljevic, 1974b) and phosphorylation (Platz and Hnilica, 1973b) of nuclear acidic proteins have been described during development. Experiments on the effect of nonhistone chromosomal proteins on the binding of heterologous RNA polymerase to sea urchin DNA suggest a role of such proteins in preventing unproductive binding of this enzyme to template (Di Mauro *et al.*, 1979). More recently Katula (1983) has also studied the so-called "high mobility group" nonhistone chromosomal proteins (HMG) extracted from eggs and embryos of *Strongylocentrotus purpuratus*. This author found that only one major HMG, named P2, can be identified in sea urchin nuclei; this is present in two forms slightly different in charge; the less basic predominates in the egg and early developmental stages, whereas the more basic predominates in the post-blastular stages.

No HMG proteins have been found by Azorin *et al.* (1983) in the sperm of *Holothuria tubulosa*, which is in line with what one would expect if they play a role in activating transcription.

The nucleosomal arrangement of sea urchin chromatin has also been investigated: Spadafora (Spadafora and Geraci, 1975, 1976; Spadafora *et al.*, 1976a) demonstrated a typical nucleosomal arrangement of chromatin. This author also described a higher resistance of sperm chromatin to nuclease digestion due to its tight package. Spadafora *et al.* (1976b) also found that, whereas the repeat DNA unit of the embryonic chromatin is of about 210 base pairs, that of the sperm is of about 240 base pairs, probably due to the higher basicity of sperm H1 histone, which binds to a longer segment of the sperm nucleosomal spacer, thus protecting it more than in the embryos from the nuclease attack (see also Cornudella and Rocha, 1979 for a comparison of the mature sperm with the male gonads). These results were essentially confirmed by Arceci and P. R. Gross (1980b).

Chambers *et al.*, (1983) also reported a higher lenght of the nucleosomal repeat in the sperm than in embryos. Their results, obtained in *Strongylocentrotus purpuratus*, indicate a repeat of 250 b.p. in the sperm, which decreases to 182 ± 2 at two blastomeres to increase to 201 ± 2 by 8 blastomeres. A further increase is observed in early development, which does not, however, reach the values of the sperm.

Keichline and Wasserman (1977, 1979), found an increased resistance of the embryonic chromatin to nuclease attack as development proceeds from morula to pluteus. Albanese *et al.*, (1980) have found a very peculiar situation in the unfertilized egg chromatin: when treated with micrococcal nuclease, this does not originate any electrophoretic pattern characteristic of nucleosomal particles, even when much more drastic hydrolytic conditions are used than those necessary to originate such a pattern with the embryonic chromatin. Interestingly, this unusual property is not shared by oocytes, sperm or fertilized eggs, thus indicating the existence of some mechanism at chromatin level which keeps all the machinery for macromolecular syntheses present in the unfertilized egg inactive. More recent work (Albanese personal communication) has demonstrated, however, that when the DNA fragments purified from micrococcal nuclease-digested egg nuclei are analyzed by electrophoresis, rather then the DNP particles, they exhibit a typical nucleosomal pattern. These studies also show the presence of some heterogeneity in the structural organization of egg chromatin which appears to consist of at least two fractions: one, more sensitive to micrococcal nuclease attack, is characterized by a longer nucleosomal repeat (206.9 $\pm$ 2.12 b.p.) while the other exhibits a shorter linked DNA size (194.7 $\pm$ 5.77 b.p.) and is more resistent to degradation. In any event, the average egg nucleosomal repeat size is significantly smaller than that of *P. lividus* sperm chromatin (233.8 $\pm$ 6.5) and rather similar to the chromatin from embryonic cells. When the chromatins of blastulae, gastrulae and plutei are compared, an increase in the resistance to the digestion with micrococcal nuclease is found as development proceeds (Arceci and P. R. Gross, 1980a).

Interestingly, Spinelli *et al.* (1982b) have demonstrated that those histone genes that are being transcribed are more sensitive to nuclease attack within the chromatin than those which are silent during that developmental stage. By using a similar approach, Bryan *et al.* (1983) have specified that it is the promoter region of the genes of the major histone repeat (the so-called h22) that is very sensitive to micrococcal nuclease and DNAase 1 at the early (128-cell) blastula stage, i.e., when these histone genes are maximally transcribed. This hypersensibility drastically decreases at hatching, i.e., when these genes are no longer transcribed. On the contrary, some sequences near the 3' end of these genes (which map just downstream the inverted DNA repeats essential for generating faithful 3' ends of the histone messengers) are very resistant to micrococcal DNAase when the genes are active, and become sensitive when they are no longer expressed.

Spinelli *et al.* (1982b) have also found that the nucleosomal pattern appears disorganized in correspondence to the histone genes which are being transcribed.

Preliminary data by Spinelli *et al.* (personal communication) indicate that the nucleosomal core particles are not randomly located along the histone genes, and that their apparent phasing is different between sperm and embryos. Experiments are in progress aimed at finding if a stage-specific pattern of phasing can be demonstrated.

The possibility of the existence of a phasing of the 5S RNA genes has been suggested also by experiments of chromatin reconstitution by Simpson and Stafford (1983), using sea urchin 5S rDNA genes and chick erythrocyte histone.

A special approach to the analysis of the potentially active genome has been followed by Di Mauro *et al.* (1980), who studied the binding of the homologous RNA polymerase II to the *Paracentrotus* 1. DNA: 4.5×10^4 stable complexes were formed, part of which with a 2000–4500 b.p. spacing and another part with up to 30,000 b.p. spacing.

The only available information about differences in chromatin organization in different embryonic territories is the work of Cognetti and Shaw (1981), who found an increased resistance of the micromere chromatin to micrococcal nuclease attack, which may correlate with the precocious determination of these cells and with their limited transcriptional complexity (Ernst *et al.*, 1980). Chambers *et al.*, however, (personal communication) call attention to the fact that this different nuclease sensitivity is abolished when the nuclei of micromeres are digested, in mixing experiments, together with the nuclei of macromeres or mesomeres, which suggests that diffusible elements are responsible for the different nuclease sensitivity.

Contradictory results have been reported in the past with regard to the chromosomal organization in sea urchins, due to the fact that chromosomes are quite small in these animals. The most recent review on the subject is from Colombera (1974), who reports that the most frequent chromosome number is 44 for a diploid set. The presence of sex chromosomes, reported in earlier papers, is not confirmed by more recent work (Colombera, 1974; Auclair, 1964; German, 1964, 1966; Delobel, 1971).

Recently, Poznanovic and Sevaljevic (1980) have isolated the nuclear matrix. The characterized proteins DNA, RNA, and phospholipids show small but significant stage-specific differences.

4.4 The Synthesis of DNA

As has already been mentioned, whereas no increase in the net mass of RNA and proteins occurs till the early pluteus stage, a very high net increase of the amount of DNA takes place during development, due to the increase in the number of nuclei per embryo. DNA synthesis, which is silent in the unfertilized egg, becomes dramatically activated following fertilization or parthenogenetic activation, for example with ammonia (Mazia and Ruby, 1974) or with procaine (Brandriff and Vacquier, 1975; Vacquier and Brandriff, 1975) or with the Ca^{2+} ionofore A 23187. The two latter might also act through a pH increase (Lopo and Vacquier, 1977).

The onset of DNA synthesis occurs, before the fusion of the pronuclei in sand dollars (Simmel and Karnofsky, 1961; Brookbank, 1970) and in *Paracentrotus* (W. A. Anderson, 1969) but not in *Strongylocentrotus* (Brookbank, 1970). Longo and Plunkett (1973) found that in *Arbacia* DNA synthesis initiated only after the fusion of the pronuclei, but that the latter is not a necessary event, since in polyspermic egg DNA synthesis initiates also in the other sperm pronuclei. Ito *et al.* (1981) reported that during the cleavage period, i.e., when cell divisions proceed at a very fast pace, virtually without any interphase, 3H-thymidine incorporation occurs in the chromosomal vesicles which form during anaphase. These chromosomal vesicles are observed in the cells of many invertebrate and fish cells which divide at a high rate, and may be interpreted as monochromosomic nuclei which have already started to swell in order to allow DNA replication to occur in the shortened cell cycle and that will then fuse to form the telophase nucleus.

Quantitative studies of the rate of DNA synthesis by the use of radioactive precursors were initiated by Nemer (1962), who was the first to show the stepwise increase of exogenous thymidine incorporation into DNA, preceding each cell division. These in vivo experiments, however, before being evaluated in quantitative terms, need to be corrected for two factors, first, the already-mentioned increase in permeability to nucleosided following fertilization (Von Ledebur-Villiger, 1975; Ord and Stocken, 1974; Nishioka and Magagna, 1981; McGwin *et al.*, 1983; J. L. Grainger and Hinegardner, 1974; see the latter authors for the false results that can be obtained with bromodeoxyuridine), second for the presence of a very large nucleotide triphosphate internal pool in the unfertilized egg, which decreases by a factor of 4 at the blastula stage, as carefully measured by Gourlie and Infante (1975) and by Mathews (1975).

Since the synthesis of DNA parallels the rate of cell division in sea urchins, it undergoes very important variation during development, i.e., a sharp start from zero level, at fertilization, a halt from the early blastula stage till the mesenchyme blastula, followed by a second increase, which in terms of DNA per embryo leads to about a doubling in the interval between mesenchyme blastula and prism, bringing the cell number from about 1000 to about 2000, but which represents only one nuclear doubling in about 15 h, whereas nuclear doubling occurs in less than 1 h during the early cleavage stage. The curve which describes the increase in cell number per embryo is therefore a typical curve of diauxic growth, as calculated by Liquori et al. (1981).

These variations of the rate of DNA synthesis have attracted the attention of many investigators, because they offer a good general model for the study of regulation of DNA synthesis. The activity of enzymes involved in such a synthesis has therefore been investigated at different developmental stages with the following results:

1. Thymidine kinase: an activation of thymidine phosphorilation in vivo has been described after fertilization (Nonaka and Terayama, 1975; Nishioka and McGwin, 1980) or activation by ammonia (Nishioka and Mazia, 1977). Nonaka and Terayama (1977) suggested that this increased phosphorylation might have been due to the liberation of the kinase from a particulate form. McGwin et al. (1983) have recently concluded that thymidine kinase and thymidylate kinase must be located in the egg cortex of Strongylocentrotus purpuratus, since a centrifugation of the egg which does not damage the cortex does not affect the increase in thymidine uptake and phosphorylation which follows fertilization or parthenogenetic activation of both nucleated or nonnucleated egg halves; whereas treatment with cytochalasin B, which disrupts the egg cortex, significantly reduces the rate of uptake and phosphorylation after fertilization. This activation is therefore not directly connected with nuclear activity. On the other hand, variations accompanying those of the rate of DNA synthesis have been described for thymidine kinase as well as for thymidilate kinase in Paracentrotus (De Petrocellis and Rossi, 1976) and in Sphaerechinus (Parisi and De Petrocellis, 1976), during embryogenesis, but not at fertilization.

2. Ribonucleotide reductase: The activity of this enzyme has been found to increase sharply after fertilization (De Petrocellis and Rossi, 1976); it cannot, however, be conceived that it is responsible for the onset of DNA synthesis, since a very high pool of deoxyribonucleotide triphosphates is already present in the unfertilized egg (Gourlie and Infante, 1975; Mathews, 1975).

3. Thymidilate synthase and dihydrofolate reductase are two key enzymes in DNA synthesis because the former methylates dUMP to dTMP by receiving the methyl group from N^5, N^{10}-methylene tetrahydrofolate, and has been found to undergo an increase just after fertilization and to undergo cyclic variations relatedto the early cleavage cycles. Dihydrofolate reductase does not undergo variations following fertilization; its experimental inhibition, however, as brought about by aminopterin, causes cleavage arrest (Yasumasu et al., 1979). Moreover, as noted by Yasumasu et al., (1982), the unfertilized egg contains an amount of long-chain acyl-CoA's high enough to inhibit dihydrofolate reductase (Mita and Yasumasu, 1981; 1982) as well

as the activity of G6P dehydrogenase and 6PG dehydrogenase (Mita and Yasumasu, 1980) enzymes which may be used in the reduction of the ribonucleotides through the production of NADPH. Following fertilization, the concentration of the long-chain acyl-CoA's decreases, while that of polyamines, known to stimulate dihydrofolate reductase, increases (Kusunoki and Yasumasu, 1976, 1980), thus providing a mechanism which might enhance the production of DNA precursors.

4. DNA polymerase: It had already been described that the activity of this enzyme does not change throughout development but is transferred from the cytoplasm to the nucleus up to the blastula stage (Mazia and Hinegardner, 1963; Loeb et al., 1967, 1969; Fansler and Loeb, 1969, 1972; Loeb and Fansler, 1970; Müller et al., 1974; Shioda et al., 1977). The existence of plurime forms of DNA polymerase in sea urchins has been suggested by I. Slater and D. W. Slater (1972) and by Siu et al., (1973); two different forms have actually been identified by De Petrocellis et al., (1976) and alpha, beta and gamma forms by others (Shioda et al., 1977; Suzuki-Hori et al., 1977; Habara et al., 1979). Several authors (Hobart and Infante, 1980; Shioda et al. 1980, 1982) have reported that it is the α-form of the DNA polymerase, that becomes transferred into the nucleus as development proceeds, in agreement with its supposed role in nuclear DNA synthesis. This enzyme increases by a factor of 2 only before the gastrular DNA replication; at this stage it increases first in the endoplasmic reticulum, then in the nucleus, while losing its ability to bind the cytoplasmic membranes, which is retained in a stage-specific fashion by the β-form. Shioda and Nagano (1983) have recently suggested that the α-DNA polymerase which is located in the endoplasmic reticulum of the unfertilized Hemicentrotus pulcherrimus egg becomes associated with the nuclear membrane at the blastula stage, which is in agreement both with the origin of the nuclear membrane from the endoplasmic reticulum and with the proposed function of the nuclear membrane as site of DNA synthesis.

The inhibition of the α-polymerase in vivo with afidicolin immediately stops DNA replication (J. Brachet and De Petrocellis, 1981). It is interesting that under these conditions RNA and protein synthesis continue, thus offering the possibility of investigating which transcriptional or translational developmental changes depend upon DNA replication. DNA polymerases of β- and γ-types have also been extracted from Hemicentrotus pulcherinus sperm, which, on the other hand, as typical of the nondividing tissue, do not contain the α-form (Habara et al., 1980).

Poccia et al., (1978) have found a DNA nicking-closing enzyme in association with chromatin, whose amount parallels the nuclear number per embryo. Proteic factors stimulating DNA synthesis have been described in sea urchin eggs and embryos (Murakami and Mano, 1973; Murakami-Murofushi and Mano, 1977; Jimenez et al., 1978). Although their mechanism of action or their in vivo role is not yet fully understood, it is interesting that at least one of them, partially purified by Shimada (1983), is present in the nuclei of Hemicentrotus pulcherrimus embryos at the cleavage stage, but not in the nuclei from unfertilized eggs.

Only a preliminary report exists to our knowledge, of the presence of an RNA-instructed DNA polymerase in Lytechinus p. embryos till the stage of 16 cells (Chilton et al., 1979).

Shimada et al. (1983) found that DNA polymerase α and γ are inhibited by long chain fatty acylCoA, whose concentration decreases by a factor of 2 following fertilization. The authors, however, warn against too simple conclusions because in the unfertilized eggs the concentration of these molecules is about 6 µm in *Hemicentrotus pulcherrimus*, i.e., still lower than that needed to give a 50% in vitro inhibition.

5. DNAse: Different forms of this enzyme have been described in sea urchin development and their intracellular distribution studied (De Petrocellis and Parisi, 1972, 1973a, b; Parisi and De Petrocellis, 1972; Gafurov and Rasskazov, 1974). No variations of its activity are observed up to the hatching blastula stage, in agreement with older observations by Mazia *et al.*, (1948); while a decrease and subsequent increase, occur after this stage, which roughly parallel the variations of DNA synthesis rate.

6. Ornithine decarboxylase: its activity has been recently found to parallel the rate of DNA synthesis during development (Pirrone *et al.*, 1983). This fact is especially interesting because DNA synthesis in sea urchin development is uncoupled from the synthesis of rRNA and proteins, thus suggesting a more specific role for this enzyme, which is known to be on the pathway of polyamine biosynthesis, and whose activity in other biological systems has been found to increase with the increase of DNA, RNA, and protein synthesis, which all together accompany cell division. That ornithine decarboxylase may play a role in the regulation of DNA synthesis in sea urchin development is also suggested by the experiments of Kusunoki and Yasumasu (1976, 1978a, b), who found that this enzyme is higher in the fertilized than in the unfertilized egg, that polyamine concentration undergoes cyclic variations correlated with the embryonic cleavage, and that α-hydrazinoornithi-

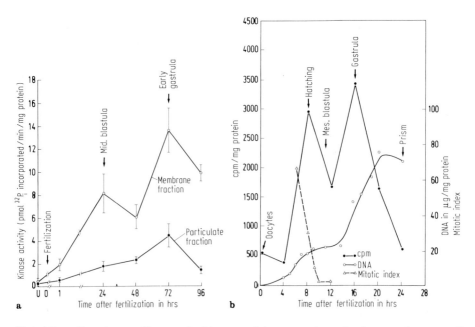

Fig. 4.7. a Tyrosine-specific protein kinase activity at various developmental stages of *Strongylocentrotus purpuratus*. (Dasgupta and Garbers, 1983). **b** ●——● Ornithine decarboxylase activity. (Pirrone et al., 1983); ○——○ DNA content, and □——□ mitotic index. (After Parisi et al., 1978) at various developmental stages of *Paracentrotus lividus*

ne, a potent competitive inhibitor of ornithine decarboxylase, inhibits the rise of polyamine level and cleavage. These authors have also described a heat-labile activator of this enzyme in the cell sap of eggs and embryos at stages with active cell division.

7. Tyrosine protein kinase: This activity has also been found to undergo variations in *Strongylocentrotus purpuratus* embryos (Dasgupta and Garbers, 1983) that positively correlate with the variations in DNA synthesis (see Fig. 4.7).

Attempts to correlate variations of concentrations of regulatory micromolecules like diadenosinetetraphosphate (Pirrone *et al.*, 1979) with the variations of DNA synthesis rate do not provide answers useful to this purpose; these authors, in fact, did not find any detectable synthesis of diadenosinetetraphosphate during the early development of *Paracentrotus lividus*. Weinman-Dorsch and F. Grummt, however, (personal communication), by means of a luciferin-luciferase assay, have found some correlation between the content of this molecule and the synthesis of DNA in sea urchins. Since the highest concentration found by these authors is in the sperm, it cannot be excluded that the destruction of these molecules participates at least in the initial burst of DNA synthesis which occurs at fertilization. Negative results had first been reported also for cyclic AMP (Nath and Rebhun, 1973; Blomquist *et al.*, 1973), but more recently Yasumasu and coworkers have provided evidence for a periodic change of cAMP concentration during the cleavage period, which is inversely correlated with changes in the activity of the phosphodiesterase, while the adenylate cyclase undergoes only an initial increase (Ishida and Yasumasu, 1982), which can be attributed to the stimulatory effect of the hexose monophosphates generated at fertilization. Amy and Rebhun (1977) have also found a four to five fold increase of the adenylate cyclase activity following fertilization in isolated egg cortices. Moreover, the stimulatory effect on the cleavage rate of cAMP and dibutyril cAMP has been directly demonstrated (Ishida and Yasumasu, 1981).

R. F. Baker (1971) and Shimada and Terayama (1976) by means of pulse and chase experiments followed by gradient analysis of the labeled fragments, first suggested that DNA synthesis in sea urchin embryos occurs according to the model implying the synthesis of small fragments, which are then assembled into longer ones. This question has also been investigated by means of electron microscope observations of the replicating DNA molecules: Howze and Van Holde (1977) have described the presence of "eye" forms in replicating DNA of *Arbacia*; Baldari *et al.*, (1978) under careful experimental conditions, found no "eye" forms but "microbubbles", which they interprete as replication sites. These are actually more frequent in the DNA of the cleavage stage than in that of gastrulae or adult tissues, i.e., where DNA replication is slower, which is in agreement with the analyses of Case *et al.*, (1974), who by means of *Aspergillus* DNAase digestion showed a decrease of the single-stranded regions of DNA from morula to gastrula (see also Wortzman and Baker, 1980, 1981). On the basis of their observations Baldari *et al.*, (1978) propose that DNA synthesis in sea urchin embryos occurs in the three steps: (1) strand separation actively operated by a DNA-unwinding protein; (2) discontinuous DNA synthesis; (3) completion and ligation of the fragments. Kurek *et al.*, (1979) find not only "bubbles" but also "eye" and "gap" forms in the embryonic DNA, but not in that of sperms, which again suggests these to represent figures of replication; since, however, their appearance does not change between 4-blastomeres and blastula, they conclude that the higher rate of DNA synthesis during the cleavage stage is not brought about by an increase in the number of replication sites.

Wortzman and Baker (1980, 1981), by analysis of the isolated single-stranded DNA regions, found that these can be divided into shorter fragments (500 b.p.) which may represent the replication forks and longer fragments (1500–3000 b.p.) which are highly enriched in sequences for histones, an intriguing observation, in view of the usual coupling of histone synthesis with DNA synthesis.

Botchan and Dayton (1982) found that the eye forms in isolated fragments *Lytechinus* rDNA are located nonrandom from the site of Kpn I restriction. From this observation they concluded that the origin of replication of the rDNA is not random. This is in our opinion based on the assumption that either it is only the origin of replication that can be visualized as an eye form, or that the replication of the rDNA genes is quite synchronous among the different genes, the different cells and the different embryos, which seems rather unlikely. From the analysis of the position of the eye forms with respect to the Kpn site, these authors also suggest that the replication origin is located within the nontranscribed spacer.

A role for the nuclear membrane as initiation site for DNA synthesis has been suggested by Infante, after pulse and chase experiments followed either by cell fractionation (Infante *et al.*, 1973; Crabb *et al.*, 1980) or autoradiography (Hobart *et al.*, 1977). The absence of repetitious sequences in the DNA associated with the nuclear membrane has been reported by Fitzmaurice and Baker (1973), which supports the idea of a nonrandom association of DNA to the latter. Also this question, however, waits for further experimental evidence for a final word, also in view of contradictory results obtained in other biological materials.

The effect of 5-bromodeoxyuridine has been studied by several authors. The effect of such an analog is much stronger on sand dollars than on sea urchin (I. M. Evans and P. R. Gross, 1978). The effect on sand dollars consists in the arrest of development at morula or gastrula, depending upon the dose, (Karnofsky and Simmel, 1963; Mazia and Gontcharoff, 1964; Tencer and J. Brachet, 1973) probably through a disturbance of chromosome separation. The effect on *Paracentrotus lividus* development is as stronger as earlier the treatment is initiated, the most sensitive time being between fertilization and 16 blastomeres (Czihak, 1978; Schreuer and Czihak, 1978). Kotzin and Baker (1972) described an inhibition of the chase of thymidine from short DNA segments into longer ones (see also Baker and Case, 1974 and Case and Baker, 1975b). The synthesis of RNA, although probably modified (Kokzin and Baker, 1972), is not inhibited by bromodeoxyuridine (Fritzmaurice and Baker, 1974a), nor is the synthesis of DNA-binding proteins (Case *et al.*, 1975).

A question of general relevance for the problem of regulation of development is whether or not changes of DNA occur during development, which may be responsible for such a regulation. It is not our purpose to go into details of the pros and cons of such an idea, but we have to mention that such a mechanism has been proposed by Scarano for sea urchins. This author has demonstrated that methylations of DNA occur during development and that they are not randomly distributed; a subsequent deamination step converts the 5-methylcytosine into thymidine, therefore ultimately converting a G into a T in one DNA strand. Upon DNA replication, one daughter cell would therefore receive a DNA molecule bearing the original G-C pair (stem cell), and the other would receive a DNA molecule bearing an A-T pair (differentiated cell) (Scarano and Augusti-Tocco, 1967; Scarano et al., 1965, 1967; Scarano, 1969, 1971; Grippo et al., 1968, 1970; Tosi et al., 1972; Tosi and Scarano, 1973). The data of Scarano and coworkers show a solid experimental background; it is more difficult to prove that these point-changes of DNA do actually play a role in differentiation. Some necessary corollaries stemming from this hypothesis have, however, been verified in other biological materials with different answers: (1) One would expect to see DNA methylation before cell differentiation: this occurs (Adams, 1973); moreover, the inhibition of DNA methylation also inhibits mouse embryo differentiation (Jones and Taylor, 1980); (2) one would expect tissue- and species specificity of DNA methylation: also this is true (Vanyushin et al., 1973; Gama-Sosa et al., 1983); (3) one would expect different compartments of DNA to become methylated in different developmental stages: no such differences have been detected at least with the employed restriction enzymes (Bird et al., 1979); neither has the ratio 5-methylcytosine: cytosine been found to change from sperm to plutei (Pollock et al., 1979), although subtle differences would escape such a kind of analysis; negative answers have been obtained also with respect to changes in methylation of the repeated DNA sequences during avian development and evolution (Sobieski and Eden, 1981).

It is pertinent to mention here that in many biological materials hypomethylation has usually been found to be a necessary, although not sufficient, condition for gene transcription (see, e.g., Hjelle et al., 1982); but negative results have been reported as well: Bower et al., (1983), for example, found no obvious correlation between undermethylation and expression of the gene for 5-crystallin in chick development.

The discovery of the somatic rearrangement of the immunoglobulin genes during lymphocyte differentiation (Hozumi and Tonegawa, 1976) has prompted new interest into the possibility that genome rearrangements play a role in embryonic development. Very little is known in this respect, but Dickinson and Baker (1978) have found differences between proportions of double- and single-stranded regions of the "hairpin" DNA sequences of different developmental stages of the sea urchins; interestingly for what we have said, these differences are especially evident if DNA has been labeled with ^{3}H-methylmethionine, and they do not appear in the presence of 5-bromodeoxyuridine (Dickinson and Baker, 1979).

It has to be recalled that Baker has also in the past, although less explicitly, reported changes of DNA during sea urchin development, e.g., with respect to the content in deoxyadenylate rich sequences (Fitzmaurice and Baker, 1974b) and to the position of polypirimidine runs (Case and Baker, 1975a).

Chapter 5

Ribonucleic Acid

5.1 Ribosomal RNA

As we have already mentioned, immediately upon fertilization the sea urchin egg undergoes a tremedously high rate of cell division not accompanied by embryonic growth, at least until the prism stage. In order to be able to do so, it has stored in its cytoplasm a number of macromolecules synthesized during oogenesis: among these, ribosomal RNA (Sconzo *et al.*, 1972; Giudice *et al.*, 1972b; Griffith *et al.*, 1981). Ribosomal RNA synthesis, therefore, is barely detectable during cleavage and becomes apparent at the gastrula stage (Nemer, 1963; P. R. Gross *et al.*, 1965; Comb *et al.*, 1965b; Giudice and Mutolo, 1967, 1969; Barros and Giudice, 1968; Sconzo *et al.*, 1970a, b; Sconzo and Giudice, 1971; Greco *et al.*, 1977).

There is general agreement in the literature about this fact, but disagreement appears when the rate of rRNA synthesis is measured per cell nucleus. The question is not irrelevant, because what it has been tried to establish is whether or not there is a regulation of the transcription of rRNA during development. Giudice and coworkers, by preloading the embryos with ^{32}P and then measuring the rate of incorporation into mature 26S rRNA under conditions of constant specific activity of the precursor nucleotide triphosphate pool, found a severalfold increase of the rate of production of mature rRNA at the mesenchyme blastula stage on a per nucleus basis (Sconzo *et al.*, 1970a, b; Sconzo and Giudice, 1971; Pirrone *et al.*, 1973).

Emerson and Humphreys (1970, 1971), measuring the incorporation of exogeneous precursors into a ribosomal RNA characterized by MAK columns on different species, denied such an increase. More recent experiments by Surrey *et al.*, (1979), based on the ratio of incorporation of methyl groups into the "cap" structure of messenger RNA versus that into rRNA, concluded a lack of activation of transcription at gastrula, leaving the possibility open of an increase of production of mature rRNA through a regulation of the precursor maturation. This result would reconcile all previous data except for those of Sconzo and Giudice (1971), who also measured methyl incorporation into ribosomal RNA, although in a much more crude way. That maturation of rRNA precursor is especially fast at the gastrula stage, at least compared to the stage of oogenesis, has indeed been found by Sconzo *et al.*, (1972) and although to a lesser extent by Griffith *et al.*, (1981); we believe, however, that some criticism can still be raised as to the identification criteria of rRNA used by Surrey *et al.*, (1979) and that the final world should come from the use of purified rDNA probes. Such a probe has already been successfully used in sea urchin (Griffith and Humphreys, 1979; Griffith *et al.*, 1981). These latter data are certainly the most accurate and lead to one conclusion on which there is general agreement, that the rate of rRNA synthesis drops from oogenesis to early cleavage, calculated on per-nucleus basis. Where the conclusions of these authors show some discrepancy with part of the previous literature is whether this low rate of rRNA synthesis ever increases before the feeding larva stage. Griffith et al., (1981) admit only a possible twofold increase; we are incline to believe that this figure should be somewhat higher and possibly multiplied by a factor of two, because we have proved that at the pluteus stage only one half of the nuclei, those of the archenteron cells, make rRNA (Roccheri et al., 1979).

The possibility remains that the different conclusion with respect to the question of the time of reactivation of rRNA synthesis might be due to the fact that different species have been used by different authors. Whatever the final outcome, one has to bear in mind that the amount of rRNA synthesized between bastula and prism stage accounts only for about 10% of the amount of rRNA that has accumulated in the unfertilized egg during oogenesis (Sconzo et al., 1970a). A much higher activation of rRNA synthesis will occur at the pluteus stage, i.e., when the feeding embryo starts real growth in terms of net mass increase, as predicted by Sconzo and Giudice (1971), and demonstrated by Tenner and Humphreys (1973).

The precursor of ribosomal RNA in sea urchin embryos was first described by Sconzo et al., (1971). Their results, including the maturation scheme reported in Fig. 5.1, have been essentially confirmed by the use of more modern techniques (Griffith and Humphreys, 1979), like rDNA cloning and electron microscope analysis (Blin et al., 1979; Bieber et al., 1981).

What regulates rRNA synthesis in sea urchin embryos? Some evidence has been provided that polyamines might play a role (Barros and Giudice, 1968), although no correlation between rRNA synthesis and ornithinedecarboxylase activity has been found by Pirrone et al., (1983). Also no correlation between rRNA synthesis and ppGpp concentration has been found (Pirrone et. al., 1976; Brandhorst and Fromson, 1976). If one looks at RNA polymerase activity (Roeder and Rutter, 1969, 1970a, b) one finds that nucleolar activity increases by about 50% following the blastula stage; a decrease is, however, observed when the activity is measured on a per-nucleus basis. Later experiments, however, show that on a per-embryo basis, the ratio RNA polymerase I:II increases as development proceeds (Morris and Rutter, 1976).

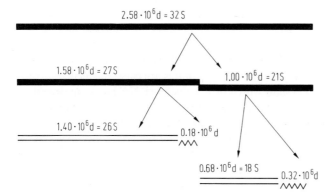

Fig. 5.1. Scheme of rRNA maturation in sea urchin oocytes and embryos. *Solid lines* indicate the immature products; *double lines* the mature products; *zig zag lines* the descarded fragments. (Giudice et al., 1973)

No developmental variations from fertilization to 5 h later have been found by Ho and Mazia (1979) for a calcium-activated ribonuclease, in *Strongylocentrotus purpuratus*.

Whatever the mechanism, the signal for the lowering of the rRNA synthetic rate seems to start from the cell membrane, since preliminary experiments of cell dissociation at the stage of 4–8 blastomeres indicate a stimulation rRNA synthesis (Arezzo and Giudice, 1983 and unpublished data).

The synthesis of the 5S ribosomal RNA has been also thoroughly investigated. Early experiments had reported that ^{32}P incorporation into such an RNA is detectable only after the gastrula stage (Comb, 1965; Comb *et al.*, 1965a; Frederiksen *et al.*, 1973). More recent experiments, however, have detected 5S RNA synthesis already during the cleavage, and if the rate of the synthesis is measured on a per-cell basis, again there is no increase, but actually a decrease (O' Melia and Villee, 1972b; O' Melia, 1979a, b, 1984) of 5S RNA synthesis as development proceeds. Nijhawan and Marzluff (1979) have calculated a rate of synthesis of 60 molecules cell^{-1} min^{-1} at the morula stage, which elevates to 230 at blastula and then decreases to 130 at gastrula and to 150 at pluteus. Lu and Stafford (1982), who have recently reviewed the subject, conclude therefore that the synthesis of 5S rRNA is probably controlled independently of nucleolar rRNA synthesis (see also O'Melia, 1979a, 1984). The most interesting comparison, however, is still lacking, i.e., that of the rate of 5S RNA synthesis during oogenesis with that of early development, i.e., of stages which differ dramatically for the nucleolar activity.

An inhibitor of RNA polimerase III activity has been purified from sea urchin egg nuclei by Morris and Marzluff (1983). Does this inhibitor play a role in vivo?

Analyses of the properties of 5S rRNA have first been reported by Sy and McCarthy (1970, 1971), and by Bellemare *et al.*, (1972).

More recently Lu et al., (1980) have studied the sequence of the 5S rRNA of *Lytechinus variegatus* embryos labeled in vivo. These data, coupled with those of 5S DNA sequencing, permitted the conclusion that the sequence of 5S RNA of *Lytechinus* differs by only 16 nucleotides from that of *Drosophila* and by only 18 nucleotides from that of *Xenopus laevis* oocytes. The secondary structure reported in Fig. 5.2 has been proposed by Hori and Osawa (1979) for sea urchin 5S rRNA. More recently, Ohama *et al.*, (1983) have analyzed the sequence of the 5S rRNA of three echinoderms, the sea urchin *Hemicentrotus pulcherrimus*, a starfish, and a sea cucumber. All three 5S rRNA's are 120-nucleotide long, and their sequences are more related to those of proterostomes (87% identity) than to those of vertebrates (82% identity) although both sea urchins and vertebrates are classified as deuterostomes. An in vitro transcription system which originates 5S rRNA has been described by Lu and Stafford (1982), who made use of a nuclear extract of *X. laevis* oocytes directed by a cloned sea urchin 5S rDNA, in which the transcribing enzyme seems to be RNA polymerase III, since it is inhibited by intermediate concentrations of α-amanitin.

5.2 Transfer RNA

Besides the early experiments based on the analysis of incorporation of labeled precursor into the RNA fractionated by sucrose gradients or by MAK columns

```
                          1         2         3        4              5
             123456789  012345  678901  2345678  9012  345678901234  5678  90123456
SEA-CUCUMBER GUUUACGAC  CAUAUC  ACGUUG  AAUAUAC  CGGU  UCUCGUCCGAUC  ACCG  AAGUCAAG
STARFISH     GUUUACGAC  CAUACU  ACGUUG  AAUACAC  CGGU  UCUCGUCCGAUC  ACCG  AAGUUAAG
SEA-URCHIN*  GUUUACGAC  CAUACC  AUGCUG  AAUAUAC  CGGU  UCUCGUCCGAUC  ACCG  AAGUCAAG
SEA-URCHIN** GCCUACGAC  CAUACC  AUGCUG  AAUAUAC  CGGU  UCUCGUCCGAUC  ACCG  AAGUCAAG
                  A      aLb      B       bLc    C       cLc'        C'    c'Lb'
```

```
                                                          1           1            1
      6        7        8           9       0           1            2
   789012  345  6789012  34567  890123456  7890  123456789  012  3456789  012345678  90
   CAGCGU  CGA  GCCUAGU  G/C AGUA  CUUGGAUGG  GUGA  CCGCCUGGG  AAU  ACUAGGU  GCCGUAGAC  UU
   CAACGU  CGG  GCUUGGU  UAGUA  CUUGGAUGG  GAGA  CCGCCUGGG  AAU  ACCAGGU  GUCGUAGGC  UU
   CAGCAU  AGG  GCCCGGU  UAGUA  CUUGGAUGG  GAGA  CCGCCUGGG  AAU  ACCGGGU  GUUGUAGGC  AU
   CAGCAU  AGG  GCUCGGU  UAGUA  CUUGGAUGG  GAGA  CCGCCUGGG  AAU  ACCGGGU  GUUGUAGGC  UU
a    B'   b'Le   E      eLd     D         dLd'    D'       d'Le'  E'       A'
```

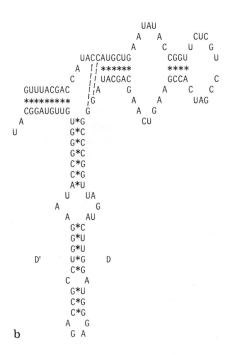

Fig. 5.2. a Sequence alignment of 5S rRNA's. *SEA-CUCUMBER Stichopus oshimae. STARFISH Asterina pectinifera. SEA URCHIN*, Hemicentrotus pulcherrimus, SEA URCHIN**, Lytechinus variegatus. Squared-off sequences* correspond to the base-paired regions in the secondary structures (*A, A', B, B',* in the lowest row). Symbols *aLb, bLc,* etc are for loop regions. The 73rd residue of the sea cucumber 5S rRNA is heterogeneous, i.e., C or G. **b** The secondary structure model of *Hemicentrotus pulcherrimus* 5S rRNA. *Dotted lines* indicate potential base pairs. (Ohama and Osawa, 1983)

followed by base ratio analysis (see Giudice, 1973, for a review), few recent experiments on the synthesis of transfer RNA during development have been reported. The sum of the early evidence indicated that the terminal turnover of the transfer RNA is very active in the stages comprised between fertilization and cleavage with little de novo synthesis, while some synthesis of transfer RNA occurs at later stages. Recently O' Melia (1979b, 1984) has made quantitative measurements of the amount of exogenous ^{3}H guanosine incorporated into the 4S peak of an acrylamide electrophoresis of *Arbacia punctulata* RNA. The results, which have been corrected for the specific activity of the GTP pool, indicate a rate of 4S RNA synthesis of g 7.62×10^{-13} h^{-1} per embryo during cleavage, of g 12.36×10^{-13} h^{-1} per embryo at early blastula, of g 15.3×10^{-13} h^{-1} per embryo at the mesenchyme blastula stage and of g 12.7×10^{-13} per embryo at the pluteus stage, i.e., with a decrease by a factor of about 2.7 of the rate of 4S synthesis from cleavage to pluteus, if the data are expressed per nucleus.

Information of the functionality of tRNA during embryogenesis will be provided in Chapter 6.

5.3 Messenger RNA

A central question in the problem of embryonic differentiation has been that of the possibility that this is operated through the synthesis of different proteins at different times of embryonic development. That this does not necessarily mean a regulation at a transcriptional level, of a bacterial type, was already suggested by the early experiments of inhibition of RNA synthesis by the use of actinomycin D (Wolsky and de Issekutz Wolsky, 1961; Lallier, 1963; P. R. Gross and Cousineau, 1963a, 1964; P. R. Gross, 1964; P. R. Gross et al., 1964; Giudice et al., 1968) or by a variety of other means as X-ray irradiation (Neifakh, 1960, 1964; Neifakh and Krigsgaber, 1968) nucleotide analogs (Crkvenjakov et al., 1970); enucleation (E. B. Harvey, 1936, 1956; Hiramoto, 1956), all showing that suppression or severe inhibition of RNA synthesis did not produce an immediate effect on development. For example, P. R. Gross and coworkers showed that inhibition of RNA synthesis since fertilization allows development to proceed till the hatched blastula. Giudice et al., (1968) produced a map of the effect on development of actinomycin administered at various development stages (Fig. 5.3). This, on one hand, demonstrated the long life of some messenger RNA's and, on the other, suggested the possibility of a post-transcriptional regulation of protein synthesis. As we shall see later, in fact, the pattern of synthesis of at least some proteins can undergo developmental changes also in the absence of RNA synthesis.

The question of developmental changes of the types of RNA synthesized at different stages was more directly addressed when the nucleic acid hybridization techniques became available. A quantitative approach to such a study has disclosed that this question is not a simple one, because while 90% of the polysomal mRNA of gastrulae deriving from single copy DNA sequences is made by a few hundreds of RNA types, each represented by a few hundreds of copies per cell, the remaining 10% is made of some 10–15,000 RNA types each represented only by 1–15 copies per cell (Galau et al., 1974).

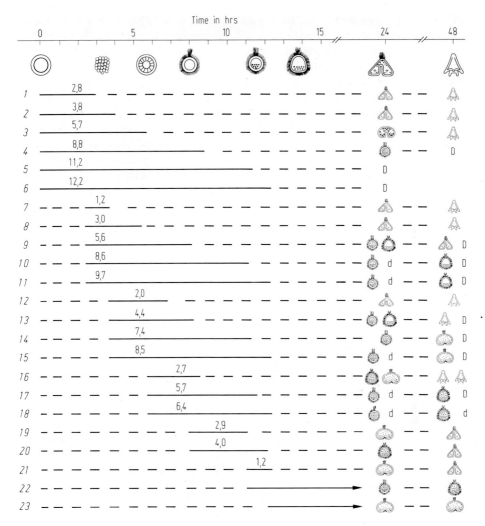

Fig. 5.3. Effect on development of "pulse" treatments with actinomycin D. *Top* the developmental stages of control embryos are indicated. *Continuous lines* the period in which the drug was present, whose length is expressed in hours by the *numbers above the lines*. In the *two last vertical columns* the stages attained at 24 and 48 h after fertilization are schematically represented. *d* degeneration of minor fraction of the embryos; *D* degeneration of a major fraction of the embryos. (Giudice et al., 1968)

This tells us how difficult it can be to look for differences of mRNA's belonging to different cells, when one thinks that a difference in the synthesis of a dozen enzymes might profoundly change the metabolic pattern of a cell. By using a single copy DNA enriched in mRNA of the gastrula stage and one lacking these sequences, it was, however, possible to establish that great differences exist among the complex mRNA's of embryonic and adult tissues and that some differences are also detectable between embryonic stages such as gastrula and pluteus (Galau *et al.*,

1976), and between embryonic territories (Bruskin et al., (1981). In a more recent analysis Lynn et al., (1983) have been able to precisely locate a family of mRNA in a group of ectodermal cells by means of in situ hybridization. It was found that a mRNA complementary to the clone pSpec1, whose abundance increases 100-fold per embryo during early development is located in a group of mor-phologically uniform ectodermal cells in the dorsal part of the pluteus. This sequence is already located in a specific group of cells at the gastrula stage, where it is found in the amount of about 1000 copies per cell, which decreases to 500 in the pluteus cells. If translated in vitro it generates a family of small acidic proteins. Furthermore, as already mentioned Shepherd et al., (1983) have found differences in the abundance of single polysomal poly(A)$^+$ species between normal and animalyzed blastulae.

One relevant question that was asked a few years ago was if the mRNA differences which arise during development are due to transcription of different DNA segments in different stages or if all nuclear transcripts are the same at all stages and only some of them reach the polysomes as mature messages at one stage and others at another stage. The latter hypothesis is theoretically possible, because the complexity of nuclear RNA of gastrulae made on single copy DNA has been found to be of about 1.7×10^8 nucleotide pairs, of which only 1.7×10^7 nucleotide pairs is found on poly-somes (Hough et al., 1975; Galau et al., 1976). When the nuclear RNA's, made on single copy DNA, of different stages were compared, at first no differences were found (Kleene and Humphreys, 1977; Wold et al., 1978) thus calling for a post-transcriptional control, but more accurate measurements, made by use of a single copy DNA tracer of adult intestine lacking the sequences present in the nuclear RNA of gastrulae, have shown a more than 10% difference between the nuclear RNA's of adults and embryos (Ernst et al., 1979). The differences between polysomal mRNA's are, however, much higher (Hough-Evans et al., 1977; McColl and Aronson, 1978; Xin et al., 1982). The analysis of the transcripts of the single copy DNA sequences has undergone further progress in recent times thanks to the use of purified gene clones. It was in such a way possible to ascertain the disappearance from the polysome, but the persistence in the nuclei, at the gastrula stage of the transcripts of a single copy gene (A. S. Lee et al., 1980). Two other sequences of 1600 and 2000 nucleotides show developmental variations of the number of copies present on polysomes, from 1000–1400 to zero, but are always represented in the same number in the nuclei (Lev et al., 1980). Furthermore, Shephered and Nemer (1980) have demonstrated that many abundant RNA classes present in the polysomes of blastulae disappear from the polysomes of gastrulae but remain present in their nuclei. Many of these findings are in agreement with a post-transcriptional regulatory model. We will report here that proposed by Davidson and Britten (1979): it had been already observed in earlier experiments (A. H. Whiteley et al., 1966, H. R. Whiteley et al., 1970; Glisin et al., 1966; Hynes and P. R. Gross, 1972; Hartman et al., 1971; Mizuno et al., 1973; H. R. Whiteley and A H. Whiteley, 1975) and then essentially confirmed by means of more quantitative analyses, that important quantitative differences exist between the nuclear transcripts made on repetitive DNA sequences of different embryonic stages (Costantini et al., 1978; Scheller et al., 1978). These RNA sequences represent about 30% of the nuclear transcripts (McColl and Aronson, 1974; M. T. Smith et al., 1974), but only a few of them, with the exception of the histone mRNA, reach the polysomes (Goldberg *et al.*, 1973). They are, however, evolutionarily conserved (G. P. Moore *et al.*, 1980). What then is their function? The proposal of Davidson and Britten is that they play a role in regulating the production of mature messages; the idea being that RNA's made on repeated sequences of DNA, which the authors call "integrating regulatory transcription units" (IRTU), when in sufficiently high concentration, make RNA–RNA duplexes with regions of the RNA's made on DNA sequences containing also the messenger RNA information, and which the authors call "constitutive transcription units" (CTU). The presence of such double helical regions in the mRNA precursor is hypothesized to be a necessary condition for the production of mature mRNA's. This model, although to be considered only as an attractive hypothesis, is consistent with all the available data; moreover, some double-stranded RNA has been found in sea urchin embryo nuclei by Kronenberg and Humphreys (1972) and recently Balmain *et al.*, (1982) have described the formation of duplexes between polyadenylated and nonpolyadenylated RNA's in the nuclei of Friend erythroleukemic cells.

One of the predictions of such a model is that of a high turnover of nuclear RNA. This has been verified and accurately measured by many authors: Brandhorst and Humphreys (1971, 1972) found that the half-life of nuclear RNA of blastulae of plutei is about 5–10 min, which is partially in agreement with earlier findings of Kjima and Witt (1969) and of Aronson *et al.*, (1972) and with the results of determinations made by the use of heavy and radioactive isotopes by R. M. Grainger and Wilt (1976). Also Aronson and Chen (1977), by measuring the rate of elongation of the HnRNA chain (i.e., 6–9 nucleotides: second, equivalent to 12–18 min for an RNA of 2×10^6 m.w.) concluded for a short half-life of nuclear RNA.

Different RNA classes, however, are synthesized at different rates, and show different half-lives, as shown by Nemer *et al.*, (1979), who reported that the longer transcripts containing the cap 1 structure are synthesized at a rate about 15 times higher than the shorter ones, which contain the cap 2 structure.

Less agreement exists on the half-life of mRNA, which was first calculated in about 70 min (Brandhorst and Humphreys 1971, 1972) but later found to be between 5 and 6 h (Nemer *et al.*, 1975; Galau *et al.*, 1977) at blastula and late gastrula. These latter measurements also permitted the conclusion that the rate of mRNA synthesis for both complex and prevalent RNA classes is of 0.13 pg m^{-1} per embryo. Brandhorst (1980) has moreover reported that the unfertilized egg synthesizes the same classes of RNA as those synthesized during oogenesis but that the former have much shorter half-lives than the latter, which, as we shall discuss later, might be due to the existence of post-transcriptional mechanisms which regulate the delayed utilization of mRNA; this is also suggested by the observation of Brandhorst that it is only the newly synthesized mRNA which is translated in the unfertilized egg and not that already present in the egg. The decay of histone mRNA or at least its functional stability has been recently measured by injecting it into *Xenopus laevis* oocytes or eggs and then seeing how long sea urchin histones are synthesized. The conclusion of a half-life of about 3 h was reached, except for a 5% fraction, which persists for weeks (Woodland and Wilt, 1980a, b).

Maxon and Wilt (1982), by measuring the precursor pool specific activity, reached the conclusion that the half-life of the histone messenger during the early cleavage is of 1.5–2 h, and suggested a regulation of the turnover of the newly synthesized histone mRNA in *Strongylocentrotus purpuratus*.

The factors governing the rate of synthesis of RNA are not thoroughly known, nor is it known if rate of synthesis and decay are different for different mRNA's. A search for the mechanism of regulation of the synthesis of individual mRNA's seems therefore a more promising approach than an investigation of the overall rate of RNA synthesis and decay. This search, as we shall see later, has already begun, in the meantime the availability of cDNA libraries has made it possible to measure the prevalence of various cytoplasmic trascripts at different developmental stages. The following conclusions have been reached: rare mRNA classes (one copy per cell) which are present in plutei are absent in adult coelomocytes; low-abundance mRNA classes (5.5 copies per cell), representing 46% of pluteus mRNA's are much less represented in adult coelomocytes; moderately abundant classes (40 copies per cell) contain some sequences present at the stages of gastrula and pluteus, but absent at 2 blastomeres, and rare in the adult coelomocytes; abundant classes (more than 100 copies per cell) undergo a decrease at gastrula while being very similar at 2 blastomeres, pluteus and in the coelomocytes (Lasky *et al.*, 1980; Xin *et al.*, 1982; Flytzanis *et al.*, 1982). Moreover, Duncan and Humphreys (1981a), have concluded by hybridization of the poly(A)$^+$ RNA to its cDNA that in the egg the maternal sequences are 55% rare, 13% moderately abundant, and 32% abundant. Developmental as well as territorial variations of 5 mRNA classes, picked up at random from a genomic library, have been described by Bruskin *et al.*, (1981); these are almost absent in the egg and increase even 100-fold during development to become 3–20 times more concentrated in a group of cells of the pluteus ectoderm (Lynn *et al.*, 1983).

An untranslated and highly (2000–3000 times) repeated sequence adjacent to the 3' termini of the genes coding for two such mRNA's has been described by Carpenter *et al.*, (1982).

The portion of genome transcribed at different developmental stages does not change appreciably at least from blastula to pluteus (Kleene and Humphreys, 1977), or even from the unfertilized egg on, according to Brandhorst (1980). What brings about a higher amount of RNA synthesis per embryo as development proceeds is essentially the increase in number of nuclei per embryo, or, to be more precise, the increase in the amount of DNA per embryo: experiments of Brookbank (1976) in fact suggest that embryos kept at low temperature, which do not cleave but replicate their DNA, increase their RNA synthesis as the control embryos.

The properties of sea urchin chromatin at different developmental stages have, however, been investigated also with the aim of finding differences in its template ability. In this respect it has to be recalled that Aoki and Koshihara (1972a, b) have suggested an inhibitory effect of an acidic polysaccharide fraction, while a stimulatory role has been proposed by Kinoshita (Kinoshita, 1974, 1976; Saiga and Kinoshita, 1976) for a heparin-like mucopolysaccharide-protein complex. Kinoshita and Yoshi (1983) have recently proposed an interesting method to study the nucleocytoplasmic interactions which may bring about structural and functional modifications of the chromatine: The embryonic cells are made permeable by treatment with glycerol. The addition to these permeable cells of proteoglycans extracted from embryos causes an accumulation of the added proteoglycans in the cell nucleus and their binding to the chromatin. It is unfortunate that the cells do not survive such a glycerol treatment. The number of RNA chain initiation sites on blastula chromatin has been estimated at 1 every 2.5×10^6 base pairs by Di Mauro et al., (1977), which is not too far from preliminary analyses performed by electron microscopy on spread nuclei (Busby and Bakken, 1977; Zdunsky et al., 1977).

No quantitative variations of the RNA polymerase activity have been found during development, but a progressive transfer from cytoplasm to nuclei has been described (Maroun, 1973; Morris and Rutter, 1976).

The best characterization of the RNA polymerase II comes from the work of Ballario et al., (1980), who, however, have measured its activity only at the gastrula stage.

Even if no final positive answer about control mechanisms is provided by the above experiments, one mechanism is excluded by the work of Duncan (Duncan and Dower, 1973; Duncan et al., 1975): RNA production, transport to cytoplasm and turnover are not linked to its translation, since they occur as normally in the presence of inhibitors of proteins synthesis.

An important role for the survival of the nuclear RNA and for its transport to the cytoplasm has been attributed to its polyadenylated 3' appendix. The early observation by D. W. Slater et al., (1972) has been repeatedly confirmed and followed by a series of works (Wilt, 1973; R. S. Wu and Wilt, 1973; I. Slater et al., 1973; Pawlawski and Rodriguez, 1974), whose most important conclusions can be summarized as follows: maternal RNA polyadenylation is activated immediately following fertilization, i.e., concomitantly with the activation of protein synthesis; $poly(A)^+$ RNA becomes progressively associated with ribosomes after fertilization; the fraction of $poly(A)^+$ RNA increases until blastula. The above observations might suggest that polyadenylation of maternal RNA plays a key role in the activation of protein synthesis following fertilization. This hypothesis, however, seems not to be acceptable, since Mescher and Humphreys (1973, 1974) have demonstrated that this activation of protein synthesis occurs also in the absence of polyadenylation, as brought about by deoxyadenosine; moreover, the $poly(A)^+$ and $poly(A)^-$ maternal RNA's have been shown to have the same biological activity at least in a wheat germ cell-free system (Fromson and Verma, 1976). One important role of polyadenylation seems to be that of permitting the transport of RNA from nucleus to cytoplasm. Sconzo and Giudice (1976) have, in fact, demonstrated that deoxyadenosine, at concentrations that do not yet inhibit RNA synthesis, causes a marked inhibition of the passage of nuclear RNA to cytoplasm and (Spieth and A. H. Whiteley, 1980) arrest of development at hatching. The transport to cytoplasm, however, does not seem to be the only function of polyadenylation, because in the fertilized egg this occurs at a cytoplasmic level (I. Slater and D. W. Slater, 1974). Moreover, polyadenylation is not a sufficient condition to prevent degradation of the RNA within the nucleus (R. S. Wu and Witt, 1974); nor is it a necessary condition for transport to the cytoplasm because at all developmental stages there is a consistent aliquot of polysomal RNA which is $poly(A)^-$, and which cannot be entirely explained as residual maternal RNA which has not been polyadenylated (Fromson and Duchastel, 1973, 1974, 1975). Much work has been devoted to the quantitation and characterization of the $poly(A)^+$ and $poly(A)^-$ fractions of RNA at different developmental stages (Nemer, 1975; Nemer and Surrey, 1975; Nemer et al., 1974, 1975; Dubroff and Nemer, 1975, 1976; Fromson *et al.*, 1977; Brandhorst *et al.*, 1979). Some important observations are that the poly(A)$^+$ and poly(A)$^-$ RNA's are transcribed partially on different genes, although essentially coding for the same proteins; that three types of HnRNA can be distinguished: α, with internal oligo(A) sequences, β, with poly(A) tail, γ with neither oligo-nor poly(A) structures; that both poly(A)$^+$ and poly(A)$^-$-mRNA can have a "cap" structure and internal methylations (Surrey and Nemer, 1976; Faust *et al.*, 1976). Duncan and Humphreys (1979, 1981a) have concluded, by experiments of hybridization of the total poly(A)$^+$-RNA to its cDNA, that each abundance class of maternal messages is composed by both poly(A)$^+$- and poly(A)$^-$-RNA in equal proportions.

A still unsolved problem is that of the role of the proteins bound to the poly(A) stretches of mRNA. C. Peters and Jeffery (1978) reported the interesting observations that they increase 2.5-fold from fertilization until 8 blastomeres. They can be resolved into two proteins of 87,000 and 130,000 molecular weight respectively. It is the lighter one that increases following fertilization: does it play a regulatory role for the utilization of maternal mRNA?

An then interesting observation has been reported by Duncan and Humphreys (1981b), that some maternal RNA sequences which are not translated contain multiple oligo(A) tracts. The authors suggest that they may represent incompletely processed mRNA precursors; they never seem to become active during development, so that a regulative hypothesis seems to have to be discarded. Costantini *et al.*, (1980) have described the presence of short repetitions sequences within the single copy poly(A)$^+$-RNA of the egg, which interestingly, are much less represented in the polysomal RNA of gastrulae. What is the function of the short oligo(A) stretches found in the nuclear RNA's? Since some oligo(U) sequences have also been found (Dubroff, 1977; Duncan and Humphreys, 1977, 1983), the possibility has been suggested that they can form intramolecular duplexes of some functional significance.

It has also been demonstrated that poly(A) is not only added to maternal RNA, but also undergoes a turnover (Wilt, 1977), so that all of it turns over within 2.5 h at polysomal level (Dolecki *et al.*, 1977). Brandhorst and Bannet (1978) have also demonstrated that the embryonic poly(A)$^+$-RNA receives the addition of a few adenylate units at its exit from the nucleus.

The mechanism through which RNA polyadenylation is activated at fertilization is not known. Hyatt (1967a, b) described a nuclear enzyme able to synthesize poly(A), and more recently Egrie and Wilt (1979) found that poly(A)-polymerase activity does not change during development, but is transferred from cytoplasm to nucleus at blastula and gastrula.

The experiments so far discussed deal with the synthesis of the bulk RNA's; detailed information about single classes of mRNA have recently started to become available; one case, that of the mRNA coding for the histone proteins, is the best studied. We will, therefore, discuss it with special care, although it is probably not representative of the majority of cases, since it belongs to the very few (still the only known one) cases of mRNA's made on repetitive DNA sequences.

We have already discussed in Chapter 4 how, following the early identification of histone mRNA by Kedes and Gross (1969a, b) and by Nemer and Lindsay (1969), a great deal of successful work has been done on this subject, leading to the isolation and partial sequencing of the histone genes, and to a good knowledge of the temporal pattern of their transcriptional activation. We shall complete this knowledge here by recalling some fundamental work more directly aimed to study the histone mRNA. K. Gross et al., (1973a, b) demonstrated that the 9S RNA of the sea urchin embryo, when translated in a cell-free system, produces histones. It was then shown also by hybridization with purified histone genes that most of this RNA is histone messenger. This represents about 9.7% of all the RNA synthesized during the cleavage stage in *Strongylocentrotus purpuratus* and 6.5% in *Lytechinus pictus*, decreasing to 0.75% in *Strongylocentrotus*, and to 1.4% in *Lytechinus*, at the mesenchyme blastula stage (Kunkel and Weinberg, 1978). This decrease of histone RNA synthesis parallels the decrease of the rate of DNA synthesis and seems to be conserved by the isolated nuclei (Shutt and Kedes, 1974). Measurements of the synthesis of the early histone messengers based on radioactive labeling and measurement of the precursor pool-specific activity have permitted Maxon and Wilt (1981) to conclude that the rate of histone mRNA synthesis is 80×10^{-15} g min^{-1} per embryo at the stage of 128 cells, followed by a decline to 12×10^{-15} g min^{-1} per embryo at the stage of about 3000 cells and that in *Strongylocentrotus p.* $\times$ *Lytechinus p.* hybrids, both the paternal and maternal histone genes are equally active in the early development. Maxon and Wilt (1982) also found that between the 16- and the 200-cell stages $7-10 \times 10^{6}$ molecules of each core histone mRNA, and 2.5×10^{5} molecules of H1 mRNA accumulate in *Strongylocentrotus*. Table 5.1 shows the time course of the synthesis of different forms of H1 during the early development of *Strongylocentrotus purpuratus* according to Harrison and Wilt (1982).

Measurements of the amount of newly synthesized H2B histone mRNA have shown that the transcription of the late H2B genes is activated already at fertilization, but that their transcription rate increases by a factor of >15 in the interval between 14 and 16 h after fertilization in *Strongylocentrotus purpuratus* (Maxson et al., 1983).

As Mauron et al., (1982) pointed out, the histone gene expression in early development "involves a complex interaction between activation of stockpiled maternal messengers, fluctuating rates of histone gene transcription and alteration of histone mRNA stability". The values reported by these authors after hybridization with probes of known specific activities are shown in Fig. 5.4. The above data compare quite well with those obtained by Weinberg et al., (1983) by three different methods: measurement of the incorporation of externally supplied radiolabeled precursors; assay of the amount of RNA by filter blot hybridization to cloned full length histone DNA repeats; assay of the translational ability in a wheat germ cell-free system. The results of these experiments indicate that the rate of histone mRNA synthesis in *Strongylocentrotus purpuratus* increases from at least 47×10^{-15} g min^{-1} per embryo at 6 h of development to 114×10^{-15} at 9 h and then drops again to 29×10^{-15} at 12 h. This means that the maximum transcriptional rate is of 2–3 molecules min^{-1} per gene copy. The estimated half life of the histone mRNA is about 1.5 h at 12 h and the maximal turnover is observed at 9 h. It follows that each species of histone mRNA increases from $6-10 \times 10^{5}$ molecules egg^{-1} to $6-10 \times 10^{6}$ molecules per embryo at 10.5 h. It was at first reported that while H3 and H4 histones are conserved during development as they are in phylogenesis (Grundstein et al., 1976), H1 undergoes developmental changes, so that at the morula stage an H1, called H1m, was distinguishable from a gastrular one, called H1g (Ruderman

Table 5.1. The program of HI synthesis in *s. purpuratus* embryos[a]

	Early		Transition				Late	
Age of Embryos (hours)	6	10	12	16	20	24	28	32
Number of cells	16-32	150	200	300	400	600		700
Embryonic Stage	Morula	Very young Blastula		Pre-Hatching Blastula	Hatching Blastula	Mesenchyme Blastula → Early Gastrula		Mid-Late Gastrula
Nature of HI-Histone Synthesis	Early HI Synthesis only (HIα$_1$ and HIα$_2$ Both Synthesized)		① ↔ HIγ Appears ↔ HIβ Appears ②		③ — — — → HIα$_1$ Disappears HIα$_2$ Disappears ④			Late HI Synthesis only (Hiγ and Hiβ Both Synthesized)

[a] The length of the arrows is an indication of the time periods over which the various changes were seen to occur; thus, the appearance of HIγ and HIβ were fairly discrete events, whereas the cessation of synthesis of HIα$_1$ and HIα$_2$ were more variable. The dashed arrow, marked (3), indicates that insufficient data were obtained to determine exactly when HIα$_2$ disappears, an event that can only be scored on acid-urea gels. Although the arrows overlap, the sequence of events in a given embryo culture was invariably 1 → 2 → 3 → 4. (From Harrison and Wilt, 1982)

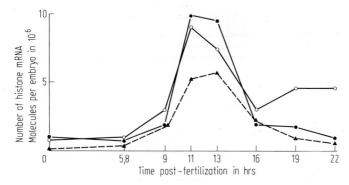

Fig. 5.4. Time course of histone mRNA accumulation in early development. The number of mRNA molecules per embryo was plotted against the time after fertilization (○) H2B mRNA; (●) H3 mRNA; (▲) H1 mRNA. (Mauron et al., 1982)

et al., 1974; Arceci *et al.*, 1976). Later work has revealed the detail that a subclass of H1m, called H1x, is synthesized between 8 and 16 blastomeres (Senger *et al.*, 1978) and a finger printing map has also shown differences arising in H4 and H3 hatching (Grundstein and Schedl, 1976; Grunstein, 1978; Hieter *et al.*, 1979). Variations of H1, H2A and H2B (Newrock *et al.*, 1977) occurring before the blastula stage have been attributed to post-transcriptional modifications, because, with the exception of the synthesis of H1x, they all occur also in the presence of actinomycin D (Ruderman and P. R. Gross, 1973, 1974; Newrock *et al.*, 1977; Senger *et al.*, 1978).

The development changes of the H1 mRNA's have been confirmed by hybridization with purified histone gene clones (Lifton and Kedes, 1976; Spinelli *et al.*, 1980); on the other hand, variations of the elctrophoretic pattern for all histone classes after the hatching blastula stage have also been shown by a variety of methods (hybridization with histone genes, translational activity in cell-free systems, partially sequencing) to be due to the activation of different gene sets (Kunkel and Weinberg, 1978; Newrock *et al.*, 1978b; Grundstein, 1978; Spinelli, *et al.* 1979; Hieter *et al.*, 1979). An analysis of Grundstein *et al.*, (1981) has revealed that also the sequences of H4 mRNA undergo developmental changes, which are minor when one compares the *maternal* H4 sequence with that of the *early* H4, i.e., that synthesized between fertilization and blastula, and become major differences when one compares the *early* with the *late*, i.e., postblastular H4 sequences; these differences are both in the translated and in the untranslated regions. Much work in this respect has been done by Kedes and coworkers (S. Levy *et al.*, 1975, 1979; Childs *et al.*, 1979a, b). The main conclusions of all this work are resumed in Fig. 5.5. It seems necessary to admit the existence of a post-transcriptional control, since mRNA's relative to late histones are present, but not translated, also at much earlier stages, while mRNA's relative to early histones persist although in small amount also at later stages, i.e., when the corresponding proteins are no longer synthesized.

The experiments of V. R. Lee and A. H. Whiteley (1982) also speak for a post-transcriptional control. These authors reported the surprising finding that, contrary to early reports, the paternal and maternal genomes of developmentally arrested

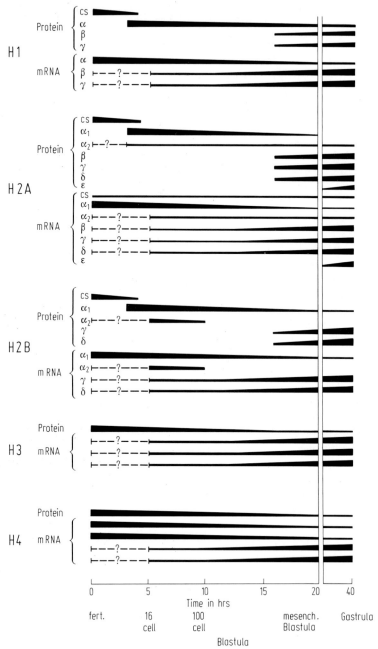

Fig. 5.5. Developmental expression of sea urchin histone genes in *Strongylocentrotus purpuratus*. *Thick line* relatively abundant component; *thin line* relatively minor component. *Broken lines* uncertainty in the synthesis of the component. *Vertical lines* the first point in development when the component has been positively identified. Data for each component in the figure is based on in vivo protein synthesis, production of a protein in the wheat germ cell-free system, or identification of a pulse-labeled radioactive mRNA species at a particular developmental stage. Histone proteins are stable and remain associated with chromatin even after its synthesis is no longer detected. *fert* time of fertilization; *mesench. Blastula* mesenchyme blastula. (Childs et al., 1979 b)

lethal hybrids are equally transcribed, as judged by DNA/RNA hybridization experiments, although it is known that maternal proteins are prevalently synthesized in the hybrid embryos (Badman and Brookbank, 1970; H. R. Whiteley and A. H. Whiteley, 1975; Ozaki, 1975; Tufaro and Brandhorst, 1982). Also Crain and Bushman (1983), looking at the mRNA specific for actin, found that the paternal and maternal genes are equally expressed and probably processed in hybrid embryos.

How is the post-transcriptional control brought about? Spirin, back in 1964, proposed a role for proteins bound to mRNA in the control of its translation (Spirin et al., 1964). He and Nemer found in sea urchin embryos some messenger-containing RNA particles which he called informosomes and which he interpreted as masked mRNA's. (Spirin and Nemer, 1965; Infante and Nemer, 1967, 1968; Nemer and Infante, 1968; Spirin, 1979). This attractive hypothesis will be discussed in detail in Chapter 6, as it will be that of cytoplasmic maturation of mRNA precursors. We shall limit ourselves in this chapter to the mechanisms of control of the synthesis of the histone mRNA, refering the reader to Chapter 6 for a discussion of a translational control, although the two subjects cannot always be separated.

A very special case has been proposed for the synthesis of histone mRNA by Spinelli, who found the synthesis of a gigantic RNA containing all the five histone sequences (Spinelli et al., 1980). This gigantic RNA has usually escaped the attention of many investigators, because it is not evident during the cleavage stage, i.e., that most used for the study of histone messenger synthesis, but it can be detected in stages when histone synthesis is slower, as at mesenchyme blastula (Kunkel et al., 1978; Spinelli et al., 1980, 1982a), and especially during oogenesis. Its synthesis can also be induced in sea urchin by special treatments, like heat shock (Spinelli et al., 1982a); chase experiments suggest that when the temperature is brought back to normal values the giant histone RNA is matured into 9S histone messengers. These data are also substantiated by the findings of Buongiorno-Nardelli et al., (1981) that the homologous RNA polymerase II binds preferentially at one single point of a DNA clone containing all five histone genes, i.e., at a sequence within the spacer between H4 and H2B, as shown by a combined analysis with the electron microscope and with restriction enzymes. The attractive hypothesis can also to be considered, that two different and developmentally regulated mechanisms of histone genes transcription exist, one by which all five genes are transcribed in the form of a polycistronic precursor, and the other by which each histone gene is separately transcribed, as described, for example, by Hentschel et al. (1980a).

One of the most important approaches to the study of the mechanism of regulation of histone transcription has been that of injecting into Xenopus laevis oocytes purified histone genes whose sequence has been experimentally modified, with the aim of ascertaining which parts of the sequence are relevant for the transcriptional processes. The series of these experiments was started by Probst et al., (1979), who showed transcription and translation of an injected Psammechinus miliaris clone containing the sequences for H2A and H2B. The fidelity of such a translation was further proved by Hentschel et al., (1980a,), Etkin and Maxon R. E. Jr. (1980) and by Bending (1981), who showed persistence of the translation of the injected genes until at least the tadpole stage. More recently, Etkin and Roberts (1983) have been able to show that a recombinant plasmid injected in the cytoplasm of Xenopus oocytes becomes integrated into the frog genome as proved by experiments of nuclear transplantation, which have generated clones of embryos containing the sea urchin histone DNA. In the meantime sequencing of the gene regions preceding the 5' end had progressed and the sequence 5'-TATAAATA-3' followed by a leader sequence equal for all five histone genes composed by 5'-PyPyATTCPu-3' had been identified (Sures et al., 1980; Hentschel et al., 1980b). Removal of the TATAAATA box sequence from the injected H2A histone genes alters the initiation of transcription with the generation of heterogenous 5' mRNA termini; an experimental substitution of this sequence with a TAGA box causes a five fold decrease of histone RNA synthesis, while the removal of another sequence preceding the TATAAATA accelerates histone RNA synthesis (Grosscheld and Birnstiel, 1980, Grosscheld et al. 1981). Grosscheld et al., (1983), in refining their analysis of the distal control elements for the synthesis of H2A mRNA, have found that the sequence comprised between nucleotides —165 and —111 upstream from the cap site of the DNA of *Psammechinus miliaris* and of *Paracentrotus lividus* (Ciaccio unpublished results) is especially important for the regulation of transcription and shows a striking resemblance to the enhancer sequences of some viruses (see Fig. 5.6).

SSV	-188	GgGCCAaagACAGATGGttcccaga
SFFV	-199	GgGCCAagAACAGATGGtccccaga
Mo-MuSV	-200	GgGCCAagAACAGATGGtccccaga
Psam, H2A	-143	GAGCCACCAACAGATGG
Para, H2A	-137	cAatCgCCAACAGAgGG
Lyt, H3	-171	tgGaagCgAACAGATGG

Fig. 5.6. Compilation of sequence homologies in far upstream regions of histone genes and viral LTR's. *Numbers* refer to the nucleotide position (counting from the mRNA cap site) of the left-most nucleotide on each line. *SSV* simian sarcoma virus; *SFFB* murine Friend spleen focus forming virus; *Mo-MuSV* Moloney murine sarcoma virus. In addition to the cases listed above the authors have found that the *Rous sarcome* virus LTR contains at position – 96 a 9 out of 10 bp homology to the H2A modulator. The RSV sequence aGGAAGgc-*AACAGAcGGN₆CATG* appears more closely related to the H3-171 sequence tGGAAGcg-*AACAGAtGGN₄CATG*. (Grosscheld et al., 1983)

For the correct 3' termination of the transcription of the injected H2A gene, a highly conserved 12 b.p. sequence immediately preceding the 3' of the transcribed region is needed, but not sufficient because another sequence within the downstream spacer appears to be also necessary (Birchmeier et al., 1982) for a length of 80 nucleotide pairs (Birchmeier et al., 1983).

Stunnenberg and Birnstiel (1982) reported the interesting finding that if a clone of *Psammechinus* DNA containing the sequences of all five histones is injected into the *Xenopus* oocyte mucleus, all five genes are capable of being transcribed with correct 5' and 3' termini, although with a different degree of faithful initiation or termination for each gene, but a read through the 3' terminus of H3 is obtained; if, however, this DNA is injected together with a factor extracted from the sea urchin chromatin at the cleavage stage, a correct termination of the transcription at the H3 3' end is observed.

The Zurigo group has then found (Galli *et al.*, 1983) that a 60-nucleotide-long RNA is able to cause the correct 3' termination of the H3 histone RNA when the DNA clone containing the histone genes is injected into the *Xenopus* oocyte. They therefore suggest that this RNA is contained in the previously described "termination factors" (see also Birchmeier *et al.*, 1984).

A new word on the problem of the mechanism of generation of the 3'-end of histone mRNA if by transcription termination or by processing of a larger precursor (as suggested by Spinelli) has been brought about by the experiments of Birchmeier et al. (1984), who injected into *Xenopus* oocytes a synthetic RNA precursor of the H2A histone messenger. This precursor was correctly processed, provided the already described dyad symmetry element was present. Also the 80 b.p. spacer sequence, especially the segment CAAGAAAGA, seems to be involved. If a synthetic precursor of the H3 mRNA is injected, processing requires the injection of the 60 b.p. sequence, which is probably the RNA component of a small nuclear ribonucleoprotein. All this unequivocally shows that the generation of the 3' termini of H2A and H3 histone messenger from a histone messenger precursor is possible.

Bendig and Hentschel (1983) have recently succeeded in transfecting HeLa cells with recombinant DNA containing the same histone gene repeat (h22). They found sea urchin histone gene transcripts with a pattern of initiation very similar to that observed when this clone was injected into *Xenopus* oocytes (Hentschel *et al.*, 1980), fertilized eggs or embryos (Bendig, 1981; Stunnenberg and Birnstiel, 1982); termination, on the other hand, was largely aberrant in HeLa cells, with the correct 3' terminus of only H2B mRNA present in significant amounts. As the authors suggest, the efficiency of correct termination may depend upon the recognition between the heterologous termination factors and the terminator sequences of the introduced genes.

The picture of the regulation of histone synthesis which we have already delineated in Chapter 4 will be completed in Chapter 6, where other mRNA's will also be taken in consideration. We shall briefly report here only the studies on the regulation of the synthesis of the m-RNA for actin by Buschman and Crain (1983). By measuring the size and the abundance of this mRNA at various developmental stages in four sea urchins (*Strongylocentrotus purpuratus, Strongylocentrotus droebachiensis, Arbacia punctulata* and *Lytechinus variegatus*) and one sand dollar (*Echinarachnius parma*), these authors obtained indications that the general pattern of developmental expression of some members of the actin gene family is conserved within the species belonging to the same genus (see Fig. 5.7). Similarities are also observed between the more distant echinoid species: the relative abundance of these sequences is low in early development and begins to rise during the late cleavage or blastula stages, with an increase of 9–35-fold by the pluteus stage in the sea urchins and of 5-fold in the sand dollar. A major sequence of 2.0–2.2 kilobases is common to all species, but a smaller species appear to lack in *Lytechinus* and *Echinaracnius*, and sharply decreases by the 16-cell stage in *S. droebachiensis*.

A careful analysis of the pattern of expression of six of the eight actin genes present in the *Strongylocentrotus purpuratus* genome has been reported by Shott *et al.*, (1984), who have recognized the transcription product of each of them by means of probes complementary to the 3' nontranslated region, which is different in the different genes. All the genes coding for the cytoskeletal actin (Cy) are expressed during early

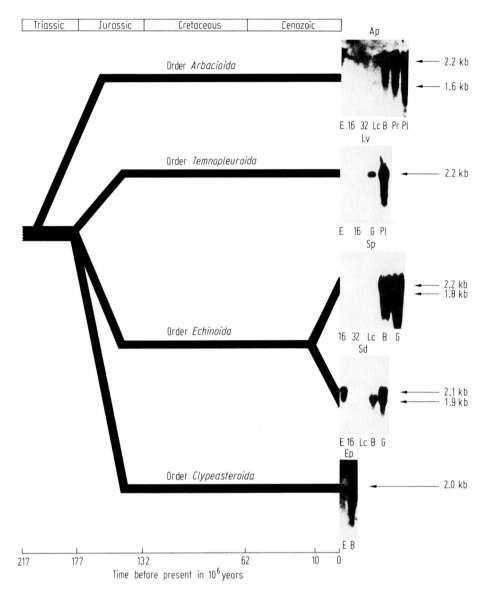

Fig. 5.7. Embryonic accumulation of actin-coding RNA's and phylogeny of five echinoderms. Total egg and embryonic RNA's were subjected to blot analysis except for *S. purpuratus*, in which cytoplasmic RNA's were analyzed. The hybridization probe was the *S. purpuratus* actin-coding clone pSpG17. RNA samples were loaded as follows: *S. droebachiensis*, *A. puctulata*, and *E. parma* 10 μg/lane; *S. purpuratus*, 30 μg/lane; *L. bariekatus*, egg-6 μg, 16-cell-10 μg, gastrula-13 μg, pluteus-14 μg. Stages of development: E, egg; 16,16-cell; 32, 32-cell; Lc, late cleavage; B, blastula; G, gastrula; Pr, prism; Pl, pluteus. Time course RNA blots were performed from four (*L. variegatus* and *E. parma*) to seven (*S. droebachiensis* and *A. punctulata*) times for each species, with a typical representative shown in the figure. (Bushman and Crain, 1983)

Table 5.2. Qualitative Estimates of Expression of Actin Genes in Sea Urchin Embryo and Adult Tissues

Gene	Transcript length (kb)	First appearance in embryogenesis	Expression in adult tissues relative to pluteus embryos						
			Testis	Ovary	Coelomocytes	Intestine	Tubefoot	Lantern muscle	Pluteus
CyI	2.2	Maternal mRNA	+++++	+++++	+++++	+++++	+++++	+++++	++++
CyIIa	2.2	Gastrula (40 hr)	−	−	+/−	+	+/−	+/−	+
CyIIb	2.1	Early blastula (14 hr)	+/−	+/−	+	+	+	+	+
CyIIIa	1.8	Maternal mRNA	−	+	−	−	−	−	+++++
CyIIIb	2.1	Early blastula (14 hr)	−	+	−	−	−	−	+
M	2.2	Early pluteus (62 hr)	+/−	+/−	−	+	+++	+++++	+

Note. Crosses indicate relative abundance of each gene transcript in different tissues. (−) Denotes undetectable (we believe this means less than two transcripts per average cell at the pluteus stage). +/− indicates detectable transcript but only on long exposure. (From Shott *et al.*, 1984)

development although at different times. The gene coding for the muscular actin (M) is expressed only from the pluteus stage on (Table 5.2).

5.4 Other RNA

Some few RNA's lie outside the classes already described. Among these are small RNA's, five of which have been described in *Paracentrotus* and *Psammechinus* (Frederiksen *et al.*, 1973; Frederiksen and Hellung-Larsen, 1974) and three in *Strongylocentrotus* (Nijhawan and Marzluff, 1979). Although they are highly represented in terms of moles per cell and it has been suggested that they play a role in RNA splicing, their function is not yet fully understood.

The genes of the small nuclear RNA's, N_1 and N_2 (corresponding to the mammalian U_1 and U_2) have been found to be organized in distinct tandemly repeating units of about 1.1–1.4 kb for a total of about 30 kb per group (Card *et al.*, 1982). It has been reported by Blin *et al.* (1983) that the sequences of U_1 and U_2 as well as the Alu-like of rat cross-hybridize with those of several species, including *Paracentrotus lividus*, thus showing their high evolutionary conservation (see Ullu *et al.*, 1982) and thus suggesting again a general role for them (see Lewin, 1982a for a comment), such·as, for example, that of participating in the correct termination of transcription, as suggested in sea urchins by the experiments of Galli *et al.* (1983) already described.

No genuine "chromosomal RNA" has been found in *Strongylocentrotus* by Alfageme and Infante (1975).

Chapter 6

Protein Synthesis

6.1 Rate of Protein Synthesis

Hultin (1950, 1952, 1953a, b, d) first proposed the idea that protein synthesis occurs at a very low level in the fertilized egg and then becomes activated at fertilization. This idea was more thoroughly investigated by Monroy and coworkers (Nakano and Monroy, 1957, 1958a, b; Monroy and Nakano, 1959; Giudice and Monroy, 1958; Monroy, 1960), who circumvented the problem of different permeability to amino acids of the fertilized eggs, by preloading the unfertilized egg with radioactive amino acids injected into the body cavity of the adult female and by observing their incorporation into proteins following fertilization.

These early results have been repeatedly confirmed later (Monroy, 1967; Epel, 1967; Mackintosh and Bell, 1967, 1969; Tyler et al., 1968; Humphreys, 1969). The most frequent figures indicated an increase of about 15-fold of the rate of protein synthesis starting about 2 min after fertilization and reaching its maximum value within about 2 h. More recent experiments in which the specific activity of the aminoacyl tRNA has been measured, give higher figures, i.e., an increase of 113-fold comparing gastrulae and unfertilized eggs (Regier and Kafatos, 1977) with an absolute rate of protein synthesis of 12.4 μg per 10^6 embryos min^{-1}, which is very close to the rate recently calculated by Goustin and Wilt (1981). Since the rate of protein synthesis at gastrula is about twice that of morula (Giudice et al., 1962; Bellemare et al., 1968; Neifakh and Krigsgaber, 1968; Ellis, 1966), this increase in the rate of synthesis after fertilization greatly exceeds the earlier reports. Although the techniques used by Regier and Kafatos are more reliable, it is unfortunate that in these very elaborate and accurate experiments the unfertilized eggs appear to have been cultured at an exceedingly high concentration, which might have made them hypooxygenated and, therefore, less active in synthesizing proteins.

The exact timing of the activation of protein synthesis following fertilization is also still being debated, but, although it should probably be set at an earlier time than that reported by Epel (1967), according to the calculations of Timourian and Watchmaker (1970), it has always to be considered as one of the "late" responses to fertilization (Epel, 1978).

The common idea that has become established for many years has been that the rate of protein synthesis, quite high during oogenesis (Ficq, 1964; Piatigorsky et al., 1967) becomes almost zero in the mature egg, to return to high levels following fertilization. This idea has to be modified according to careful measurements of Rinaldi and Parente (1976), based on the assumption of a dilution of the exogenous radioactive amino acids into an internal noncompartimentalized pool. These authors have shown that the rate of protein synthesis is comparable in oocytes and unfertilized eggs and becomes highly accelerated following fertilization. This seems reasonable in view of the fact that in order to synthesize all the proteins of the egg the oocyte has a time period measurable in weeks, whereas the fertilized egg, within 2 days, synthesizes (by transforming the stored raw material) an amount of proteins equivalent to about one half of those synthesized during the entire oogenesis (Fry and P. R. Gross, 1970b). There is also a selective pressure for this, because the oocyte does not need to rush, being well protected within the spiny maternal body, whereas the fertilized egg needs to build as soon as possible a swimming larva which is able to feed and has a higher chance of escaping predation than the immotile egg; moreover, there is evidence, as will be mentioned later, that most of the yolk proteins are synthesized out of the oocyte and then transported into this one.

What is the mechanism that brings about such an acceleration at fertilization? It has become progressively clearer that this is a multistep process. The first trigger is not necessarily the sperm, because parthenogenetic activation can do the same in all respects (Nakano et al., 1958; Giudice and Monroy, 1958; Hultin, 1961; P. R. Gross et al., 1963; T. Brachet et al., 1963; Denny and Tyler, 1964; Baltus et al., 1965; Tyler, 1966; Sargent and Raff, 1976). Here again the agents that can cause a partial parthenogenetic activation have been exploited in order to discriminate between those physiological events occurring at fertilization, which are necessary for the acceleration of protein synthesis from those which are not. In such a way it was found that the activation by ammonia induces no cortical exocytosis, membrane depolarization, respiratory burst, and, at low ammonia, no chromosome condensation, but still causes the increase of protein synthesis (Epel et al., 1974), thus permitting exclusion of the former events as a cause of the latter; also the inhibition of K^+ conductance does not absolish the increase of protein synthesis (see also Chap. 1).

The natural trigger for the acceleration of protein synthesis at fertilization seems to be the increase in the intracellular pH, because this has been found to parallel the variations of protein synthesis rate during the cell cycle, and because experimental modification of the intracellular pH are accompanied by parallel variations of the protein synthesis rate (J. L. Grainger et al., 1979). Moreover, it has been found (Winkler et al., 1979, Winkler and Steinhardt, 1981) that raising the pH from 6.9 to 7.4 stimulates the activity of a cell-free system for protein synthesis derived from Lytechinus pictus eggs. The increase in pH is caused by the known increase in the free calcium. According to Winkler et al., (1980) Ca^{2+} has also a direct effect on protein synthesis activation, as indicated by the fact that the enhancement of the protein synthesis rate brought about by the mere pH increase, as as by the addition of NH_3, is higher if external calcium is present. Dubé and Guerrier (1983), on the other hand, have reinvestigated this question on Sphaerechinus granularis eggs and failed to confirm that an external Ca^{2+} influx causes a further stimulation of the protein synthesis rate activated by NH_4Cl. T. Evans et al. (1983), as already mentioned, found that ammonia activates the synthesis of most, but not all, of the abundant proteins whose synthesis is activated by fertilization of by the Ca^{2+} ionophore A 23187. Changes in the cell surface properties responsible for the rise of the intracellular pH through calcium release, as previously described, may therefore be involved in the activation of protein synthesis at fertilization. This is also suggested by experiments of Johnson and Epel (1975) showing that, following fertilization or ammonia activation, there is a release of some surface proteins, which if added back to the egg, inhibit proteins synthesis (see also Shaprio, 1975 and Ribot et al., 1983 for the release of surface proteins following fertilization). Later results, however, E. J. Carroll and Epel, 1981) failed to confirm the inhibitory effect of the proteins released after ammonia activation.

Dissection of the mechanism of acceleration of protein synthesis at fertilization began as early as the cell-free systems for protein synthesis became available; thus Hultin and Bergstrand (1960) and Giudice (1962b) found that the differences of protein synthesis rate were also observed in cell-free systems. The various component elements of the cell-free systems have therefore been analyzed: early experiments (Hultin and Bergstrand, 1960; Hultin, 1961) showed that ribosomes of unfertilized eggs were inactive in protein synthesis, and it was found by Monroy and coworkers that these ribosomes could be made active by mild trypsin treatment (Monroy et al., 1965). The stimulatory activity of proteases on unfertilized egg ribosomes was confirmed later (Grossman et al., 1971), and since it is known that proteases become activated at fertilization (Lundblad, 1954; Lundblad and Runnström, 1962; Lundblad and Falksveden, 1964; Lundblad and Schilling, 1968; Lundblad et al., 1966; Mano, 1966; Maggio, 1957; Grossman et al., 1971; Kirsten et al., 1973; Grossman et al., 1973a, b), their role in the activation of protein synthesis seems very probable. But do they act directly on ribosomes? Probably at least part of the effect found by Monroy is due to the presence of some protein-bound mRNA in his ribosome preparations, since as we shall see later, the unfertilized egg mRNA can be made inactive by the attachment of proteins, and since Monroy obtained stimulation of endogenous activity upon trypsin treatment of ribosomes; part, however, of the trypsin action is probably due to an effect on ribosomes them selves, since they have been shown to contain a protein inhibitor of protein synthesis (Metafora et al., 1971; Hille, 1974) and, since trypsin treatment reverts the low dissociability into subunits of unfertilized egg ribosomes at low Mg^{2+} concentration (Maggio et al., 1968), which also might be due to the presence of some proteic inhibitor of dissociation, that may be removed by sucrose gradient purification of ribosomes (Infante and Graves, 1971). Monoribosomes from unfertilized eggs and from embryos are equally dissociated by increasing the KCl concentration (Infante and Graves, 1971). Differences between unfertilized and fertilized egg ribosomes were absent or very little when poly(U) was used as a stimulator of polypeptide synthesis (Stavy and P. R. Gross, 1967, 1969a; Monroy et al., 1965; Castaneda, 1969; Nemer and Bard, 1963; Vittorelli et al., 1969).

Clegg and Denny (1974) failed to find differences in the activity of 80S ribosomes or microsomes purified from Lytechinus pictus unfertilized eggs or zygote when tested in a cell-free system from Krebs II ascites cells supplemented with rabbit globin mRNA.

That ribosomes from unfertilized eggs are less active than those from embryos has been, however, more recently confirmed by Danilchick and Hille (1981) by incubating them in a reticulocyte cell-free system. Under these conditions, indeed, the unfertilized egg ribosomes become progressively activated especially at pH's higher than 7.4. The question can then be asked again as to what activates the unfertilized egg ribosomes.

One factor, as already mentioned, is the high pH itself, another may be the protease activation and a third has been proposed more recently, i.e., the phosphorylation of one protein of the 40S ribosomal subunit of 31,000 molecular weight, corresponding to the mammalian S6 protein (Ballinger et al., 1979, 1984; Ballinger and Hunt, 1980,

1981) accompanied by a decrease in activity of a specific phosphatase. Complex changes in the pattern of phosphorylation of ribosomal proteins within 5 min following fertilization or activation by A 23187, but not by ammonia, have also been described by C. Keller *et al.*, (1980) in *Strongylocentrotus purpuratus* (this latter being confirmed by Ballinger *et al.*, 1984 in *Arbacia*). Phosphorylation of 40S proteins has been found to accompany frequently, but not always (Leader *et al.*, 1981) the increases in protein synthesis rate in other biological systems. As far as sea urchins are concerned, the increased phosphorylation of the 31,000 ribosomal protein 10 min after ferti- lization has been confirmed by Ward *et al.*, (1983) for the eggs of *Arbacia puctulata*, but not for those of *Lytechinus pictus* or of *Strongylocentrotus purpuratus*, therefore showing that also for these animals phosphorylation of this protein does not always accompany the increase in protein synthesis rate (see also Ballinger *et al.*, 1984). Takeshima and Nakano (1983) have described changes in the electrophoretic be- havior of five ribosomal proteins (S7, S16, L19 and L31) which occur within 30 min after fertilization, in three (except for L31) Japanese sea urchin species. No proof has been provided yet that these changes cause ribosomal activation, although the hypo- thesis is appealing on our opinion.

It was very soon discovered (Monroy and Tyler, 1963), and repeatedly confirmed later (Hultin, 1964; Wilt, 1963; Piatigorsky, 1968; Infante and Nemer, 1967; Rinaldi and Monroy, 1969; Stafford *et al.*, 1964; Mangia *et al.*, 1973; Ruzdijic and Glisin, 1972; Goustin and Wilt, 1981; Martin and Miller, 1983), that ribosomes are mostly in the form of monoribosomes in the unfertilized eggs and that they become more and more associated in the form of polyribosomes beginning already 2 min (Rinaldi and Monroy, 1969) after fertilization and reaching levels even 30-fold higher than in the unfertilized eggs (Humphreys, 1971).

Apart from what, has already been said on ribosomes, the main reason why polyribosomes are almost absent in the unfertilized egg seems to have to be sought in the low availability of mRNA. It was very erly suggested that this can be due to the presence of some inhibitory proteins bound to mRNA which serve the purpose of storing it is an inactive state in the form of "informosomes" (Spirin and Nemer, 1965; Spirin, 1966; Nemer and Infante, 1968; Mano, 1966; Rurdijic *et al.*, 1973). The presence of messenger-containing ribonucleoprotein (RNP) particles in the post- mitochondrial supernatant of unfertilized eggs has, in fact, been demonstrated (Skoultchi and Gross, 1973). The mRNA for histones has been directly demonstrated within these by nucleic acid hybridization experiments and studies of their template activity (K. Gross *et al.*, 1973 a, b); and it has been shown that almost all the histone messenger can be found in these particles in the unfertilized eggs and that it becomes progressively transferred to polysomes following fertilization (Woods and Fitschen, 1978; Young and Raff, 1979). These RNP particles have been shown to have little template activity in cell-free systems, whereas their RNA, when purified from proteins in quite active (Cummins and Hunt, 1977), at least under certain ionic conditions like 3 mM Mg^{2+} (Ilan and Ilan, 1978) or 5 mM Mg^{2+}. Jenkins *et al.* (1978) have also reportest that in 0.35 M KCl the protein content of RNP particles is higher and their activity is lower than that of their RNA. All this suggests a regulatory role for the messenger-bound proteins of the RNP's present in the unfertilized eggs, and, as we shall see later in this chapter, RNP's seem to be involved not merely in sequestering the bulk messenger RNA before fertilization, but also in permitting an appropriately timed availability for translation of single messenger classes, thus realizing a post-transcriptional control of protein synthesis (Shepherd and Nemer, 1980; Infante and Heilman, 1981; E. J. Baker and Infante, 1982).

An electrophoretic analysis of the proteins contained in the inactive RNP's of the unfertilized egg has been performed by Moon et al., (1980), who reported the interesting finding that three of these proteins, of 140,000, 67,000 and 22,000 molecular weight, disappear shortly after fertilization.

More recently, however, Moon (1983) has resolved by electrophoresis 5–10 major and many minor proteins associated with the RNP particles of unfertilized eggs of *Strongylocentrotus purpuratus* and found at least no qualitative differences between these and those present in the RNP's associated with the polysomes of embryos.

These results did not change when obtained with eggs of *Arbacia punctulata* irradiated with UV in order to cross-link RNA and proteins.

Moon et al., (1982), moreover, have shown that the messenger RNA's purified from the RNP particles of unfertilized egg is at most only twofold more active in stimulating a cell-free system for protein synthesis than the RNP's themselves, which, in the authors' opinion, makes the hypothesis of RNA masking by protein less attractive. It has to be considered, on the other hand, that no quantitative comparison between the *in vivo* and *in vitro* activities can be made.

The genesis and fate of these RNP's are difficult to study, due to the difficulty of labeling them during oogenesis and then of obtaining in vitro maturation of the oocytes. When the unfertilized egg has been labeled in spite of its low permeability to precursors, it has been found that 80% of the radioactive RNA leaving the nucleus is incorporated into free RNP particles (Dworkin and Infante, 1978); when the metabolism of these RNP particles has been studied by labeling the embryos after fertilization, the kinetics of labeling does not favor the hypothesis of a precursor product relationship between the mRNA of the free RNP's and that of the polysomes (Dworkin and Infante, 1976; Enger and Hanners, 1978). This does not necessarily hold true for the RNP particles present in the unfertilized eggs, whose protein content, studied under proper ionic conditions, was found to be higher than that of RNP particles of later stages, thus reinforcing the idea of their role in messenger storage (Kaumeyer et al., 1978; Young and Raff, 1979). The higher protein content of the RNP particles containing maternal messenger has been found to hold true for those containing the histone mRNA of maternal origin (Gordon and Infante, 1983); Simplicistic conclusions cannot, however, be drawn, because these RNP's are translated in a cell-free system with the same efficiency as those containing embryonic mRNA.

Even if it has been repeatedly demonstrated that mRNA purified from unfertilized eggs is active as a template in cell-free systems (Maggio et al., 1964; D. W. Slater and Spiegelman, 1966, 1968; Jenkins et al., 1973, 1978; K. Gross et al., 1973b; Tonelli and Hunt, 1977), some later experiments have clearly shown that also the purified mRNA, if extracted from unfertilized eggs, is less active by a factor of 2 or 3 than if extracted from fertilized eggs or embryos (Pirrone et al., 1977; Ruderman and Pardue, 1977; Murray and Sosnowski, 1980), partially because of its lower ability to bind ribosomes (Dworkin et al., 1977; Rudensey and Infante, 1979). Why this RNA is less active is not known, but it has been excluded that this is due to the lack of the 5′ terminal "cap" because, although the sea urchin embryo cell-free systems show an absolute requirement for the "cap" in order to translate mRNA's (Winkler et al., 1983), the mRNA's of the unfertilized eggs have repeatedly been shown to be capped (Hickey et al., 1976; Pirrone et al., 1977; Sconzo et al., 1977; Rudensey and Infante, 1979). Neither this lower activity of the unfertilized egg mRNA can be attributed to the lack of poly(A) because also the poly(A)⁻-RNA from embryos is more active than that from unfertilized eggs. Recently, however, Duncan and Humphreys (1983) have described a decrease following fertilization in the content of the oligo (U) sequences present in the maternal RNA, and suggested that this might participate in the processes which activate maternal mRNA's; one theoretical possibility being that the oligo (U) sequences produce a more compact structure for mRNA.

Another step of the protein synthetic pathway that has been investigated has been that of amino acid activation. Early experiments carried out with crude cell sap fractions failed to show dramatic changes following fertilization, when the biological activity in stimulating amino acid incorporation into ribosomes was studied (Hultin and Bergstrand, 1960; Candelas and Iverson, 1966; Castaneda, 1969; Yasumasu and Koshihara, 1963; Stavy and P. R. Gross, 1967, 1969; Kedes and Stavy, 1969). Also when the ability of the egg cell sap to activate the carboxylic group of amino acids (Scarano and Maggio, 1957, 1959; Maggio and Catalano, 1963) or to load them on transfer RNA (Ceccarini et al., 1967; Ceccarini and Maggio, 1969; Zeikus et al., 1969) were directly investigated, again no dramatic quantitative changes were detected, although differences of a factor of two cannot be excluded when unfertilized eggs and embryos are compared.

Also the ability to load amino acids on the different isoaccepting transfer RNA's, as resolved by MAK chromatography, does not change appreciably, except for lysin and methionine, when the enzymes of unfertilized eggs and embryos are compared (Ceccarini and Maggio, 1969). A change for leucyl-tRNA-synthetase occurring at fertilization was described by Spadafora et al., (1973).

To conclude, no major role in the activation of protein synthesis at fertilization seems to have to be attributed to changes of the activity of the aminocyl tRNA synthetases. The same can be said for transfer RNA itself, as already suggested by experiments of Nemer and Bard (1963). When changes in the codon recognition properties of single tRNA's were investigated, no important changes were detected immediately following fertilization, although some changes appeared later in development (Molinaro and Mozzi, 1969; Molinaro and Farace, 1972; Zeikus et al., 1969; Yang and Comb, 1968; Spadafora et al., 1973).

A doubling of the activity of the elongation factor T1 has been described as occurring 2 min after fertilization (Felicetti et al., 1972), while an inhibitor of the initiation disappears from ribosomes (Metafora et al., 1971; Gambino et al., 1973). Yablonka-Reuveni and Hille (1983) have recently purified the elongation factor EF2 from eggs and embryos of Strongylocentrotus purpuratus, and found that although its total amount is about the same in unfertilized eggs and embryos, its intracellular distribution undergoes important changes following fertilization, which may be relevant in the increase of the protein synthesis rate.

Another question that has been asked has been whether the "transit time" of ribosomes along messenger RNA changes also at fertilization. This calculation has been made several times, with a negative answer in earlier experiments

(Humphreys, 1969; Mackintosh and Bell, 1969) and with a positive answer in the last ones (Brandis and Raff, 1978, 1979; Hille and Albers, 1979), which after very accurate measurements conclude that the "transit time" is decreased by a factor of 2 or 3 soon after fertilization, being about 40 min for the unfertilized eggs of *Strongylocentrotus purpuratus* at 17 °C. The problem of the measurement of the peptide elongation rate has been studied by a variety of approaches. Humphreys (1969) has exposed the embryos of *Lytechinus pictus* to radioactive amino acids and measured the time required for an amount of radioactivity equal to that present on polysomes (as growing peptide chains) to be transferred to the soluble cytoplasm. The result is 1.16 min for the embryo and 0.61 min for the unfertilized egg at 18 °C, which, assuming an average peptide chain of molecular weights 35–50,000, means an elongation rate of 2–3.3 amino acids s^{-1} in the embryo (or 1.4–2.3 at 15 °C). Mackintosh and Bell (1969), using a similar approach, concluded a time of growth of an average peptide chain of 6.1 min in the unfertilized eggs of *Arbacia punctulata* at 22°–24 °C and of 5.9 min for the fertilized eggs, which corresponds to about 0.39–0.44 amino acids s^{-1}. Brandis and Raff (1978, 1979) and Hille and Albers, comparing the curves for total incorporation and for polysomal incorporation, and also measuring the incorporation of exogenous amino acids into the proteins of various size classes at different time intervals (Brandis and Raff, 1978), concluded for a time of 43 min to complete an average length peptide (50,000 M.W.) in unfertilized eggs of *Strongylocentrotus purpuratus* at 16.5 °C (i.e., about 0.09 amino acid s^{-1}) and of 17 min for the fertilized eggs (i.e., about 1.94 amino acids s^{-1}). If the embryos were cultured at 12 °C, the values were 60 min for unfertilized eggs and 20–29 min for fertilized eggs or embryos (Hille and Alberts, 1979). Goustin and Wilt (1981) have exposed *Strongylocentrotus purpuratus* embryos to radioactive lysine, and measured the specific activity of the amino acid pool and therefore the absolute rate of protein synthesis. This referred to the number of ribosomes at work gives a value of elongation rate of 1.8 amino acids s^{-1} in embryos cultured at 14°–16 °C. The same authors (Goustin and Wilt, 1982) have also measured the elongation rate of two specific proteins, histones H2A and H1 α in cleaving *Strongylocentrotus* embryos by measuring the time lag incorporation of labeled amino acids into two known positions of the peptide chain. The results indicate an elongation rate of 0.69 and 0.80 amino acids s^{-1} respectively, at 15 °C. To sum up, different methods have provided different answers; the experiments of Martin and Miller (1983) have therefore been welcome because these authors, unlike all the others, do not make use of radioactive amino acids. They have, in fact, used a modified chromatin-spreading technique to study the structure of polysomes of *Lytechinus pictus* eggs and embryos in the electron microscope. Since they found some tailed polysomes, i.e., not yet fully loaded with ribosomes during the time of ribosome recruitment on the mRNA shortly after fertilization, they were able, by measuring the increase in number of ribosomes on these incompletely loaded polysomes at close time intervals, to calculate the rate of ribosome movement in the messenger. These experiments lead to a value of an elongation rate of 1.5–1.8 amino acids s^{-1}, at 16 °C, that is very close to the values reported by Goustin and Wilt (1981).

    In conclusion, there is general agreement on the existence of a marked acceleration of protein synthesis following fertilization, most of which can be explained by the increase in the number of ribosomes interacting with messenger to form polysomes, and for a small aliquot by the faster movement of ribosomes along the messenger. The increase in the number of ribosomes per unit length of mRNA, which might have been suggested by the data of Humphreys (1969) and of Rinaldi and Monroy (1969), has to be excluded according to the electron microscopie measurements of Martin and Miller (1983), who actually reported a decrease in the average size of the zygote polysomes (7.14 ribosomes) with respect to those of the unfertilized egg (11.96 ribosomes).

Why there is such an increase in polysome formation at fertilization can be explained either with the unmasking of the mRNA from the RNP particles, possibly accompanied by an intrinsic (conformational or maturational) change of RNA, or by an increased initiation rate due to the removal of the inhibitor from ribosomes. Obviously both these mechanisms might be at work, and both might be caused by an activation of proteases. The second one, however, has to be excluded, according to Hille *et al.* (1981), because if the elongation is experimentally inhibited with emetine there is no crowding of the ribosomes along the messenger, thus indicating that initiation is not rate-limiting in the unfertilized egg and until the 16-blastomere.

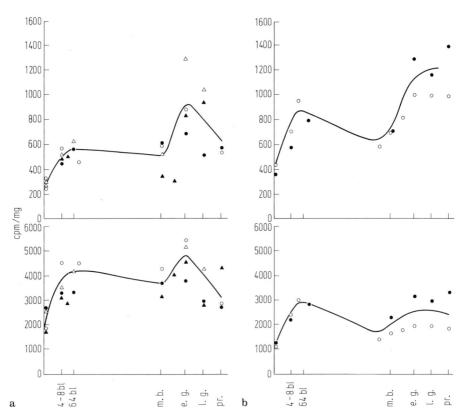

Fig. 6.1. Rate of uptake of 35S-methionine **a**, and ^{14}C-leucine **b**, in the total eggs and embryos (lower curves) and in the TCA insoluble fraction (upper curves). Vertical lines indicate stages of development. *4–8 bl.* 4–8 cell stage; *64 bl.* 64-cell stage; *m.b.* mesenchyme blastula; *e.g.* early gastrula; *l.g.* late gastrula; *pr* prism; *pl.* pluteus. (Giudice et al., 1962)

Interestingly, it becomes rate limiting at blastula. The faster movement of ribosomes along mRNA can be explained by the increased activity of the elongation factor T1. A computer-simulated model by Raff *et al.* (1981), for example, brings about the conclusion that the observed increase or protein synthesis rate at fertilization is brought about by an increased availability of messenger RNA and by an increase in the elongation rate.

Few variations of the overall rate of protein synthesis occur throughout development till early pluteus, following the initial burst, except for a second acceleration at the gastrula stage, which, as already said, brings the rate of protein synthesis to a level 50–100% higher than that of the cleavage stage (Giudice *et al.*, 1962; Berg, 1965; Neifakh and Krigsgaber, 1968; Ellis, 1966; Bellmare *et al.*, 1968) (see Fig. 6.1).

A word of caution has to be given in the evaluation of data obtained by the use of exogenous labeled methionine as a tracer, since Ilan and Ilan (1981) have shown that ^{35}S-methionine exogenously supplied becomes incorporated into the embryonic proteins through a route different from that of the internal pool, which, however, does not happen for exogenous leucine. In experiments, however, where measurements of the specific activity of the pool were made, the early data were essentially confirmed (see Fig. 6.2), with an increase of the protein synthesis rate to 240 pg h^{-1} embryo^{-1} from 2 blastomeres until 8 blastomeres, followed by a minor increase to 500 pg, by the blastula stage (Goustin and Wilt, 1981). These figures have been confirmed by the same authors when measuring the rate of histone synthesis by a method involving the cyanogen bromide cleavage of the histones.

It is important to recall once again that no net increase in the total amount of proteins is observed until the early pluteus stage, because all this active protein synthesis is at the expense of the yolk proteins stored into the egg cytoplasm during oogenesis, which become transformed into other cellular proteins after fertilization.

Not many authors have worked on the characterization of the yolk proteins.

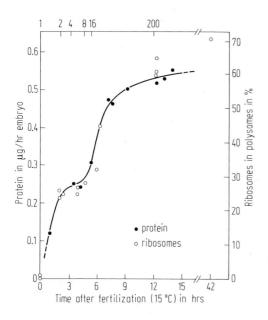

Fig. 6.2. Changing absolute rates of protein synthesis and levels of polysomes during development. Measurements of absolute rate of protein synthesis were conducted on embryos of *S. purpuratus* grown at 15 °C to various stages. The numerals on the top, of the figure indicate the number of cells embryo. Embryos were labeled with ^{3}H lysine at 15 °C and rates of protein synthesis calculated, utilizing measurements of lysine pool specific radioactivities. The fraction of ribosomes in polysomes at various developmental stages was determined from absorbance profiles. *Curve* through the points is an attempt to hand-draw a reasonable fit to the data on protein synthesis. (Goustin and Wilt, 1981)

Some good data are, however, available, which have been reviewed by Cognetti (1982): Kari and Rotman (1980) have indicated the existence of about 15 glycoproteins as components of the yolk granules, which undergo some relative developmental changes especially with respect to the major one, of 200,000 molecular weight, which disappears at blastula. Also Harrington and Easton (1980) found the disappearance of a major yolk glycoprotein at blastula. Ozaki (1980, 1982) and Harrington and Easton (1982) also reported an electrophoretic analysis of the proteins contained in the isolated yolk granules. Ozaki's conclusions are that these glycoproteins can be separated into three main components: a high molecular weight glycoprotein sedimenting as a 14S particle and two proteins immunologically unrelated to the glycoprotein; this latter contains neutral sugars, mannose and lipids, plus 62% uncharged, 24% negatively charged and 14% positively charged, amino acids. As to the site of origin of the yolk proteins, early observations (Chatlynne, 1969; Tsukahara and Sugiyama, 1969) have suggested an accumulation in the ovary accessory cells of some material which is later transferred into the growing oocytes. Ozaki (1982) and Harrington and Easton (1982) have demonstrated that this material has biochemical and immunological similarities with the egg yolk glycoproteins. According to Ozaki (1982), these glycoproteins are primarily synthesized by the coelomocytes present within the body cavity of the adult female, then transferred to the accessory cells and from these to the oocytes. Most recent data of Ozaki (communicated at the 2nd Symposium on Developmental Biology of the Sea Urchin, 1983, Santa Barbara, California) show that the yolk proteins are synthesized by those coelomocytes called trephocytes, and represent one half of the coelomocyte population. This is also suggested by studies on in vitro cultures of the coelomocytes, which continue to synthesize a putative precursor of a large yolk protein for at least 24 h. This precursor, of 200,000 molecular weight, is converted into a 93,000 protein when translocated into the oocyte yolk granules of the sand dollar *Dendraster excentricus*. This molecular weight reduction does not occur in three sea urchins investigated. It was again confirmed that none of these yolk proteins contains phosphates in detectable amounts.

Previous experiments of Ichio *et al.* (1978) had, on the other hand, shown that 40% of the material extractable from *Hemicentrotus pulcherrimus* is made of lipoproteins. These were divided into three main components on the basis of density, each component containing glyceride as a major lipid. These lipoproteins, however, also contain large amounts of carbohydrates.

No conclusive work is available about the mechanism of yolk protein breakdown. That this must occur is substantiated by the data of Fry and P. R. Gross (1970b), already mentioned, in which it has been calculated that about one half of the amount of stored yolk protein must be broken down from fertilization to the pluteus stage in order to balance as much synthesized protein during the same period without any major change in the net amount of protein per embryo. Several authors have described lytic enzymes in association with the yolk granules such as, for example, proteases and phosphatases. We refer the reader to Giudice (1973) for a review of the older data, and will mention here only the new data of Yokota and Nakano (1979), who have reinvestigated the nitrophenyl phosphatase found in association with the yolk granules of several sea urchin embryos, which can be divided in to four different forms with different pH optima. These authors, however, conclude that it is unlikely that these enzymes are involved in yolk breakdown, since they are not able to hydrolyze phosphoserine and phosphoproteins, which in any case are not the major components of the yolk. Several lysosomal enzymes have been found within the yolk granules of *Strongylocentrotus purpuratus* eggs by Schuel *et al.* (1975b).

Since the first cell divisions following fertilization are fairly synchronous in cultures of sea urchin eggs, these have been used as a naturally synchronized cell culture in order to investigate the question of variations of the protein synthesis rate during the cell cycle. The results have been conflicting: negative for P. R. Gross and Fry (1966) and Fry and P. R. Gross (1970a), variable for Timourian (1966) and positive for Meeker and Iverson (1971) and for Mano. The latter author has published a series of papers devoted to the problem such as studied in Japanese sea urchins (Mano, 1968, 1969, 1970, 1971a, b, c, 1977; Hirama and Mano, 1976; Mano and Kano, 1977a, b), and concluded that the cyclic variations of GSH, inversely correlated to those of GSSG, may be responsible for the cyclic stimulation of that part of the ribosome population which is bound to the endoplasmic reticulum.

As already discussed in Chapter 5, the possibility of the existence of a post-transcriptional regulation of protein synthesis in sea urchin embryos has been repeatedly suggested. The crucial argument in support of this hypothesis has been that of the possibility of obtaining developmental changes of the type of protein synthesized, in the absence of RNA synthesis, by taking advantage of the fact that development can proceed up to blastula also when RNA synthesis has been severely inhibited. The best evidence for this would be the appearance at a certain stage of specific proteins under such conditions. Since, however, individual proteins have been characterized only recently, most of the past evidence rests upon studies of changes of the overall pattern of protein synthesis. We shall therefore first describe these and then mention the studies on the synthesis of some specific proteins. Among the early experiments which made use of the monodimensional electrophoresis, those of M. Spiegel *et al.*, (1970) are worth mentioning, showing the appearance of a new band at about the 4–8-cell stage and those of Terman (Terman and Gross, 1965; Terman, 1970), showing that changes of the electrophoretic profile occur between fertilization and blastula also in the presence of actinomycin D. Two-dimensional electrophoreses (Brandhorst, 1975, 1976; Brandhorst *et al.*, 1979), resolving up to 400 spots, have demonstrated that only few changes are detectable between fertilization and blastula, but that up to 20% of proteins are changed by the pluteus stage. With the same technique it was shown that some proteins are characteristically synthesized only during oogenesis and that no major differences exist between those coded for by the poly (A)$^+$- and

Table 6.1. Quantitation of Changes in Relative Rates of Protein Synthesis

Developmental period	Number of changes observed[a]		
	Total	10 ×	100 ×
Zygote-mesenchyme blastula	101 (70)	61 (43)	9 (4)
Mesenchyme blastula-pluteus	76 (50)	37 (23)	0 (0)

(From Bédard and Brandhorst, 1983)

poly(A)⁻-RNA's. It would be interesting to know if the few changes observed between fertilization and blastula occur also in the presence of actinomycin D.

Extending these observation by a two-dimensional electrophoresis which resolves over 800 polypeptides Bédard and Brandhorst (1983) have essentially confirmed them, and specified that during development there is a substantial change in the rate of synthesis of about 20% of the 900 proteins investigated. For one half of them the change was of over ten fold (see Table 6.1). The qualitative changes detected are reported in Fig. 6.3.

It is of interest that T. Evans *et al.*, (1983) found that the synthesis of three major proteins is switched off after fertilization, while the synthesis of the other three major proteins is switched on.

The sensitivity of such a kind of analysis is markedly increased when the pattern of protein synthesis of a specific cell line is followed throughout development. This is what Harkey and A. H. Whiteley (1982b, 1983) have done by isolating the micromeres and then culturing them in sea water with the addition of horse serum, which permits them to differentiate and to form spicules as their in situ counterparts. By

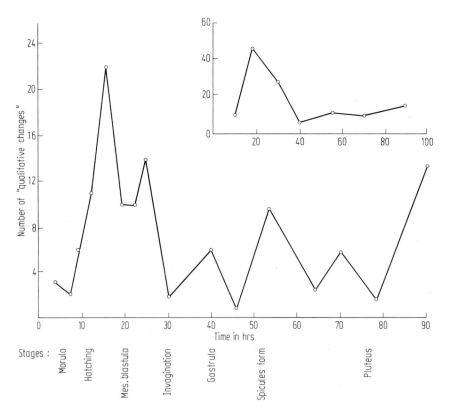

Fig. 6.3. Qualitative changes in protein synthesis during development. Each *point* represents the number of qualitative changes detected at a particular labeling interval when compared with the preceding labeling interval. *Inset* shows the sum of changes detected every 10 to 15 h of development. (Bédard and Brandhorst, 1983)

pulse-labeling these cells with ^{3}H-valine and then examining the labeled proteins by two-dimensional electrophoresis, it was found that more than one half of these proteins underwent quantitative or qualitative changes in the rate of labelig. The greatest number of changes occurs at the time when the mesenchyme cells glide into the blastocoel in the control embryos (Fig. 6.4). That the observed changes of the pattern of proteins synthesis occur also in the entire embryos is strongly suggested by control experiments in which the mesenchyme cells were isolated from prisms or early plutei and immediately labeled.

Early experiments based on immunological methods had revealed at least six new antigens appearing after the mesenchyme blastula stage (Westin and Perlman, 1972a), while others had been observed to change cell compartment (Westin and Perlmaan, 1972b; Lundgren and Westin, 1974; Westin, 1976). Since Harkey and A. H. Whiteley (1980, 1982a, b) have, however, been able to separate epithelial cells from mesenchyme cells of early gastrulae, the interesting finding was reported that out of 454 labeled proteins resolved by two-dimensional electrophoresis, 69 were labeled exclusively in one or the other cell type, thus indicating profound synthetic differences in different embryonic territories already at such an early stage.

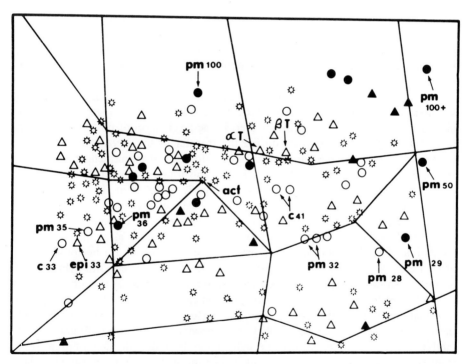

Fig. 6.4. Comparison of the protein synthesized by primary mesenchyme cells and epithelial cells of the pluteus larva. The electrophoretic positions of the labeled spots included in the analysis are indicated. Each spot is classified as described in the text as: mesenchyme-specific (●), mesenchyme-enriched (○), epithelium-specific (▲), epithelium-enriched (△), or cell-common (*). (Harkey and A. H. Whiteley, 1983)

6.2 Synthesis of Specific Proteins

The time course of the synthesis of the various histone forms has already been described in Chapters 4 and 5. We shall add here the surprising findings of Harrison and Wilt (1982), that the time of synthesis of the late H2 forms appears to be controlled by a chronological clock, because it is not altered by experimental disruption of the cell cycle, cessation of DNA synthesis, or alteration of nucleocytoplasmic ratios.

We have already described under the RNA section how transcriptional regulation occurs for the synthesis of different histone types during development. That a mechanism of post-transcriptional regulation of histone synthesis is, however, also at work in sea urchin embryos has been recently suggested again by several authors: Brandhorst (1980) found that the newly synthesized histone mRNA is preferentially translated versus the oogenetic one in the unfertilized egg, while the second is preferentially utilized following fertilization. Wells *et al.* (1981) observed that the RNA for H3 histone remains stored and untraslated in the embryonic cytoplasm for at least 90 min after fertilization and then becomes progressively transferred into polysomes, while that for α-tubulin is translated within 30 min of fertilization. E. J. Baker and Infante (1982) found that the RNA for the different histones is not only progressively but also differentially mobilitated from RNP's (again individuated as a place for mRNA storage) and transferred to polysomes at different development times. Goustin (1981), by accurate measurements of several parameters, concluded that in *Strongylo-centrotus purpuratus* during the first 6 h of development is the aliquot of histone mRNA, which becomes bound to polysomes to regulate the rate of histone synthesis, which is thereafter regulated by the amount of histone messenger present in the embryo.

Again, therefore, it is concluded for a post-transcriptional regulation of histone synthesis during the cleavage, followed by a transcriptional regulation in later phases. As already mentioned in Chapter 5, one mechanism of post-transcriptional regulation may also be at work during the transfer of the RNA from nucleus to cytoplasm. That such a mechanism may operate for the histone messenger is suggested by the experiments of Venezky *et al.* (1981), who found by in situ hybridization a concentration of histone messenger for the early histones in the pronucleus of *Strongylocentrotus purpuratus* egg 20–50 times exceedingly that of the cytoplasm (about 0.36 pg egg^{-1}), this RNA becoming transferred to the cytoplasm at the 2-cell stage. The hypothesis of a role of pronuclear breakdown as a mechanism for regulation of mRNA utilization has been reinforced again by further observations of De Leon *et al.* (1983), made by means of a specially sensitive in situ hybridization technique, showing that most of the early variant histone mRNA is contained in the pronucleus of the *Strongylocentrotus* unfertilized egg and then released into the cytoplasm at the time of pronuclear breakdown (see also Angerer *et al.*, 1984). On the other hand, it should be recalled that Showman *et al.* (1982) found that the inhibition of nuclear breakdown, brought about by 6-dimethylaminopurine treatment, does not inhibit the recruitment of histone mRNA onto polysomes. These authors, however, also found by cell fractionation a special accumulation of histone mRNA in the nucleated part of the egg, and by subfractionating the nucleated halves were eventually able to show that all the α-subtype histone mRNA was contained in the egg pronucleus (Showman *et al.* 1983).

It is of interst in this context to recall that Moon *et al.* (1983) found that eggs of *Arbacia, Lytechinus* and *Strongylocentrotus* contain a Triton X 100 insoluble cytoskeleton to which the RNP's of the unfertilized eggs, but not the polysomal mRNA of the embryos, are anchored. Does it play a role in sequestering the maternal messenger?

Herlands *et al.* (1982) find that the messenger RNA for the α-variants of histone is present in the unfertilized egg of *Strongylocentrotus purpuratus*, but is translated only after fertilization, whereas the messenger of the Cs histones are translated also in the unfertilized egg, which again calls for a post-transcriptional control.

The general picture therefore clearly emerges that both transcriptional and post-transcriptional control mechanisms are at work in the regulation of histone synthesis, and that the first one seems to predominate in the late development, i.e., in post-blastular stages, whereas the second seems to predominate in the early development.

It is also interesting to mention that Arceci and P. R. Gross (1977) have shown that the synthesis of histones does not coincide with that of DNA during the cell cycle of the early cleavage, which is in agreement with the findings of Cognetti *et al.*, (1974, 1977a) that sea urchin oocytes synthesize histones while not synthesizing DNA; part of the histones utilized for the nuclei are therefore derived from a maternal store, and part are newly synthesized. One of the most recent and accurate measurements of such a synthesis is due to Goustin (1981), who found that in *Strongylocentrotus purpuratus* the rate of histone synthesis increases at least 48 times within the first 6 h of development, i.e., from 0.5 pg embryo^{-1} h^{-1} to 24 pg; and to 182 pg in the following 6 h. The estimates of the amount of maternal histones agree on an amount of histones enough for about 100 nuclei, as estimated by measurements of the rate of histone synthesis during oogenesis (Cognetti *et al.*, 1974, 1977a) or by direct measurements of the egg histone pool (Salik *et al.*, 1981).

The protein whose synthesis has been studied in the greatest detail after histones in sea urchins development is tubulin. This important protein is needed in a great amount in the early stages because of the high number of mitotic spindles to be formed during the cleavage period, and at the blastula stage because of ciliogenesis. Meeker and Iverson (1971) were among the first to directly demonstrate synthesis of tubulin as identified by vinblastine precipitation and by binding of colchicine, during the cleavage stage. Raff *et al.*, (1971, 1972), Brandis and Raff (1978), and Raff (1975) provided the same demonstration by electrophorectic identification of tubulins α and β; they also indirectly proved that there is a storage of messenger RNA for such a protein in the unfertilized egg, and suggested that there is also a storage of the protein itself in the egg cytoplasm. This has been repeatedly confirmed by the studies of Bibring and Baxandall (1974, 1977, 1981), by the direct demonstration of tubulin synthesis during oogenesis (Cognetti *et al.*, 1977b), and by the possibility of causing crystallization of the tubulin stored in the unfertilized egg by treatment of the latter with vinblastine (Bryan, 1970, 1972). Finally, Detrich and Wilson (1983) have purified tubulin from *Strongylocentrotus purpuratus* eggs and found that it contains two major α-tubulin and a single β-tubulin species, whose electrophoretic behavior differs slightly from that of the tubulins of bovine brain.

Many studies have been carried out on the mechanism of production of the cilia, whose main protein, as already mentioned, is tubulin. Already in 1966 Auclair and

Siegel (1966a, b) reported that a storage of ciliary proteins must exist in the blastula cells because cilia can be regenerated up to four times, also under conditions of severe inhibition of protein synthesis, after they have been stripped off by treatment of the swimming blastulae with hypertonic sea water. Essentially the same results were obtained by Iwaikawa (1967) and Stephens (1972b, 1977). The latter author has shown that a pool of tubulin and dynein exists in the blastulae (see also Pratt, 1980), but not of other minor ciliary proteins like nexin and ribbon component 20, which might play a regulatory role in ciliogenesis. Studies on the kinetics of cilia regeneration have been carried out by Burns (1973, 1977, 1979), who has reported the interesting observation that the cilia stripped off from embryos animalized by trypsin grow longer again than in the normal embryos (see also Riederer-Henderson and Rosenbaum, 1975) unless protein synthesis has been inhibited, in which case they grow only to an approximate normal length. Finally, as will be described later, Ruderman and Alexandraki (1983) have demonstrated both a transcriptional and post-transcriptional regulation of the synthesis of tubulins.

Another stage-specific protein whose synthesis has been studied in sea urchin development is collagen. After the early suggestion by Ellis and Wintex (1967) that collagen might be synthesized in connection with spiculae formation, Pucci-Minafra et al. (Pucci-Minafra et al., 1972, 1975, 1978, 1980; Minafra et al., 1975, 1980) proved that collagen is synthesized by sea urchin gastrulae and is used for the formation of the spiculae. Pucci-Minafra has also made the interesting suggestion that the embryonic collagen might be synthesized on genes different from those of the adult animal. Golob et al., (1974) have confirmed the synthesis of collagen at the gastrula stage, and Gould and Benson (1978) have interpreted the fact that minimal concentrations of actinomycin D allow the formation of plutei apparently normal, but without skeleton (Giudice et al., 1968, Peltz and Giudice, 1967) with a selective inhibition of the synthesis of collagen; Benson and Sessions (1980) have also shown that the sea urchin embryos contain a prolyl-hydroxylase whose activity increases seven fold by the prism stage. Mizoguchi and Yasumasu (1982b) by stimulating the prolyne hydroxylase have provided indirect evidence that proto-collagen might be produced also at stages earlier than gastrula, to be hydroxylated later on.

Nakano and Iwata (1982) have described the synthesis of three collagen-binding proteins of about 88,000 molecular weight. This synthesis begins within 1 of fertilization and ceases at blastula.

Last, but not least, the synthesis of actin has been studied during development. A progressive increase of the mRNA for actin during development was detected by Merlino et al. (1980, 1981), both by measurements of its biological activity and of its hybridization to cloned DNA. Beside the transcriptional control a translational control of actin synthesis has also been described in sea urchins: Infante and Heilmann (1981) found a stage-specific and differential transfer of the messenger RNA for the different actin forms from RNP's to polysomes during development. At first two but then three, forms of actin (α, β and γ) have been described in sea urchins (Merlino et al. 1980, 1981; Kabat-Zinn and Singer, 1981; Poenie and Fromson, 1979; Infante and Heilmann, 1981). Bédard and Brandhorst (1983) found that an actin variant comigrating with the β-actin of mammalian cells is synthesized at a very low rate in the unfertilized egg, and its synthesis rate increase over 100-fold during embryogenesis; the major increase occurring between the 64-cell stage and blastula in *Strongylocentrotus purpuratus.*

Studies the structure of F-actin needles from sea urchin eggs have been reported by De Rosier and Censullo (1981).

No recent study on the synthesis of ribosomal proteins exists. Takeshima and Nakano (1982a, b), however, have reported a detailed electrophoretic analysis of those present in the cytoplasmic ribosome and found that they are in the number of 30 (plus two acidic ones) ranging from 11,900 to 51,700 molecular weight in the 40S subunit, and in the number of 42 (plus an acidic one), ranging from 12,700 to 76,300 in the 60S subunit. These authors have also found some species-specific differences between the ribosomal proteins of the large ribosomal subunit which can be revealed by a two-dimensional slab gel electrophoresis of the ribosomal proteins of *Pseudocentrotus depressus, Hemicentrotus pulcherrimus* and *Anthocidaris crassispina.*

Indirect evidence for the de novo synthesis of an enzyme, the ribonucleotide reductase, shortly after fertilization, has been provided by Noronha *et al.* (1972). Interestingly, the template for such an enzyme is already present in the unfertilized egg, since actinomycin D does not inhibit its synthesis.

One of the latest proteins whose synthesis has been investigated is "cyclin", a protein so named by T. Evans *et al.* (1983). These authors found that if *Arbacia punctulata* eggs are fertilized or parthenogenetically activated, three abundant new proteins start to be synthesized while the synthesis of the other three abundant proteins ceases. One of the new proteins, cyclin, is periodically destroyed at a particular point of each cell cycle, while it is continuously synthesized in the early cleavage. Cyclin disruption is correlated not merely in a temporal way with the cell cycle, since it does not occur in the activated eggs, which do not divide, and it is slowed down by inhibitors of cell division. But perhaps the most interesting feature of cyclin synthesis is the fact that its messenger is present in the unfertilized eggs, as directly shown by its translation in a cell-free system, but it is not used before egg fertilization or activation, which calls again for a post-transcriptional selective control of protein synthesis. Cyclin has also been detected but in the form of two proteins in *Lytechinus pictus* and *Spisula solidissima.*

A preliminary report has also been published (Nemer *et al.*, 1983) on the existence of synthesis of metallothionein in sea urchins, whose messenger RNA, characterized by the authors, changes in amount in various developmental stages, and accumulates preferentially in the ectoderm of the pluteus. Treatment of the embryos with Zn^{2+} ions, which is known to cause animalization, produces a 25-fold increase of the mRNA for metallothionein.

A promising approach to the study of the mechanisms of regulation of specific proteins stems from the work of Giudice *et al.* (1980), who have found that upon heating at 30°–32 °C, the embryos of *Paracentrotus lividus* or of *Arbacia lixula* greatly reduce the synthesis of the bulk proteins, while the synthesis of two proteins of the molecular weight of about 70,000 is initiated or greatly enhanced together with that of five minor proteins. Interestingly enough, these "heat shock" proteins are synthesized only when the embryos are submitted to heating after the hatching blastula stage (see Figure 6.6), whereas no such answer is obtained if they are heated at earlier stages. What is of interest for a possible biological role of the heat shock

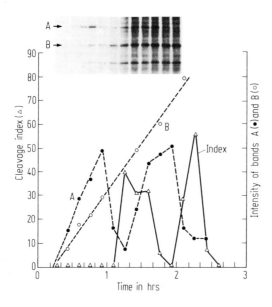

Fig. 6.5. Correlation of the level of cyclin with the cell division cycle. A suspension of eggs was fertilized, and after 6 min, ^{35}S-methionine was added to a final concentration of 25 μCi ml^{-1}. Samples were taken for analysis on gels at 10-min intervals, starting at 16 min after fertilization. Samples were taken 20–30 s later into 1% glutaraldehyde in calcium-free artificial sea water for later microscopic examination; the cleavage index is shown thus: $\triangle$———$\triangle$ The autoradiograph shown as an inset was scanned to yield the data plotted thus: cyclin, ●---●; protein B, ○---○. (T. Evans et al., 1983)

proteins is that if embryos are subjected to heat shock at stages when they are unable to produce the heat shock proteins, i.e., till the hatching blastula, their development is immediately arrested and they eventually degenerate; whereas if they are subjected to heat shock from the hatching blastula on, they produce the heat shock proteins and their development proceeds normally (Roccheri et al., 1981a, b, 1982b). The synthesis of the heat shock proteins is transcriptionally regulated, since it is prevented by actinomycin D and is linked to the appearance of specific mRNA's; the reversal to the normal pattern of protein synthesis seems, on the other hand, to be post-transcriptionally regulated, as judged by its insensitivity to actinomycin D (Roccheri et al., 1982a). The intracellular distribution of the h.s.p. indicates a preferential localization in the soluble cytoplasm, where they accumulate and survive for long time. A fugacious apprearance in the nucleus has, however, also been observed (Roccheri et al., 1981a), which might be of functional relevance (Fig. 6.6).

Many of the reported facts strongly suggest that sea urchin eggs also after fertilization are able store messenger RNA's made somehow inactive and to call them into action at specific developmental stages. What are the mechanisms that allow such a storage? The main theory that has been put forward, as already mentioned, and not disproved yet is that of "informosomes" (Spirin, 1966). The observation that giant RNA's are found also in the cytoplasm of sea urchin embryos (Giudice and Mutolo, 1969; Giudice et al., 1972a; Hogan and P. R. Gross, 1972; Brandhorst and Humphreys, 1972; Kung, 1974; Giudice et al., 1974; Sconzo et al., 1974; Rinaldi et al., 1974) has led Giudice and coworkers to formulate the hypothesis that mRNA precursors are stored in a nonactive form in the egg cytoplasm and matured into active mRNA's at specific developmental stages (Rinaldi et al., 1974). This is only a hypothesis to date; what has been ascertained is that the presence of these giant RNA's is not due to artifactual nuclear leakage because it is also

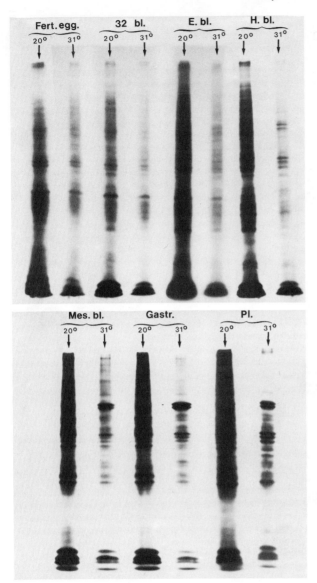

Fig. 6.6. Effect on the pattern of protein synthesis of *Paracentrotus lividus* embryos of heating at 31 ëC at different developmental stages. *Fert. egg.* fertilized eggs; *32 bl.* 32 blastomeres; *E.bl.* early blastula; *H.bl.* hatched blastula; *Mes.bl.* mesenchyme blastula; *Gastr.* gastrula; *Pl.* pluteus. (Roccheri et al., 1981 b)

found in microsurgically enucleated oocytes, as shown in Figures 6.7 and 6.8 (Giudice *et al.*, 1974), and many arguments strongly suggest that they do not derive from an artifactual molecular aggregation (Figs. 6.7, 6.8).

The findings of A. H. Whiteley and Mizuno (1981) of giant polysomes in sea urchin embryos (see Figure 6.9) are interesting to mention in this context: These

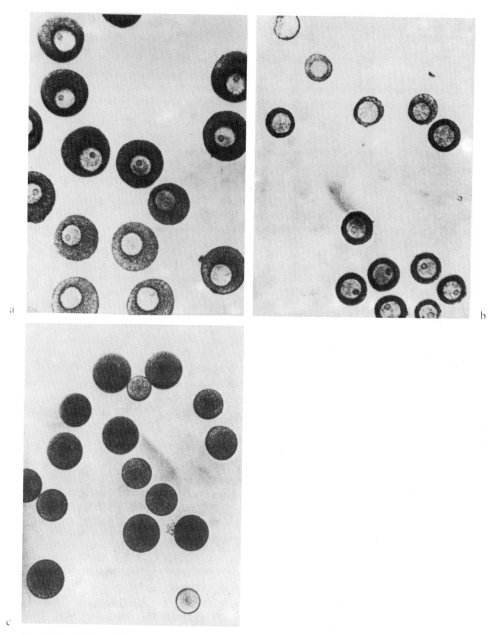

Fig. 6.7. a Isolated *Paracentrotus lividus* fully grown oocytes. (×150). b Nucleated halves (×150). c Non-nucleated halves (×150). (Giudice et al., 1974)

authors also suppose that they may have a regulatory significance, because they are formed also in the presence of actinomycin D or in enucleated eggs, but not if these are not activated by fertilization or parthenogenesis (Fig. 6.9). Also Martin and Miller (1983), with a new method of spreading, found that 14% of the polysomes of the

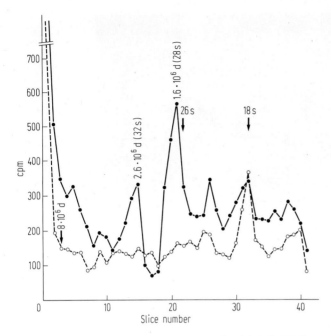

Fig. 6.8. Electrophoretical analysis of the labeled RNA extracted from the nucleated (●———●) and non-nucleated (○――○) halves of about 300 oocytes. (Giudice et al., 1974)

unfertilized egg of *Lytechinus pictus*, observed in the electron microscope, contain more than 20 ribosomes, some of them reaching up to 85 ribosomes. These authors find that the polysome size decreases after fertilization and speculate that this reflects the lower transcriptional efficiency of the eggs versus the embryos. It is worth mentioning that the hypothesis of a role of cytoplasmic processing in the control of the utilization of maternal information has been again put forward most recently by Davidson *et al.*, (1982), based on the observation of the presence of many interspersed repeats in very long trascripts found in the egg. The study of such a long transcript, carried out by the use of a cloned DNA probe called gene S88, has shown a great similarity and probably a colinearity of this 9.5 kb transcript stored in the egg and an RNA found in the nuclei of blastulae and gastrulae (Thomas *et al.*, 1982). The single copy part of this long transcript is found in the form of three smaller RNA's on the polysomes of embryos at the 16-cell stage, thus reinforcing the hypothesis of a storage in the egg of nonprocessed RNA.

Furthermore, Ruderman and Alexandraki (1983) have recently found that the RNA coding for α-tubulin is primarily in the form of transcripts of 2.5–2.8 kb, i.e., larger than the mature mRNA (which is 1.75–2.2 kb) before fertilization and becomes converted to the mature form following fertilization. In contrast, the RNA for the β-tubulin is in the mature form already before fertilization; the overall level of tubulin mRNA's rises from cleavage through pluteus. This shows that the synthesis of tubulins is both transcriptionally and post-transcriptionally regulated.

Another example of possible translational regulation of protein synthesis has been recently provided by Lau and Lennarz (1983) for two glycoproteins of 65,000 and

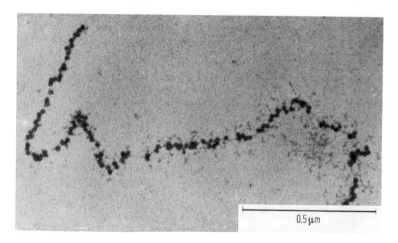

Fig. 6.9. Polysome from lysate of *Strongylocentrotus purpuratus* gastrula, showing a single expanding array of fibrous material believed to be nascent peptide chain. The polysome contains about 148 ribosomes and is 3.1 μm long. Stain = PTA; scale = 0.5 μm. (A. H. Whiteley and Mizuno, 1981)

51,000 molecular weight respectively. Their messenger RNA's are present already in the unfertilized egg of *Strongylocentrotus purpuratus*, but become associated with the membrane fraction, where the glycosylation should occur, only at the gastrula stage. It is interesting to note, however, that the regulation of the synthesis of other two glycoproteins studied by the same authors seems to be operated at a transcriptional level, since their mRNA's appear only at the gastrula stage, at least in a translatable form.

The conclusion is therefore reinforced that both translational and post-translational controls are at work during embryogenesis, and that the first one is predominant in the post-blastular stages, while the second predominates in the earlier stages. The reason for this should be sought in the fact that the embryo, in order to reach hatching, and to swim around freely, thus probably increasing its chances of survival in the environment, needs to build up at least 1000 cells; it therefore has to make the first cleavages at a very rapid pace, about every half an hour. This high division rate is especially necessary when the number of cells to be doubled at each division is still low; because once the early blastula is reached, about 500 new cells are made with only one synchronous cleavage and therefore at blastula the cleavage rate slows down (see Chap. 4). Therefore the strategy adopted by the animal has been that of making up a store of macromolecules and organelles during the oogenesis, i.e., inside the safe maternal body, so that the main macromolecule it has to synthesize up to hatching is only DNA. Since messenger RNA's are among the stored macromolecules to be used up to the blastula stage, it needs a mechanism of post-translational control to call them to work to synthesize the appropriate proteins at appropriate times during the interval between fertilization and hatching. A transcriptional control comes into operation subsequently as a predominant pathway.

Chapter 7

Nucleomitochondrial Interactions

Among the various materials that the sea urchin egg accumulates during oogenesis, a very important place is occupied by mitochondria. They are indeed produced during oogenesis (Matsumoto *et al.*, 1973, 1974; Enesco and Man, 1974) in great amounts, so that their DNA in the mature egg is more abundant that the nuclear DNA (Pikò and Tyler, 1965; Pikò, 1969; Pikò *et al.*, 1967, 1968; Kaneko and Terayama, 1974, 1975). When the oocyte becomes a mature egg, mitochondrial DNA synthesis is halted. Following fertilization, an intense nuclear DNA synthesis is started, as already described, but no mitochondrial DNA synthesis is resumed at least until the prism stage (Matsumoto *et al.*, 1974; Bresch, 1978; Rinaldi *et al.*, 1979b). This poses an interesting problem, that of the mechanism of uncoupling

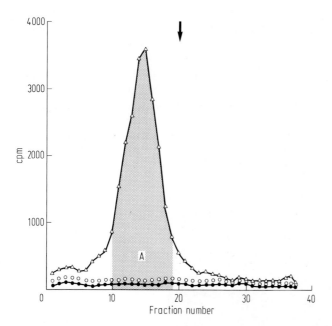

Fig. 7.1. CsCl-ethidium bromide gradient of mitochondrial DNA from activated eggs (●), activated enucleated halves (△) or fertilized enucleated halves (○). At 1 h after fertilization or parthenogenate activation, the eggs and the enucleated halves were incubated for 6 h with ^{3}H thymidine, the mitochondria were isolated and the DNA was purified. *Arrow* position of purified nuclear DNA. (Rinaldi et al., 1979)

of the nuclear DNA synthesis from that of mitochondrial DNA, i.e., of two functions that are always coupled in the somatic cells.

What is it that prevents mitochondrial DNA synthesis from occurring after fertilization in the presence of an active nuclear DNA synthesis, accompanied by cell division? This problem has been studied by Rinaldi et al. (1977, 1979a, b), taking advantage of the fact that mitochondria can easily be separated from the nucleus by centrifuging the unfertilized egg, according to the technique first described in detail by E. B. Harvey (1956) and then modified by several authors (see, e.g., Wilt, 1973). By doing so, a nucleated half is obtained which practically lacks mitochondria and a non-nucleated one, which contains almost all the mitochondria (see Fig. 2.7) in the species used (De Leo et al., 1979). Both halves are viable and can be fertilized or parthenogenetically activated. The experiments of Rinaldi et al. (1977, 1979, a, b), clearly demonstrate that it is the nucleus that prevents mitochondrial DNA synthesis. In fact, if one removes the nucleus through centrifugation and activates the non-nucleated half by parthenogenesis with butiric acid, then mitochondrial DNA synthesis is started, as demonstrated by ^{3}H-thymidine incorporation into mitochondrial DNA, and by the finding of replicative figures of mitochondrial DNA upon electron microscopic analysis. (Fig. 7.1 and 7.2) As a further proof that it is the nucleus that inhibits mitochondrial DNA synthesis, if one introduces a male pronucleus into the non-nucleated half instead of parthenogenetically activating it, then mitochondrial DNA synthesis is not activated. Removal of the nucleus not only activates mitochondrial DNA synthesis, but also enhances mitochondrial RNA and protein (Rinaldi et al., 1983a) synthesis, which becomes several-fold higher than the level already present in the fertilized eggs. Also mitochondrial replication is started, as suggested by figures of dividing mitochondria observed in the non-nucleated activated halves and not in the fertilized eggs (Fig. 7.3).

What is the mechanism by which the nucleus prevents mitochondrial replication following fertilization? This is not known, at present, but if this nuclear control is mediated through a molecule, this must be a short-lived one because it disappears

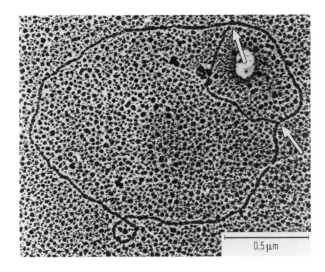

Fig. 7.2. A mitochondrial DNA replicative molecule from an activated enucleated half of *Paracentrotus lividus* egg. *Arrows* D loop. (Rinaldi et al., 1979b)

0.5 μm

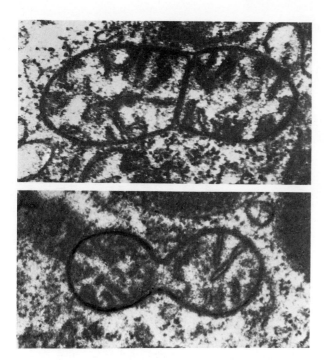

Fig. 7.3. Replicating mitochondria in non-nucleated
activated halves of *Paracentrotus lividus* eggs. (Rinaldi
et al., 1979a)

as the nucleus is removed and reappears as it is reintroduced. Another theoretical
possibility is that nuclei and mitochondria compete for one of the factors for DNA
synthesis.

There are no hints at present as to the question of what, at a certain point
during the later development, reverts the situation from the embryonic to that of the
somatic cells, i.e., that of coupling of nuclear division with mitochondrial division.
It can be conceived that the signal for such a reversal is represented by a critical
ratio between the nuclear and mitochondrial masses within the cells. This hypothesis
can be experimentally tested, because, if true, nucleated *Paracentrotus* egg halves, which
contain very few mitochondria, if fertilized and allowed to develop, should start
mitochondrial replication earlier than the embryos developing from entire eggs.

Many studies have been done in the past on mitochondrial RNA synthesis in
sea urchin eggs, by taking advantage of the fact that the nucleus can be removed
by centrifugation and therefore nuclear RNA synthesis does not interfere with that
of mitochondria (J. P. Chamberlain, 1968, 1970; J. P. Chamberlein and Metz, 1972;
Craig, 1970; Selvig et al., 1970; Craig and Piatigorsky, 1971). Mitochondrial rRNA
synthesis has therefore been demonstrated (Selvig et al., 1972; Pepe et al., 1976; Greco
et al., 1977; Innis and Graig, 1978; Craig and Innis, 1978; Rinaldi et al., 1977) as well
as 4S RNA synthesis (J. P. Chamberlain, 1970; Selvig et al., 1972; Rinaldi et al., 1977),
and messenger, or at least poly $(A)^+$-RNA, synthesis (Devlin, 1976; Devlin and
Swanson, 1975). Many of these experiments, however, need to be reinterpretated
at least from a quantitative point of view, in view of the findings of Rinaldi et al.
(1977, 1979b), that the removal of the nucleus activates mitochondrial nucleic acid
synthesis.

Table 7.1. Cytoplasmic Prevalence and Turnover Rates of Polyadenylated Transcripts of Mitochondrial Genes in Early Gastrula Stage Sea Urchin Embryos

Gene	Clone[a]	Approximate number of transcripts[b]			Kinetic measurements (blastula-gastrula stage)		
		Per gastrula embryo	Per average cell	Per egg	$t_{1/2}$ (hr)[c]	Apparent entry rate (poly(A) RNA) (molecules/embryo·hr)[d]	Apparent entry rate (total embryo RNA) (molecules/embryo·hr)[d]
16S rRNA	SpP389	1.6×10^6	2500	2.2×10^6	>17, 13	6.1×10^3	1.8×10^5
Cytochrome oxidase I	SpG30	1.1×10^5	180	4.5×10^5	10.5, 15.4 17.0, 11.5 ave. 13.6	4.0×10^4, 2.4×10^4 ave. 3.2×10^4	3.0×10^4, 5.8×10^4 ave. 4.4×10^4
URF-1	SpG41	2.4×10^4	40	ND	3.5, 4.6 ave. 4.0	ND[e]	ND[e]

(From Cabrera et al., 1983)

Note. ND, not determined.

[a] These clones were generated as described by Lasky et al. (1980) by addition of poly(dA) tails to hybrids of gastrula cDNA and poly(A) RNA (SpG clones), or of pluteus cDNA and poly(A) RNA (SpP clones). After tailing, the cDNA-poly(A) RNA hybrids were inserted into poly(dT) tailed pBR322 that had been cut at the *Bam*HI and *Eco*RI sites.

[b] Independent prevalence measurements of SpP389 and SpG30 were made by labeling and strand separating insert fragments and using them to titrate complementary sequences in the poly(A) RNA. Values for SpP389 are from Flytzanis et al. (1982), and for SpG30 from Lasky et al. (1980). The value for SpG41 is from the fraction of ^{125}I-poly(A) RNA (labeled *in vitro*) hybridizing with filters containing SpG41 DNA. The amount of ^{125}I-poly(A) RNA hybridizing was normalized to the hybridization obtained with standard filters that were present in the same reaction containing two other clones, SpG30 and SpG6, for both of which titration values of transcript prevalence are available (Lasky et al., 1980). Thus comparisons with the kinetic parameters must be made using the poly(A) RNA rather than total RNA entry rate estimates.

[c] $t_{1/2}$ is ln $2/k_d$, where k_d (see legend to Fig. 3) is the decay rate obtained in the least-squares solution to the labeling and pool-specific activity data.

[d] Apparent entry rate is the synthesis rate, k_s (see legend to Fig. 3), obtained in the least-squares solution for the labeling and pool-specific activity data. The apparent entry rate would be the actual entry rate into the poly(A) RNA compartment if at all times during the labeling experiment the mitochondrial GTP pool-specific activity is the same as the cellular GTP pool-specific activity (which may well be the case) (Gelfand and Attardi, 1981).

[e] Entry rates for this sequence were uncertain due to lack of precise knowledge regarding the length of insert DNA hybridizing.

The search for specific mitochondrial transcripts has now received a new hint from the use of cDNA clones (Cabrera *et al.*, 1983), which has revealed that the 16S mitochondrial rRNA [part of which is poly(A)$^+$] is the most prevalent among these transcripts in *Strongylocentrotus purpuratus* gastrulae (see Table 7.1).

The formation of mitochondrial polyribosomes following fertilization has been demonstrated by Innis and Craig (Innis and Craig, 1978; Craig and Innis, 1978). Harpold and Craig (1975) have proposed a role for the ratio ATP/ADP in the regulation of mitochondrial RNA synthesis, on the basis of the observation that it increases following fertilization and that its decrease inhibits mitochondrial RNA polymerase activity in a cell-free system.

The sensitivity of RNA synthesis by isolated mitochondria to various drugs has been studied by Cantatore *et al.*, (1974).

Variations of DNA polymerase activity in extracts of mitochondria isolated from different developmental stages have been reported by Gadaleta *et al.* (1977), which, however, do not parallel the in vivo mitochondrial DNA synthesis.

Once again, through the combination of techniques of the old experimental embryology, like that of enucleation by centrifugation, with those of the modern molecular biology, the sea urchin eggs have provided one of the best biological materials for the study of problems of central interest to cell biologists, like that of nucleomitochondrial interactions.

It is actually probable that the observed phenomenon is not restricted to sea urchins but extends to other eggs whith a storage of mitochondria. Rinaldi *et al.* (1981) have in fact provided preliminary evidence that mitochondrial DNA synthesis is activated also in *Xenopus laevis* oocytes if the nucleus is microsurgically removed.

We cannot conclude this survey of sea urchin mitochondria without mentioning the analyses of mitochondrial DNA recently carried out by Roberts *et al.* (1983) on isolated clones of full length mitochondrial DNA obtained from a *Strongylocentrotus franciscanus* DNA library. These analyses have shown that genes for 16S rRNA and for the subunit 1 of cytochrone oxydase are adjacent in sea urchins, but are separated in mammals, although gene polarity appears to have been conserved between these different phyla.

Again sea urchins promise to be a helpful material to solve the important problem of inheritance of mitochondrial DNA, and to study the fate of the paternal DNA during development. As already mentioned, Rinaldi *et al.* (1983b) have in fact. developed a method which allows isolation in bulk of mitochondria from *Paracentrotus lividus* sperm, from which good yelds of sperm mitochondrial DNA can be easily obtained.

Addendum

The papers discussed in this section were published, or came to my attention, after the present book had been completed. They are therefore reported here and listed corresponding to the order of the chapters and paragraphs of the book. The references are given in a separated list.

1 Fertilization

1.1 The Egg Surface

Nomura *et al.* (1983) have isolated also from *Anthocidaris crassispina* jelly coat three other sperm-activating peptides, which stimulate sperm respiration 20–30-fold at pH 6.8. The sequence of two of them has also been determined.

1.1.2 The Vitelline Layer

The study of the vitelline layer of *Strongylocentrotus purpuratus* eggs by means of monoclonal antibodies has made further progress (Niman *et al.*, 1984), which mate it possible to isolate the genes of the 25 different proteins identified by these authors by means of bidimensional electrophoresis.

Further research in the isolation and analysis of a sperm receptor glycoprotein from *Strongylocentrotus purpuratus* and *Arbacia punctulata* eggs has been presented by Rossignol *et al.* (1984c) and by Kubota and Carroll (1984).

1.1.4 The Cortical Granules and the Formation of the Fertilization Membrane

An interesting review has been published by Guraya (1982), giving a comparative view of the cortical granules of the eggs of several animal species. Extending his previous studies, Sasaki (1984) has provided evidence that chaotropic anions are responsible for differential Ca^{2+} sensitivity of the cortical reaction in isolated egg cortices and that a proteic factor of 100,000 m.w. is responsible for the Ca^{2+} sensitivity of the cortical granule discharge. Rossignol *et al.* (1984b) have made an accurate study of the well-known effect of an electric current on cortical granule discharge, and concluded that a high voltage pulsation of short duration causes the formation of functionally reversible pores in the egg plasmamembrane, through which Ca^{2+} flows into the cortical granules, causing a localized discharge, with a partial elevation of the fertilization membrane.

Turner *et al.* (1984) have found that cortical granule exocytosis in *Strongylocentrotus purpuratus* is accompanied by an increase in triphosphoinositide and diphosphoinositide (40 % and 20 %, respectively) and therefore suggest that these nucleotides may play a role in cortical granule exocytosis, perhaps by eliciting calcium liberation.

The method developed by Vacquier, allowing the egg cortex to be attached to a culture dish with its cortical granules still functioning has been employed by Moy *et al.* (1983) for studying the Ca^{2+}-induced release of glucanase from the cortical granules. It may be relevant, however, to keep in mind the effect described by Crabb and Jackson (1984) of the polycations used by this technique on the Ca^{2+} threshold.

The idea that the protease contained in the cortical granules contributes to the block to polyspermy has been further substantiated by the observation that the inhibitory activity of a wide series of protease inhibitors correlates quite well with their ability to induce polyspermy when added to *Arbacia* eggs (Alliegro and Schuel, 1983, 1984).

Further evidence that a trypsine-like enzyme contributes to the elevation of the vitelline membrane at fertilization has been provided by Sawada *et al.* (1984) and also by Justice and Carroll (1984).

In agreement with the idea that the H_2O_2 excreted by the fertilized egg plays a role in inactivating the sperms in order to prevent polyspermy, Blum *et al.* (1983) have observed that superoxide dismutase biomimetic compounds, which are able to penetrate the egg and to catalyze H_2O_2 formation, inhibit fertilization.

In line with this hypothesis are also the observations by Moss *et al.* (1983) that an antagonist of leucotriene (which is a lipoxygenase product of the arachidonic acid) causes polyspermy (see also Schuel *et al.*, 1984).

Details of the mechanism of ovoperoxidase action in the production of the fertilization membrane have been presented by Kay *et al.* (1984) and by Turner and Shapiro (1984). S. A. M. Chambers *et al.* (1984a, b) have followed the cellular location of the β-glucanase throughout development. This enzyme, which is first located within the cortical granules, moves to the hyaline layer and the fertilization membrane after fertilization, and becomes intracellular at the late gastrula stage.

1.1.5 The Block to Polyspermy

An interesting review on this controversial subject has been published by Nuccitelli and Grey (1984), who made an accurate comparison of most (perhaps all) of the available data. Their main conclusions are that the very rapid egg membrane depolarization brought about the "fertilization potential" represents a fast block to polyspermy which is temporary and also partial, since it can be overcome by high sperm concentration. This is the case for at least five sea urchin species.

It is relevant in this respect that Lynn and Chambers (1984), by electrode clamping eggs of *Lytechinus variegatus*, have found that the artificial suppression of the depolarization which normally accompanies fertilization inhibits sperm entry, although the eggs become activated, thus suggesting again a role of the electrical field across the plasma membrane in preventing sperm penetration. Similar observations have been reported on oocytes with intact germinal vesicle (Lynn *et al.* 1984; Mc Culloch *et al.* 1984, E. L. Chambers *et al.* 1984). Further evidence for the fast electrical block to polyspermy comes from experiments with anesthetics (Hinkley and Wright, 1984a, b).

1.1.6 The Hyaline Layer

The presence of antigens of the hyaline layer within the cortical granules has been confirmed by S. A. M. Chambers *et al.* (1984a), who have identified β-glucanase in both locations.

1.2 The Sperm Surface and the Acrosomal Reaction

The temperature dependence of the fertilization of some Japanese eggs, such as those of *Anthocidaris crassispina*, has been explained by Mita *et al.* (1984) with temperature dependence of the acrosomal reaction.

The role of Ca^{2+} influx through the sperm plasmamembrane in eliciting the acrosomal reaction has been further detailed by Darszon *et al.* (1984) by the use of selected sperm membrane compounds incorporated into artificial membrane systems, and by Kazazoglu *et al.* (1984) by means of Ca^{2+} antagonists. Trimmer and Vacquier (1984) have characterized, by means of monoclonal antibodies, a sperm glycoprotein of 210 kd involved in the jelly-activated acrosomal reaction. This glycoprotein binds the wheat germ agglutinin in a species-specific fashion (Podell and Vacquier, 1984a, b). Among the experiments aimed at demonstrating a role for sperm proteases in fertilization, those of Hoshi *et al.* (1979) are worth mentioning. These authors, by studying the effect of various protease inhibitors, suggest that a chymotrypsin-like activity is necessary for the penetration of the sperm through the vitelline layer.

1.3 Sperm-Egg Interaction

1.3.1 Sperm Motility

Sale *et al.* (1984), following high resolution electron microscope studies, have suggested a two-headed from for the 21S flagellar dynein. Therashita *et al.* (1983a, b) have studied the pre-steady-state kinetics of the 21S dynein and dynein subunits of *Pseudocentrotus depressus* sperms. The formation of the reaction intermediates was also analyzed in detail by these authors.

In line with the proposed role for neurotransmitters in sperm motility, Parisi *et al.* (1984) have shown that metergoline, a serotonine antagonist, depresses sperm motility.

More evidence has been added by Christen *et al.* (1983) to their previous results, suggesting that it is the low internal pH which inhibits motility of sperm within the gonad, by inhibiting the dynein ATP-ase. This is rate-limiting for the respiration of tightly coupled mitochondria. When the internal pH is alkalinized, the ATP-ase activity is increased, the respiration rate increases up to 50-fold, while mitochondria undergo transition to state 3. Bibring *et al.* (1984) have stressed the importance of the extracellular Na^+ in this chain of events and add that sodium requirement persists for the active sperms, since it is needed to balance the respiration-dependent production of H^+.

Further studies on the Na^+/H^+ exchange on isolated sperm flagella have been reported by Lee (1984).

An interesting hypothesis, substantiated by a series of facts, has been presented by Tombes *et al.* (1984): the distribution of energy from the mitochondrial site of production to the flagellar site of utilization does not occur through ATP, but through a creatine phosphate shuttle system.

1.3.2 Sperm Binding

To the question of the persistence of the sperm surface proteins after fertilization, Gunderson and Shapiro (1984) have added the observation that six sperm-specific proteins persist undergraded within the embryo at least till the gastrula stage. Longo (1984b) using cationized ferritin staining, has found that soon after fertilization the components of egg and sperm plasmamembranes intermix, which may explain the apparent contradiction in the early results.

H. A. Farach et al. (1984) have suggested a role for metalloproteases in the still poorly understood process of fusion of the sperm membrane with that of the egg. A role for bindin in this fusion process has been also suggested by experiments with mixed-phase vesicle whose fusion is elicited in vitro by bindin (Glabe, 1984).

1.3.3 Sperm Internalization and Pronuclear Movement

In the effort to clarify which mechanisms are responsible for the transformation of the sperm nucleus into a male pronucleus, Longo (1984a) has inseminated Arbacia zygotes and two- to eight-cell stage embryos during different phases of the cell cycle. Although pronuclei were never formed, in support of the idea that changes occur in the egg cytoplasm following fertilization which make it unable to support the transformation of sperm nuclei into pronuclei, it is interesting to note that the chromatin of sperms incorporated into prometaphase-anaphase embryos dispersed and then condensed into chromosome-like masses, whereas the chromatin of sperms incorporated into embryos just prior to prometaphase and at telophase failed to decondense. Sperm asters developed only in embryos inseminated during prophase to anaphase.

1.4 Some Physiological Changes that Occur at Fertilization or Following Parthenogenetic Activation

1.4.1 Ionic Movements

Ciapa et al. (1984a, b) have studied the biochemical and kinetic characteristics of the Na^+-K^+ exchange which occurs following fertilization, mainly by measuring the ^{86}Rb uptake.

Their main conclusions are the following: (1) the egg plasma membrane contains a Na^+-K^+ transport system which is stimulated by fertilization; (2) this stimulation depends upon the acid efflux; (3) the intensity of the Na^+-K^+ exchange is determined by the amount of Na^+ entering the egg at fertilization.

1.4.2 pH Changes

The uptake and efflux of amines has been further investigated by Christen (1983), who found that fluorescent amines and acridine orange added to the eggs of

Paracentrotus or *Strongylocentrotus* are concentrated in small intracellular compartments, probably vesicles, upon fertilization.

1.4.4 Other Permeability Changes

Allemand *et al.* (1984) have studied the mechanism of valine uptake before and after fertilization and concluded that about 5 min after fertilization a new mechanism of valine uptake is formed which is Na^+-sensitive and responsible for about 90% of the valine transport. This follows the Michaelis-Menten kinetics with a 15-fold increase in V and no change in K_m. Therefore the unfertilized egg valine transport of *Paracentrotus* resembles the L-system in the unfertilized egg and the ASC system in the fertilized egg.

The temporal sequence of the early changes which follow fertilization has been carefully reinvestigated on single *Arbacia* eggs by Eisen *et al.* (1984). These authors have shown that the first two events are a wave of surface contraction and the membrane depolarization; Ca^{2+} release occurs ca. 23 s later; then (at ca. 51 s after insemination) NADPH starts to increase and the fertilization membrane becomes elevated. These results, which are similar if *Lytechinus* eggs are used, essentially confirm the current views on the subject.

2 Embryonic Morphogenesis

2.1 General Description

Studies on metamorphosis might receive a new hint from the discovery of Burke (1984) that larvae of *Dendraster excentricus* can be induced to undergo metamorphosis by a pheromone releases by the adults.

A beautiful ultrastructural description of the process of primary mesenchyme cell invagination in *Lytechinus pictus* has been published by Katow and Solursh (1980). These authors suggest a relaxation of the basal cytoplasm of these cells followed by their active elongation and by the compression of the adjacent cells.

Ishimoda-Tagagi *et al.* (1984) have provided evidence that muscle tropomyosin is involved in the contraction of the pluteus coelom-esophagus complex.

2.3 Embryo Polarity

2.3.1 Animal-Vegetal Axis

With the aim of finding physiological differences between the egg poles, Peters *et al.* (1984) have measured the lipid and protein diffusion coefficients in the egg membrane with negative results. They were found to be 0.8 $\mu m^2 s^{-1}$ and 0.04 $\mu m^2 s^{-1}$ respectively at either pole in *Paracentrotus* eggs, and were not modified by Zn^{2+} or Li^+ treatments.

Ransick and Pollock (1984) have confirmed and extended their observation of the existence of particles of 137 Å located inside the plasmamembrane of the external surface of macro- and mesomeres, which are rare in micromeres.

O'Melia (1983) has failed to find differences between the rates of tRNA and 5RNA synthesis of normal and animalized embryos, which is at a variance with what it is known for the synthesis of ribosomal RNA. Leaf et al. (1984) have shown that clones of embryonic genes which are expressed by the differentiated mesenchyme cells are silent in micromeres; their RNA is also not yet present at this stage (see also Harkey, 1983 for a review on micromere determination and differentiation).

S. A. M. Chambers et al. (1984b) extending their previous observation on the different sensitivity to nuclease of the chromatin from micromeres or other blasto-meres, concluded that this is modulated by diffusible nuclear factors.

2.3.4 The Egg Cortex

A thorough study of the egg cortices of Strongylocentrotus purpuratus eggs isolated by means of moderate shear forces has been performed by Sardet (1984), who described an elaborate network of rough endoplasmic reticulum which connects the cortex to the cytoplasmic organelles.

The cortex was found to contain, in addition to the cortical granules, also some acidic vesicles, short filaments of actin, and plaques located in the inside face of the plasma membrane. Further studies of the isolated egg cortices have been presented by Chandler (1984), who followed the process of exocytosis by means of rotatory palladium shadowing.

2.3.5 The Mitotic Apparatus and the Mechanism of Cleavage

A method for isolating sperm asters from Lytechinus eggs has been described by Hinkley and Wright (1984b), by which also pronuclei can be isolated.

Further work has been carried out on the cytoplasmic dynein of sea urchin eggs: Sholey et al. (1984) have prepared microtubules from a variety of sea urchin species by the use of taxol. The dynein-like MgATPase associated with these microtubules has a molecular weight of about 350 kd. Further experiments with monoclonal antibodies (Piperno, 1984) confirm the existence in mitotic apparati of Strongylo-centrotus purpuratus of two dynein subunits, of 330 kd and 124 kd, comparable to those of the sperm tail. Hisanaga and Pratt (1984) have described a three- to eight-fold stimulation of the cytoplasmic dynein-like MgATPase by calmodulin, which suggests a role for calmodulin in the regulation of mitotic dynein activity. In keeping with the known role of Ca^{2+} in mitosis are the observations of Silver (1983) that antibodies against a protein with the function of a Ca^{2+} pump, associated with the mitotic apparatus vesicles inhibit mitosis, when microinjected into blastomeres of sand dollars or sea urchins (Silver et al. 1984).

Silver et al. (1983) had previously demonstrated the existence within the vesicles of mitotic apparati isolated from Strongylocentrotus purpuratus zygotes of a creatine kinase, which is able to generate ATP and therefore to catalyze the ATP-dependent Ca^{2+} sequestration. Details of the free Ca fluctuations during the mitotic cycle have been described by Suprynowicz and Mazia (1984) and by Poenie et al. (1984), using different approaches.

Further studies on the cycling of tubulin and actin polymerization have been carried out by Coffe *et al.* (1984) on eggs activated by procain. Their conclusion is that the polymerization of both proteins is responsible for the cyclic increase of cytoplasmic cohesiveness, but that the cycling is regulated by tubulin polymerization and depolymerization.

Hosoya and Mabuchi (1984) and Wang and Spudich (1984) have described in *Hemicentrotus pulcherrimus* and *Strongylocentrotus purpuratus* respectively a 45 kd protein which binds actin and modulates actin polymerization in a Ca^{2+}-dependent manner, increasing the steady-state concentration of the nonfilamentous actin. A similar protein has also been found by Sedlar and Bryan (1984) in *Tripneustes gracilis*.

Zimmerman *et al.* (1984) suggest that the ordered state of the water molecules is independent of the state of tubulin polymerization in vivo.

Two new inhibitors of sea urchin mitosis have been described, one, named stypoldione, is an orthoquinone which seems to act by disorganizing the cytoskeleton (O'Brien *et al.*, 1984); the other, named latrunculin, is a toxin which disrupts microfilament organization without necessarily disassembling microfilaments (Schatten *et al.*, 1984a, b).

Paweltz *et al.* (1984) have revived the idea of the existence of a *centrosome* as a physical entity, defined as a "definite body, changing its form in relation to processes where by one mitotic pole makes two mitotic poles". These authors pay a fair tribute to the old ideas of Boveri, which in their opinion should find better acknowledgement in the present literature. The observations of Paweltz *et al.* are based on both electron and optical microscopy and take advantage of the fact that a treatment with mercaptoethanol inhibits the chromosome cycle, while the centrosome cycle continues, which makes it possible to dissect phenomena connected with these two events. The conclusions are that the centrosomes are compact spherical bodies at the time of formation of the mitotic apparatus, which then become thin, flat plates, perpendicular to the spindle axis, to which (and not to centrioles) microtubules point. Each flat plate divides into two halves and becomes more compact, defining two separate poles.

Experiments by Sluder *et al.* (1984), in which microtubule assembly has been transiently inhibited by colcemid, further substantiate the hypothesis that the centrosomes provided by the sperm are the elements which organize a bipolar spindle.

The nuclear "lamins" and "perichromin-like" proteins have been studied by G. Schatten *et al.* (1984b). It is interesting that lamins are restricted from the nucleus of the polar bodies.

2.4 Cell Interactions

2.4.1 The Process of Gastrulation

The composition and the formation of the basal lamina have been studied by means of immunofluorescence by Wessel *et al.* (1984). These authors have prepared polyclonal antibodies against a series of proteins known to be part of the extracellular matrix of vertebrates, i.e., fibronectin, heparan sulfate proteoglycan, laminin, and collagen types I, III, and IV. Most of these antigens are found already in the unfertilized eggs of *Lytechinus variegatus* in granules of 0.5–2.0 μm, distinct from the cortical granules. After fertilization, these antigens migrate to the basal lamina of the developing embryos. In addition to these, two other components of the basal lamina appear to be synthesized de novo, as revealed by monoclonal antibodies.

Grant and Lennarz (1984) have found that a great deal of polymannose protein are also already present in the zigote of *Strongylocentrotus purpuratus*, but that some glycoproteins are synthesized at specific development stages (over 30 species per stage).

Extending their previous observations, Iwata and Nakano (1983) have characterized the fibronectin which can be extracted by the ovaries of *Pseudocentrotus depressus*.

In an ultrastructural description of the blastocoelic content, Katow and Solursh (1979) report the existence of granules of 30 nm diameter whose appearance is impaired by the absence of sulfate. At the gastrula stage some fibers add to the granules, to form a rough network lining the blastocoel wall. Katow *et al.* (1982) have also detected, by indirect immunofluorescence, fibronectin on the surface of the primary mesenchyme cells only at the stage when they migrate.

It is interesting that Katow has communicated (*Symposium on Cell Interaction and During Early Development* 1984) that fibronectin stimulates the migratory activty of the primary mesenchyme cells.

Venkatasubramanian and Solursh (1984) have also found that isolated primary mesenchyme cells spread and migrate in vitro on fibronectin, but not on the other substrates tested. Matranga *et al.* (1984) have found that fibronectin becomes stimulatory of cell reaggregation only at the blastula stage. All this suggests a role for fibronectin in the mesenchyme cell-specific migration.

Anstrom *et al.* (1984) have identified, by immunoelectron microscopy, an antigen which appears specifically on the surface of the primary mesenchyme cells at the time of their migration inside the blastocoel and seems to be already present in the laminar portion of some of the cortical granules.

Further work has been carried out by the group of Lennarz on the metabolism of dolichol and dolichyl phosphate in view of its proposed role in gastrulation. Rossignol *et al.* (1984a) have provided indications that the phosphate group of dolichyl phosphate turns over rather rapidly and at a rate which increases as development proceeds in *Strongylocentrotus purpuratus*, apparently by the increase in activity of a dolichyl phosphate phosphatase.

2.4.2 Cell Dissociation-Reaggregation Studies

Using a quantitative approach, Fujisawa and Amemiya (1982) have found that the strength of intercellular adhesion keeps increasing after the blastula stage in *Anthocidaris crassispina*. This increase, however, fails to occur in the presence of zinc ions.

Another piece of information has been added by Spiegel and Spiegel (1980) to the question of whether dissociated cells differentiate and reaggregate at random. The answer which comes from the observations on the differentiation of aggregates deriving from cells which have been dissociated one or two times, suggests again that cells do not dedifferentiate and that they do not aggregate at random; in fact the time required for these cells to make up a pluteus-like structure is not appreciably longer than for the nondissociated embryo.

A good approach to the solution of such a question can come from the experiments of Yazaki (1984). This author has succeeded in distinguishing, by means of fluorescent antibodies, the surfaces of the embryonic cells facing inward from those facing outward. It will therefore be theoretically possible, by means of this method, to establish whether or not the cells reverse their position during the dissociation reaggregation process.

Further work has been carried out by McCarthy and Spiegl (1983a) to analyze the hyaline layer, which has been indicated by Spiegel as the main responsible for cell adhesion in sea urchins. Using I^{125} surface labeling, these authors established a species- and stage-specific electrophoretic pattern of the hyaline proteins. Cell dissociation removes from the surface eight of these proteins, which are replaced during reaggregation. These authors (1983b) also recommend caution when the effect of various "reaggregation factors" is studied on cell reaggregation, since this can be stimulated also by nonspecific proteins like bovine serum albumin, as already stated in Chapter 2.4.2. Special attention is dedicated in this paper to the "butanol factor", which is shown to contain also many nonsurface iodinable proteins. It is my opinion that the "butanol factor" should not be regarded as "the reaggregation factor" but as something that "contains reaggregation factors". One of the best proofs of this is that univalent antibodies against the "butanol factor" dissociate the embryos in a species-specific manner, and that the "butanol factor" promotes aggregation in a stage-specific way (Cervello et al. 1984). Noll (personal communication) has found that the main component of the "butanol factor" is a 22S glycoprotein complex which is initially located within the yolk granules, then undergoes developmentally regulated processing into six subunits ranging from 70 to 160 kd. The 22S complex strongly binds concanavalin A. Noll proposes that the 22S complex binds from one side to the cell surface in a way which is not species-specific and from the other side to an identical molecule, thus bridging two cells together in a species-specific way.

The role of the fibronectin extractable from *Hemicentrotus pulcherrimus* ovaries in enhancing cell-to-cell and cell-to-substratum adhesion of micromere-derived cells has been stressed by Miyachi et al. (1984).

The Nagoya group has also tested the stimulating effect of some plant lectins on the ability of dissociated cells of *Pseudocentrotus depressus* embryos to adhere to substratum. Iwata and Nakano (1984) have also observed that Con A stimulates the release by the dissociated cells into the medium of some preexisting fibronectin and of two newly synthesized proteins of 88 and 140 kd respectively. These fibronectin-associated proteins (FAP's) seem to exert a negative control on spiculae formation, since microinjection of anti FAP's antisera into the embryonic blastocoel causes the formation of a higher number of spiculae (Iwata and Nakano, personal communication).

2.4.3 Metabolism and Cell Interactions

Following the observation that dCMP-aminhydrolase activity fails to undergo the decrease which accompanies normal development if the embryo has been dissociated into cells. De Petrocellis *et al.* (1984) have tried the effect of several inhibitors of macromolecular syntheses on this enzymatic activity in nondissociated embryos. The conclusion, which stems mainly from the effect of actinomycin D, is that the synthesis of a repressor-type molecule is needed for this activity to decrease.

3 Energy Metabolism

3.2 Carbohydrate Metabolism

Okabayashi and Nakano (1983) have carried out new studies on the cytochrome system of eggs and embryos of *Hemicentrotus pulcherrimus*.

Their main conclusions are: the unfertilized egg contains a complete cytochrome system. Only the level of cytochromes $C + C_1$ increases with development (1.5-fold at gastrulation). Antimycin A-sensitive and -insensitive NADH cytochrome c reductase activities increase during development. No cytochrome oxidase inhibitor was detected in the unfertilized eggs of this species.

Two types of acid phosphatases have been isolated from eggs and embryos by Yokota and Nakano (1984), one showing a preferential activity for flavin mononucleotide and the other for ADP and ATP.

Swezey and Epel (1984) have reinvestigated the question of the distribution of glucose-6 phosphate dehydrogenase within the unfertilized and fertilized egg, and concluded that the enzyme is distributed throughout the unfertilized egg cell, but is excluded from the cortex following fertilization. The enzyme is bound to a cytoskeletol element which is not soluble in Triton X 100.

4 Deoxyribonucleic Acid

4.1 Organization of the Genome

It is interesting that Moore (1984) has demonstrated that the DNA complementary to the RNA stored in the unfertilized egg is evolutionarily more conserved than the average single-copy DNA, thus stressing a role for the maternal RNA.

The procedure by Kaneko and Terayama (1974) is worth mentioning if one wants to make accurate colorimetric measurements of the total DNA content of sea urchin eggs or early embryos, without the interference of other chromogenic materials.

4.2 Some Specific Loci

Hindenach and Stafford (1984) have determined the nucleotide sequence of a region of the *Lytechinus variegatus* ribosomal DNA genes which spans from 232 nucleotides before the 3' terminus of the 18S region to 338 nucleotides after the 5' end of the 26S region, and comprises the two internal transcribed spacers and the 5.8 rDNA. The internal transcribed spacers were found to be GC-rich, as in vertebrates, whereas the 5.8 and 5' 26S regions recall the yeast rDNA.

Extending their previous studies, Maxon *et al.* (1983) have cloned the genes for the late H2 B histones of *Strongylocentrotus purpuratus*. They are in the range of 5–12 copies for genome and, unlike the early counterparts, lie scattered and irregularly arranged. Their transcription products are detectable at the 16-cell stage and rapidly accumulate in the interval between 14 and 16 h after fertilization.

Holt and Childs (1984) have found in *Lytechinus pictus* a third class of histone early repeat gene family (LpE). The study of such a family, which is also tandemly repetitive, shows good homogeneity within the members of this family in a fashion, which suggests a concerted evolution,

Brookbank (1984) has found that the H1 of adult sea urchin tissues has an electrophoretic behavior different from that of H1 of sperms or embryos.

Other genes which have recently been studied are: that of β-glucanase (S. A. M. Chambers *et al.*, 1984a); and those coding for transcripts prevalently expressed in primary mesenchyme cells (Leaf *et al.* 1984), which have been isolated in both cases from cDNA libraries.

4.3 Chromatin

A thorough investigation of the various histone forms of embryos of *Parechinus angolosus* has been carried out by Schwager *et al.* (1983) by several types of electrophoresis, analysis of the amino acid composition and partial sequencing. The results show that at least 24 different histone types are synthesized by the developing embryos.

Richards and Share (1984) have shown that the nucleosomal population of *Strongylocentrotus purpuratus* embryos becomes increasingly heterogenous as development proceeds, which correlates with an increasing heterogeneneity of the nucleosomal proteins. Nucleosomal heterogeneity seems to be brought about by variations of the core histone subtypes and by changes in some nucleosome-associated nonhistone proteins.

Poccia *et al.* (1984) have found that the decrease of the nucleosomal repeat length which occurs after fertilization is independent of DNA and protein synthesis. On the other hand, Poccia *et al.* (1984) found that under these inhibitory conditions the substitution of the male histone variants with the CS H2A and H2B also occurs almost normally.

4.4 DNA Synthesis

Thymidine Kinase.

Nishioka et al. (1984) have found that uptake and phosphorylation of externally supplied thymidine are fully stimulated by fertilization also if DNA synthesis has been experimentally inhibited with aphidicolin.

Morioka and Shimada (1984), using tritiated nucleositic precursors and higher embryonic concentrations than those used by Pirrone et al. (1979), have also been able to detect some synthesis of diadenosinetetraphosphate in morulae and prisms of *Hemicentrotus pulcherrimus*. No correlation with the rate of DNA synthesis has, however, been investigated yet.

5 Ribonucleic Acid

5.1 Ribosomal RNA

Steinert et al. (1984), using a specific rDNA radioactive probe, have demonstrated the presence of ribosomal RNA within the so-called heavy bodies of the eggs.

5.3 Messenger RNA's

More data about the ribonucleoside triphosphate pool have recently been produced by Killian and Nishioka (1984), who have measured the GTP pool before and after fertilization of *Strongylocentrotus* eggs.

Uzman and Wilt (1984) have assayed the RNA polymerase activity by incubating nuclei isolated from the 4- to 600-cell stages and found that, in such a "run-off" type assay, the in vitro activity closely parallels that in vivo. If, however, part of the nuclear proteins is removed by sarkosyl, the nuclei of later stages are more stimulated, but do not synthesize early histone mRNA's, indicating that the relative genes are inactive at these stages because they do not possess any initiated polymerases.

Preliminary work on isolated nuclei by Shupe and Weinberg (1983) leads also to the conclusion of a transcriptional regulation of the synthesis of the late histones, but adds the important point that there might also be a decrease in the stability of the early histone mRNA's as development proceeds.

Cox et al. (1984) have further improved their in situ hybridization technique by the use of asymmetric probes for mRNA, reaching a resolution power of 20–75 molecules of mRNA cell^{-1}. They came to the important conclusion that while histone mRNA molecules are uniformily distributed in the early developmental stages, they show a characteristic pattern of distribution at later stages. These findings may open a new way of looking at the embryo, i.e., no longer as a homogenous system, which is the limit of the usual molecular biology techniques, but as a mosaic of different cells.

Krieg and Molton (1984), injecting chicken H2B histone synthetic free-mRNA into *Xenopus* oocytes, have also found that correct 3' ends can be formed by processing of a longer precursor.

Vitelli *et al.* (1984) have now succeeded in microinjecting cloned sea urchin histone genes into *Paracentrotus lividus* eggs. These injected genes undergo replication and faithful transcription.

Raff *et al.* (1984) have made the interesting observation that whereas cidaroids, sea urchins, and other classes of echinoderms, such as starfish and sea cucumbers, all have α-type histone genes, only the modern sea urchins, split from the older ones between 190 and 200 million years ago, accumulate α-type mRNA during oogenesis and store it in the egg cytoplasm.

The following information should be added to the list of results obtained by Crain in his analysis of actin mRNA synthesis: The rate of actin synthesis parallels the amount of actin mRNA at all the studied stages (Crain *et al.* 1981).

6 Protein Synthesis

6.1 Rate of Protein Synthesis

Dube *et al.* (1984) have found that the rise of the intracellular pH, although necessary to trigger the acceleration of proteic synthesis at fertilization, is later on not required to maintain this high level of synthesis.

The experiments of Branning and Winkler (1984) and of Richter and Winkler (1984) support the hypothesis of mRNA "unmasking" following fertilization. These authors have found a change in the affinity to poly(A) and poly(U) of the ribosome and RNP-associated proteins following fertilization, and that if only one ribosome is allowed to bind per mRNA chain, by the use of anisomycin in vivo (which blocks elongation), less mRNA-ribosome complexes are formed following fertilization.

6.2 Synthesis of Specific Proteins

A preliminary study on the synthesis of various histone subtypes in *Arbacia punctulata*, carried out by Kuwabara *et al.* (1983), has indicated that some synthesis of histones occurs also before fertilization.

A mRNA coding for a 41 kd protein which binds to anti-tubulin affinity columns has already been found in the unfertilized egg (Bray and Hunt, 1983; George *et al.*, 1983). This RNA, however, starts to be translated at a high rate only after fertilization. It is interesting that the size of this RNA is considerably larger than that required for the protein it codes for.

Benson and coworkers (Benson *et al.* 1983; Blaukenship and Benson, 1984; Benson and Benson, 1984) have questioned the idea that the organic matrix of the spicule is made of collagen. They suggest rather that collagen is the main component of an extracellular matrix closely associated with spicule and necessary for their formation. These conclusions are based on electron microscopic studies of the spiculae and on the observation that inhibition of collagen synthesis inhibits the formation of spicule by isolated micromeres, but that spicule are formed also in the absence of collagen synthesis if the micromeres are cultured on a collagenous substratum.

It is clear therefore that collagen is associated with spicule and is necessary for their formation; the point which needs further clarification is whether it remains outside the spicule, as these experiments suggest, or becomes part of their organic matrix.

The well-known role of an increased Ca^{2+} uptake for the formation of spiculae has been further substantiated by the experiments of Farach and Carson (1984), in which spiculogenesis is inhibited by monoclonal antibodies against a surface protein of the mesenchyme cells, which blocks Ca^{2+} uptake.

It was just by chance that I reported on Chapter 6 of this book the paper of Nakano et al. (1972) concerning the indirect evidence of synthesis of ribonucleotide reductase on preformed templates, next to the new discovery of Hunt and coworkers of "cyclin". I was therefore very pleased when in April this year I received a letter from Dr. Hunt telling me that one of the two other major proteins which are actively synthesized on maternal templates together with cyclin following fertilization, i.e., the B band, has been identified as the H2 subunit of the ribonucleotide reductase (N. Standart, S. Bray, T. Hunt and J. V. Ruderman, unpublished results).

Continuing his studies on cyclin, Hunt has found that the inhibition of DNA synthesis also prevents cyclin degradation, and that cyclin turns over much more slowly when cell division is prevented by colchicine, taxol, or cytochalasin. Cyclin synthesis is activated if eggs are parthenogenetically activated with 10 mM NH_4Cl, which activates protein synthesis but does not allow cleavage to occur. In this case, too, there is no degradation of cyclin (Cornall et al., 1983).

The discovery of cyclin may explain the observations of Wagenaar and Mazia (1978) and of Watanabe and Shimada (1983) that the inhibition of protein synthesis, brought about by emetine, inhibits the cleavage of fertilized eggs.

Nemer and coworkers have now published an extended paper on the synthesis of metallothionein in sea urchin embryos (Nemer et al., 1984). The results confirm and extend what was already mentioned in Chapter 6.2, leading to the conclusion that the level of metallothionein mRNA is kept low in tissues other than ectoderm, in which, on the other hand, it can be induced to rise by Zn^{2+} treatment. Its level is constitutively high in the ectoderm.

The synthesis and the characteristics of two proteins, one specific for the mesenchyme cells and the other specific for the endodermal cells, have been studied by Wessel and McClay (1984).

All the result on the synthesis of heat shock proteins in sea urchins have been essentially confirmed for other species by Howlett et al. (1983) and by Maglott (1983, 1984). The latter author has also found that treatment with diamide stimulates the synthesis of proteins which in two-dimensional polyacrylamide gels migrate like the heat shock proteins.

The Caltech group has provided further evidence (Posakony et al., 1983) that the bulk of cytoplasmic RNA of the sea urchin eggs and early embryos is more similar to the unprocessed nuclear RNA than to the mRNA which can usually be observed in somatic cells. Does this play a role in the regulation of protein synthesis?

7 Nucleomitochondrial Interactions

Jacobs *et al.* (1983) have found that a sequence homologous to the 3′ end of mitochondrial 16S rRNA gene flanked by what seems to be the sequence coding for cytochrome oxidase subunit 1, is present in the nucleus of *Strongylocentrotus purpuratus* embryos. The authors suggest that the results are indicative of an evolutionary germ-line transposition of a fragment of mitochondrial DNA into nuclear DNA followed by some rearrangements and single nucleotide substitutions.

References

Adams R L P (973) Delayed mothylation of DNA in developing sea urchin embryos. Nature (London) New Biol 344: 37–39

Afzelius B (1959) Electron microscopy of the sperm tail. Results obrained with a new fixative. J Biophys Biochem Cytol 5: 269–278

Atello F, Maggio R (1961) The effect of sea water on cytochrome oxidase and oxidative phosphorylation. Experientia 17: 390–392

Akasaka K, Terayama H (1983) Sulfated glycan present in the EDTA extract of *Hemicentrotus* embryos (mid gastrula). Exp Cell Res 146: 177–185

Akasaka K, Terayama H (1984) A proteoglycan fraction isolated from the EDTA extract of sea urchin (*Hemicentrotus pulcherrimus*) gastrulae stimulates reaggregation of dissociated embryonic cells. Exp Cell Res 150: 226–233

Akasaka K, Amemija S, Terajama H (1980) Scanning electron microscopical study on the inside of sea urchin embryos (Pseudocentrotus depressus). Effect of aryl-B-xyloside, tunicamycin and deprivation of sulfate ions. Exp Cell Res 129: 1–14

Aketa K (1967a) On the sperm-egg bonding as the initial step of fertilization in the sea urchin. *Embryologia* 9: 238–245

Aketa E (1967b) I Isolation of the plasma membrane from sea urchin egg. Exp Cell Res 48: 222–224

Aketa K (1973) Physiological studies on the sperm surface component responsible for sperm-egg bonding in sea urchin fertilization. I Effect of sperm-binding protein on the fertilizing capacity of sperm. Exp Cell Res 80: 439–441

Aketa K (1975) Physiological studies on the sperm surface component responsible for sperm-egg bonding in sea urchin fertilization. II Effect of concanavalin A on the fertilizing capacity of sperm. Exp Cell Res 90: 56–62

Aketa K, Ohta T (1977) When do sperms of the sea urchin *Pseudocentrotus depressus*, undergo acrosome reaction at fertilization? Dev Biol 61: 366–372

Aketa K, Ohta T (1979) Possible causal relation between the acrosome reaction and cross-fertilization in the sea urchins *Hemicentrotus pulcherrimus* and *Pseudocentrotus depressus*. Gamete Res 2: 15–23

Aketa K, Onitake K (1969) Effect on fertilization of antiserum against sperm-binding protein from homo- and heterologous sea urchin egg surfaces. Exp Cell Res 56: 84–86

Aketa K, Tomita H (1958) Oxidative phosphorylation by a cell-free particulate system from unfertilized eggs of the sea urchin, *Hemicentrotus pulcherrimus*. Bull Mar Biol Stn Asamushi 9: 57–58

Aketa K, Tsuzuki H (1968) Sperm-binding capacity of the S–S reduced protein of the vitellino membrane of the sea urchin egg. Exp Cell Res 50: 675–676

Aketa I, Bianchetti R, Marré E, Monroy A (1964) Hexose monophosphate level as a limiting factor for respiration in unfertilized sea-urchin eggs. Biochim Biophys Acta 86: 211–215

Aketa K, Tsuzuki H, Onitake H (1968) Characterization of the sperm-binding protein from sea urchin egg surface. Exp Cell Res 50: 676–679

Aketa K, Onitake K, Tsuzuki H (1972) Tryptic disruption of sperm-binding site of sea urchin egg surface. Exp Cell Res 71: 27–32

Aketa K, Miyazaki S, Yoshida M, Tsuzuki H (1978) A sperm factor as the counter part to the sperm-binding factor of the homologous eggs. Biochem Biophys Res Commun 80: 917–922

Aketa K, Yoshida M, Miyazaki S, Ohta T (1979) Sperm banding to an egg model composed of agarose beads. Exp Cell Res 123: 281–284

Albanese I, Di Liegro I, Cognetti G (1980) Unusual properties of sea urchin unfertilized egg chromatin. Cell Biol Intern Rep 4: 201–210

Alexandraki D, Ruderman J V (1981) Sequence heterogeneity multiplicity and genimic organization of α and β tubulin genes in sea urchins. Mol Cell Biol 1: 1125–1137

Alfageme 'C R, Infante A A (1975) Nuclear RNA in sea urchin embryos. II. Absence of "cRNA". Exp Cell Res 96: 263–270

Allen R D (1954) Fertilization and activation of sea urchin eggs in glass capillaries. I Membrane elevation and nuclear movements in totally and partially fertilized eggs. Exp Cell Res 6: 403–431

Amemiya S (1971) Relationship between cilia formation and cell association in sea urchin embryos. Exp Cell Res 64: 227–230

Amemiya S, Akasaki K, Terayama H (1979) Reversal of polarity in ciliated cells of the isolated sea urchin pluteus gut. J Exp Zool 210: 177–182

Amemiya S, Akasaki K, Terayama H (1982a) Scanning electron microscopy of gastrulation in a sea urchin (*Anthocidaris crassispina*). J Embryol Exp Morphol 67: 27–35

Amemiya S, Akasaki K, Terayama H (1982b) Scanning electron microscopical observations on early morphogenetic processes in developing sea urchin embryos. Cell Differ 11: 291–293

Amy C M, Rebhun L I (1977) Properties of adenylate cyclase activity during early sea urchin development. Exp Cell Res 104: 399–410

Anderson D M, Scheller R H, Posakonj J W, Mc Allister L B, Trabert S G, Beall C, Britten R J, Davidson E (1981) Repetitive sequences of the sea urchin genome. Distribution of members of specific repetitive families. J Mol Biol 145: 5–28

Anderson E (1968) Oocyte differentiation in the sea urchin *Arbacia punctulata*, with particular reference to the origin of cortical granules and their participation in the cortical reaction. J Cell Biol 37: 514–539

Anderson W A (1969) Nuclear and cytoplasmic DNA synthesis during early embryogenesis of *Paracentrotus lividus*. J Ultrastruct Res 36: 95–110

Anderson W A, Perotti M B (1975) An ultracytochemical study of the respiratory potency, integrity, and fate of the sea urchin sperm mitochondria during early embryogenesis. J Cell Biol 66: 367–376

Andreuccetti P, Filosa A, Monroy A, Parisi E (1982) Cell-cell interactions and the role of micromeres in the control of mitotic pattern in sea urchin embryos. In: Embryonic Development, part B. Liss, New York, pp 21–29

Angerer L M, Angerer R C (1981) Detection of poly A$^+$ RNA in sea urchin eggs and embryos by quantitative in situ hybridization. Nucleic Acid Res 9: 2819–2840

Angerer L M, Levine M J, Angerer R C (1980) Detection of mRNA in sea urchin embryos by in situ hybridization. J Cell Biol 87: 141 (a)

Angerer L M, De Leon D Y, Angerer R C, Showman R M, Wells D E, Raff R A (1984) Delayed accumulation of maternal histone mRNA during sea urchin oogenesis. Dev Biol 101: 477–484

Aoki Y, Koshihara H (1972a) Inhibitory effects of acid polysaccharides from sea urchin embryos in RNA synthesis in vitro. Exp Cell Res 70: 431–436

Aoki Y, Koshihara H (1972b) Inhibitory effects of acid polysaccharides from sea urchin embryos on RNA polymerase activity. Biochem Biophys Acta 272: 33–43

Arceci R J, Gross P R (1977) Noncoincidence of histone and DNA synthesis in cleavage cycles of early development. Proc Natl Acad Sci USA 74: 5016–5020

Arceci R J, Gross P R (1980a) Histone variants and chromatin structure during sea urchin development Dev Biol 80: 186–209

Arceci R J, Gross P R (1980b) Sea urchin sperm chromatin structure as probed by pancreatic DNAase I: Evidence for a novel cutting periodicity. Dev Biol 80: 210–224

Arceci R J, Gross P R (1980c) Histone gene expression progeny of isolated early blastomeres in culture make the same change as in the embryo. Science 209: 607–609

Arceci R J, Senger D R, Gross P R (1976) The programmed switch, in lysine-rich histone synthesis at gastrulation. Cell 9: 171–178

Arezzo F, Giudice G (1983) The lack of cell contact causes genomic activation in early sea urchin embryos. Cell Biol Int Rep 7: 5–10

Aronson A I, Chen K (1977) Rates of RNA chain growth in developing sea urchin embryos. Dev Biol 59: 39–48

Aronson A I, Wilt F H, Wartiovaara J (1972) Characterization of pulse-labeled nuclear RNA in sea urchin embryos. Exp Cell Res 72: 309–324

Asai D J, Brokaw C J (1980) Effects of antibodies against tubulin on the movement of reactivated sea urchin sperm flagella J Cell Biol 87: 114–123

Asai D J, Wilson L (1983) A latent activity cytoplasmic dynein from sea urchin egg. J Cell Biol 97: 205 (abstr)

Agao H I, Oppenheimer S B (1979) Inhibition of cell aggregation by specific carbohydrates. Exp Cell Res 120: 101–110

Asnes C F, Schroeder T E (1979) Cell cloavage-Ultrastructural evidence against equatorial stimulation by aster microtubules. Exp Cell Res 122: 327–338

Astorino A, McClay D (1979) Appearance and localization of hyalin during sea urchin development. J Cell Biol 33: 211 abstr

Auclair W (1964) The chromosome number of *Arbacia punctulata*. Biol Bull 127: 359 (Abstr)

Auclair W, Siegel B W (1966a) Cilia regeneration in sea urchin embryos Biol Bull 131: 379 (Abstr)

Auclair W, Siegel B W (1966b) Cilia regeneration in the sea urchin embryo: Evidence for a pool of ciliary proteins. Science 154: 913–915

Azarnia R, Chambers E L (1970) Effect of fertilization on the calcium and magnesium content of the eggs of *Arbacia punctulata*. Biol Bull 139: 413–414

Azorîn F, Olivares C, Jordán A, Pérez-Gran L, Cornudella L, Subirana J A (1983) Heterogeneity of the histone-containing chromatin of the sea cucumber spermatozoa. Distribution of the basic protein ♀♂ and absence of non-histone proteins. Exp Cell Res 148: 331–344

Baccetti B, Burrini A G (1983) The pericentriolar region in sea urchin spermatozoa. J Submicrosc Cytol 15: 115–120

Baccetti B, Burrini A G, Dallai R, Pallini V (1979) The dynein electrophoretic bonds in axonemes naturally lacking the inner or the outer arm. J Cell Biol 80: 334–340

Badman W S, Brookbank J W (1970) Serological studies of two hybrid sea urchin. Dev Biol 21: 243–256

Bäckström S, Hultin K, Hultin T (1960) Pathways of glucose metabolism in early sea urchin development. Exp Cell Res 19: 634–635

Baginski R M, McBlain P J, Carrol E J Jr (1982) Novel procedures for collection of sea urchin cortical granule exudate: Partial characterization and evidence for post-secretion processing. Gam Res 6: 39–52

Baker E J, Infante A A (1982) Non-random distribution of histone mRNAs into polysomes and nonpolysomal ribonucleoprotein particles in sea urchin embryos. Proc Natl Acad Sci USA 79: 2455–2459

Baker P P, Whitaker M J (1978) Influence of ATP and calcium on the cortical reaction in sea urchin eggs. Nature (London) 276: 513–515

Baker R F (1971) Changing size pattern of newly synthesized nuclear DNA during early development of the sea urchin. Biochem Biophys Res Commun 43: 1415–1420

Baker R F, Case S T (1974) Effect of 5-bromodeoxyuridine on the size distribution of DNAs isolated from sea urchine embryos. Nature (London) 249: 350–352

Baldari C T, Amaldi F, Buongiorno-Nardelli M (1978) Electron microscopic analysis of replicating DNA of sea urchin embryos. Cell 15: 1095–1107

Ballario P, Di Mauro E, Giuliani C, Pedone F (1980) Purification of sea-urchin RNA polymerase II Characterization by template requirement and sensitivity to inhibitors. Eur J Bioch 105: 225–234

Ballinger D, Hunt T (1980) Further observation on the phosphorylation of ribosomal proteins after fertilization of *Arbacia punctulata* eggs. Biol Bull 159: 473 (Abstr)

Ballinger D G, Hunt T (1981) Fertilization in sea urchin eggs is accompanied by 40S ribosomal subunit phosphorylation. Dev Biol 87: 277–285

Ballinger D, Peterson S, Hunt T (1979) Phosphorylation of the 40S ribosomal subunit after fertilization of *Arbacia punctulata* eggs. Biol Bull 157: 357 (Abstr)

Ballinger D G, Bray S J, Hunt T (1984) Studies of the kinetics and ionic requirements for the phosphorylation of ribosomal proteins S6 after fertilization of *Arbacia punctulata* eggs. Dev Biol 101: 192–200

Balmain A, Frew L, Cole G, Krumlauf R, Ritchie R, Birnie G D (1982) Transcription of repeated sequences of the mouse B1 family in Friend erythroleukaemic cells. Intermolecular RNA–RNA duplex formation between polydenylated and non-polydenylated nuclear RNAs. J Mol Biol 160: 163–180

Baltus H, Quertier J, Ficq A, Brachet J (1965) Biochemical studies of nucleate and anucleate fragments isolated from sea urchin eggs. A comparison between fertilization and parthenogenetic activation. Biochim Biophys Acta 95: 408–417

Baltzer F, Harding C, Lehman H E, Bopp P (1954): Über die Entwicklungshemmungen der Seeigelbastarde Paracentrotus ♀ X Arbacia ♂ und Psammechinus ♀ X Arbacia ♂ Revue Suisse Zool 61 402–416

Baltzer F, Bernhard M (1955) Weitere Beobachtungen über Letalität und Vererbungsrichtung beim Seeigelbastard Paracentrotus ♀ X Arbacia ♂ Exptl Cell Res Suppl 3 16–26

Baltzer F, Chen P S (1960) Cytological behavior and the synthesis of nucleic acids in sea urchin hybrids Paracentrotus ♀ X Arbacia ♂ and Paracentrotus ♀ X Sphaerechinus ♂ Rev Suisse Zool 67 183–194

Baltzer F, Chen P S, Tardant P (1961) Embryonalentwicklung, DNA-synthese und Respiration des Bastards Arbacia ♀ X Paracentrotus ♂, mit Vergleichen zu anderen Seeigelbastarden Arch Klaus Stift Vererb Forsch 36 126–135

Banjhal W C, Warren R H, McClay D R (1980) Cortical reaction following fertilization of sea urchin eggs: Sensitivity to cytochalasir B. Dev Biol 80: 506–515

Banzhaf C, Warren R H, McClay D R (1980) Cortical reaction following fertilization of sea urchin eggs. Sensitivity to cytochalin B. Dev Biol 80: 506–515

Barber M L, Foy J E (1973) An enzymatic comparison of sea urchin egg ghosts prepared before and after fertilization. J Exp Zool 184: 157–165

Barber M I, Mead J F (1973) Lipids in sea urchin egg surface ghosts. J Cell Biol 59: No 2 16a

Barber M L, Mead J P (1975) Comparison of lipids of sea urchin egg ghosts prepared before and after fertilization. Wilhelm Roux' Arch Dev Biol 177: 19–28

Barrett D, Angelo G M (1969) Maternal characteristics of hatching enzymes in hybrid sea urchin embryos. Exptl Cell Res 57 159–166

Barros C, Giudice G (1968) Effect of polyamines on ribosomal RNA synthesis during sea urchin development. Exp Cell Res 50: 671–674

Bédard P A, Brandhorst B P (1983) Patterns of protein synthesis and metabolism during sea urchin embryogenesis. Dev. Biol 96: 74–83

Begg D A, Rebhun L I (1979) pH regulates the polymerization of actin in the sea urchin egg cortex. J Cell Biol 83: 241–248

Begg D A, Morell R C, Rebhun L J (1977) Studies of actin in two model systems from sea urchin eggs. J Cell Biol 75: 256a

Begg D A, Rebhun L I, Hyatt H (1982) Structural organization of actin in the sea urchin egg cortex: Microvillar elongation in the absence of actin bundle formation. J Cell Biol 93: 24–32

Begg D A, Salmon E D, Hyatt H A (1983) The changes in structural organization of actin in the sea urchin egg cortex in response to hydrostatic pressure. J Cell Biol 97: 1795–1805

Bellemare G, Pinard J, Aubin A, and Cousineau, G H (1968) Uptake and incorporation of leucine and thymidine in developing sea urchin eggs. The effect hexahomoserine. Exp Cell Res 51: 406–412

Bellemare G, Cedergren R J, Cousineau G H (1972) Comparison of the physical and optical properties of *Escherichia coli* and sea urchin 5S ribosomal RNA's. J Mol Biol 68: 445–454

Bellet N F, Vacquier J P, Vacquier V D (1977a) Characterization and comparison of "bindin" isolated from sperm of two species of sea urchins. Biochem Biophys Res Commun 79: 159–165

Bellet N F, Glabe C G, Vacquier V D (1977b) Species specific agglutination of sea urchin eggs by bindin: biochemical comparison of two bindins. J Cell Biol 75: 165a

Bendig M M (1981) Persistence and expression of histone genes injected into *Xenopus* eggs in early development. Nature (London) 292: 65–67

Bendig M M, Hentschel C C (1983) Transcription of histone genes in HeLa cells. Nucleic Acid Res 11: 2337–2346

Bennett J, Mazia D (1981a) Interspecific fusion of sea urchin eggs. Surface events and cytoplasmic mixing. Exp Cell Res 131: 197–207

Bennett J, Mazia D (1981b) Fusion of fertilized and infertilized sea urchin eggs. Maintenance of cell surface integrity. Exp Cell Res 134: 494–498

Benson S C, Sessions A (1980) Prolyl: hydroxylase activity during sea urchin development. Exp Cell Res 130: 467–470

Benson S, Jones E M E, Crise-Benson N, Wilt F (1983) Morphology of the organic matrix of the spiculae of the sea urchin larva. Exp Cell Res 148: 249–253

Benttinen L C, Comb D G (1971) Early and late histones during sea urchin development. J Mol Biol 57: 355–358

Berg W E (1965) Rates of protein synthesis in whole and half embryos of the sea urchin. Exp Cell Res 40: 469–489

Berg W B (1970) Further studies on the kinetics of incorporation of valine in the sea urchin embryo. Exp Cell Res 60: 210–217

Berg W E, Long N D (1964) Regional differences of mitochondrial size in the sea urchin embryo. Exp Cell Res 33: 423–437

Berg W E, Mertes D H (1970) Rates of synthesis and degradation of proteins in the sea urchin embryos. Exp Cell Res 60: 218–224

Berg W E, Taylor D A, Humphreys W J (1962) Distribution of mitochondria in echinoderm embryos as determined by electron microscopy. Dev Biol 5: 165–176

Bergami M, Mansour T B, Scarano E (1968) Properties of glycogen phosphorylase before and after fertilization in the sea urchin eggs. Exp Cell Res 49: 650–655

Bestor T H, Schatten G (1980) Immunofluorescence microscopy of the microtubules involved in the pronuclear movements of sea urchin fertilization. J Cell Biol 87: 136 (Abstr)

Bestor T H, Schatten G (1981) Anti-tubulin immunofluorescence microscopy of microtubules present during the pronuclear movements of sea urchin fertilization. Dev Biol 88: 80–91

Bestor T H, Schatten G (1982) Configurations of microtubules in artificially activated eggs of the sea urchin *Lytechinus variegatus*. Exp Cell Res 141: 71–78

Bibring T, Baxandall J (1971) Selective extraction of isolated mitotic apparatus. Evidence that typical microtubule protein is extracted by organic mercurial. J Cell Biol 48: 324–339

Bibring T, Baxandall J (1974) Tubulins 1 and 2: Failure of quantitation in polyacrylamide gel electrophoresis may influence their identification. Exp Cell Res 86: 120–126

Bibring T, Baxandall J (1977) Tubulin synthesis in sea urchin embryos: Almost all tubulin of the first cleavage mitotic apparatus derives from the unfertilized egg. Dev Biol 55: 191–195

Bibring T, Baxandall J (1981) Tubulin synthesis in sea urchin embryos II. Ciliary A tubulin derives from the unfertilized egg. Dev Biol 83: 122–126

Bibring T, Baxandall J, Denslow S, Walker B (1976) Heterogeneity of the alpha subunit of tubulin and the variability of tubulin within a single organisms. J Cell Biol 69: 301–312

Bieber D, Blin N, Stafford D W (1981) The region of transcription initiation in *Lytechinus variegatus* rRNA genes. Biochim Biophys Acta 655: 366–373

Binder L I, Rosenbaum J L (1973) Directionality of assembly of chick brain tubulin onto sea urchin flagella microtubules Biol Bull 145: 424 (Abstr)

Binder L I, Rosenbaum J L (1979) The in vitro assembly of flagellar outer doublet tubulin. J Cell Biol 79: 500–515

Binder L I, Deutler W L, Rosenbaum J L (1975) Assembly of chick brain tubulin onto flagellar microtubules from Chlamydomonas and sea urchin sperm. Proc Natl Acad Sci USA 72: 1122–1126

Birchmeier C, Grosschedl R, Birnstiel M L (1982) Generation of authentic 3' tremini of an H2A mRNA in vivo is dependent on a short inverted DNA repeat and on spacer sequences. Cell 28: 739–745

Birchmeier C, Folk W, Birnstiel M L (1983) The terminal RNA stem-loop structure and 80 bp of spacer DNA are required for the formation of 3′ termini of sea urchin H2A mRNA. Cell 35: 433–440

Birchmeier C, Schümperli D, Sconzo G, Birnstiel M L (1984) Sequence requirements and involvement of a 60 nucleotide RNA in the maturation of histone mRNA precursors. Proc Natl Acad Sci USA 81: 1057–1061

Bird A P, Taggart M H, Smith B A (1979) Methylated and umethylated DNA compartments in the sea urchin genome. Cell 17: 889–902

Birnstiel M, Telford J, Weinberg E, Stafford D (1974) Isolation and some propertied of the genes coding for histone proteins. Proc Natl Acad Sci USA 71: 2900–2904

Birnstiel M, Schaffner W, Smith H O (1977) DNA sequences coding for the H2B histone of Psammechinus miliaris. Nature (London) 366: 603–606

Blanchard K C (1935) The nucleic acid of the eggs of Arbacia punctulata. J Biol Chem 108: 251–256

Blaukenship J W, Benson S C (1980) In vitro differentiation of sea urchin micromeres. J Cell Biol 87: 135 (Abstr)

Blin N, Sferrazza J M, Wilson F E, Bieber D G, Michel F S, Stafford D W (1979) Organization of the ribosomal gene cluster in Lytechinus variegatus. J Biol Chem 254: 2716–2721

Blin N, Weber T, Alonzo A (1983) Cross-reaction of snRNA and an Alu I-like sequence from rat with DNAs from different eucaryotic species. Nucleic Acid Res 11: 1375–1388

Blomquist C H, Haddox M K, O'Dea R F, Hadden J W, Goldberg N D (1973) Cyclic GMP and cyclic AMP in sea urchin and frog gametes. J Cell Biol 59: N 2-part 2 27a

Bloom G S, Lucar F C, Vallee R B (1983) Identification of microtubule-associated proteins (MAPs) in purified sea urchin egg microtubules and in the mitotic spindle of dividing sea urchin eggs. J Cell Biol 97: 202 (Abstr)

Bohus-Jensen A A (1953) The effect of trypsin on the cross fertilizability of sea urchin eggs. Exp Cell Res 5: 325–328

Boldt J, Schuel H, Schuel R, Dandekar P, Troll W (1980) Sperm peroxidase mediated block to polyspermy in sea urchins. J Cell Biol 27: 134 (Abstr)

Borei H (1948) Respiration of oocytes, unfertilized eggs from Psammechinus and Asterias. Biol Bull 95: 124

Borei H G, Björklund U (1953) Oxidation through the cytochrome system of substituated phenylenediamines. Biochem J 54: 357–362

Borisy G G, Olmstead J B, Klugman R A (1972) In vitro aggregation of cytoplasmic microtubule subunit. Proc Natl Acad Sci USA 69: 2890–2894

Botcahn P M, Dayton A I (1982) A specific replication origin in the chromosomal rDNA of Lytechinus variegatus. Nature (London) 299: 453–456

Boveri Th (1883) Über partielle Befruchtung. Ber Naturforsch Ges Freiburg im Breisgau 4: 64–72

Bower D J, Errington L H, Cooper D N, Morris S, Clayton R M (1983) Chicken lens S-crystallin gene expression and methylation in several non lens tissues. Nucleic Acid Res 11: 2513–2528

Brachet A (1910) La polyspermie expérimentale comme moyen d'analyse de la fécondation. Wilhelm Roux' Arch Entwicklungsmech Org 30: 261–303

Brachet J, De Petrocellis B (1981) The effects of aphidicolin, an inhibitor of DNA replication, on sea urchin development. Exp Cell Res 135: 179–199

Brachet J, Ficq A, Tencer R (1963) Amino acid incorporation into proteins of nucleate and anucleate fragments of sea urchin egg: effect of parthenogenetic activation. Exp Cell Res 32: 168–170

Brandhorst B P (1975) Proteins synthesized before and after fertilization of sea urchin eggs. J Cell Biol 67: 41 (Abstr)

Brandhorst B P (1976) Two-dimensional gel patterns of protein synthesis before and after fertilization in sea urchin eggs. Dev Biol 52: 310–317

Brandhorst B P (1980) Simultaneous synthesis translation, and storage of mRNA including histone mRNA in sea urchin eggs. Dev Biol 79: 139–148

174   References

Brandhorst B P, Bannet M (1978) Terminal completion of poly (A) synthesis in sea urchin embryos. Dev Biol 63: 421–431

Brandhorst B P, Fromson D (1976) Lack of accumulation of ppGpp in sea urchin embryos. Dev Biol 48: 458–460

Brandhorst B P, Humphreys T (1971) Synthesis and decay rutes of major classes of DNA-like RNA in sea urchin embryos. Biochemistry 10: 877–881

Brandhorst B P, Humphreys T (1972) Stabilities of nuclear and messenger RNA molecules in sea urchin embryos. J Cell Biol 53: 474–482

Brandhorst B P, Verma D B S, Fromson D (1979) Polyadenylated and non-polyadenylated messenger RNA fractions from sea urchin embryos code for the same abundant proteins. Dev Biol 71: 128–141

Brandis J W, Raff R A (1978) Translation of oogenetic mRNA in sea urchin eggs and early embryos. Dev Biol 67: 99–113

Brandis J W, Raff R A (1979) Elevation of protein synthesis is a complex response to fertilization. Nature (London) 278: 467–469

Brandriff B, Vacquier V D (1975) Reversible activation of DNA synthesis and the chromosome cycle in unfertilized sea urchin eggs treated with procaine. J Cell Biol 67: 42 (Abstr)

Brandt W F, Strickland W N, Strickland M, Carlisle L, Woods D, von Holt C (1979) A histone program during the life cycle of sea urchin. Eur J Biochem 94: 1–10

Bresch H (1978) Mitochondrial profile densities and areas in different developmental stages of the sea urchin Sphaerechinus granularis. Exp Cell Res 111: 205–209

Britten R J, Kone D E (1968) Repeated sequences in DNA. Science 161: 529–540

Britten R J, Cetta A, Davidson E H (1978) The single-copy DNA sequence polymorphism of the sea urchin Strongylocentrotus purpuratus. Cell 15: 1175–1186

Brokaw C J (1977) Multiple modes of oscillation of reactivated sea urchin sperm flagella following digestion by trypsin. J Cell Biol 75: 373 (Abstr)

Brokaw C J (1979) Calcium-induced asymmetrical beating of Triton-demembranated sea urchin sperm flagella. J Cell Biol 82: 401–411

Brokaw C J, Simonick T F (1977) Mechanochemical coupling in flagella: effects of viscosity on movement and ATP-dephosphorylation of Triton-demembranated sea-urchin spermatozoa. J Cell Biol 23: 227–242

Brokaw C J, Josslin R, Barrow L (1974) Calcium ion regulation of flagellar beat symmetry in reactivated sea urchin spermatozoa. Biochem Biophys Res Commun 58: 795–800

Brookbank J W (1970) DNA synthesis and development in reciprocal interordinal hybrids of a sea urchin and a sand dollar. Dev Biol 21: 29–47

Brookbank J W (1976) DNA and RNA synthesis by fertilized, cleavage arrested sea urchin eggs. Differentiation 6: 33–39

Brookbank J W (1978) Histone synthesis by cleavage arrested sea urchin eggs. Cell Differ 7: 153–158

Brookbank J W (1980) Effects of condycepin and cell dissociation on the synthesis of Hl histone by sea urchin embryos. Cell Differ 9: 315–321

Brooks S C (1943) Intake and loss of ions by living cells. I. Eggs and larvae of Arbacia punctulata and Asterias forbesi exposed to phosphate and sodium ions. Biol Bull 84: 213–225

Brown A E, Bosman H B (1978) Glycoprotein synthesis in developing sea urchin embryos. Biochem Biophys Res Commun 80: 833–840

Bruskin A M, Tyner A L, Wells D E, Showman R M, Klein W H (1981) Accumulation in embryogenesis of five mRNAs enriched in the ectoderm of sea urchin pluteus. Dev Biol 87: 308–318

Bryan J (1970) On the reconstitution of the crystalline components of the sea urchin fertilization membrane. J Cell Biol 45: 606–614

Bryan J (1972) Vinblastine and microtubules. II Caracterization of two protein subunits from the isolated crystals. J Mol Biol 66: 157–168

Bryan J (1982) Redistribution of actin and fascin during fertilization in sea urchin eggs. Cell Differ 11: 279–280

Bryan J, Kane R E (1977) Separation and interaction of the components of sea archin actin gel. J Cell Biol 75: 268 (Abstr)

Bryan J, Kane R E (1978) Separation and interaction of the major components of sea urchin actin gel. J Mol Biol 125: 207–224

Bryan P N, Olah J, Birnstiel M L (1983) Major changes in the 5′ and 3′ chromatin structure of sea urchin histone genes accompany their activation and inactivation in development. Cell 33: 843–848

Buongiorno-Nardelli M, Ballario P, DiMauro E (1981) Binding of sea urchin RNA polymerase II on homologous histone genes. Eur J Biochem 116: 171–176

Burdick C J, Taylor B A (1976) Histone acetylation during early stages of sea urchin (*Arbacia punctulata*) development Exp Cell Res) 100: 428–433

Burgess D R, Schroeder T B (1977) Polarized bundles of actin filamente within microvilli of fertilized sea urchin eggs. J Cell Biol 74: 1032–1037

Burke R D (1983) Neural control of metamorphosis in *Dendraster excentricus*. Dev Biol 164: 176–188

Burns R G (1973) Kinetics of regeneration of sea urchin cilia. J Cell Sci 13: 55–67

Burns R G (1975) Comparison of the properties of sea urchin egg ATPhase and Sperm tail dynein. J Cell Biol 67: 48 (Abstr)

Burns R G (1977) The regeneration of cilia of Arbacia punctulata blastulae. Biol Bull 153: 417–418 (Abstr)

Burns R G (1979) Kinetics of the regeneration of sea-urchin cilia, II Regeneration of animalized cilia. J Cell Sci 37: 205–215

Busby S J, Bakken A H (1977) A quantitative analysis of sea urchin embryo transcription. J Cell Biol 75: 150 (Abstr)

Bushman F D, Crain W R Jr (1983) Conserved pattern of embryonic actin gene expression in several sea urchins and a sand dollar. Dev Biol 98: 429–436

Busslinger M, Portman R, Birnstiel. (1979) A regulatory sequence near the 3′end of sea urchin histone gene. Nucleic Acid Res 6: 2997–3008

Busslinger M, Rusconi S, Birnstiel M L (1982) An unusual evolutionary behaviour of sea urchin histone gene cluster. EMBO J 1: 27–33

Buznikow G A, Chudakova I V, Zvezdina N D (1964) The role of neurohumours in early embryogenesis. I Serotonin content of developing embryos of sea urchin and loach. J Embryol Exp Morphol 12: 563–573

Buznikov G A, Chudakova J V, Berdysheva L V, Vyazmina N M (1968) The role of neurohumours in erly embryogenesis. II Acetylcholine and catecholamine content in developing embryos of sea-urchin eggs. J Embryol Exp Morphol 20: 119–128

Buznikov G A, Sakharova A V, Manukhin B N, Markova L N (1972) The role of neurohumours in early embryogenesis IV Fluorimetric and histochemical study of serotonin in cleaving eggs and larvae of sea urchins. J Embryol Exp Morphol 27: 339–351

Byrd B W Jr (1975a) The block to polyspery in sea urchin fertilization: an analysis with agents affecting the cell surface. Biol Bull 149: 423 (Abstr)

Byrd B W Jr (1975b) Phospholipid metabolism following fertilization in sea urchin eggs and embryos. Dev Biol 46: 309–316

Byrd B W Jr, Collins P D (1975a) Reexamination of the blocks to polysmery in sea urchin eggs and mechanisms of polyspermic agents. J Cell Biol 67: 52a

Byrd B W Jr, Collins F D (1975b) Absence of fast block to polyspermy in eggs of sea urchin *Strongylocentrotus purpuratus*. Nature (London) 257: 675–677

Byrd W, Perry G (1980) Cytocholasin B blocks sperm incorporation but allows activation of the sea urchin egg. Exp Cell Res 126: 333–342

Byrd W, Perry G, Weidnen E (1977) Role of the egg cortex and actin in fertilization of the sea urchin egg. J Cell Biol 75: 267 (Abstr)

Cabrera C C, Jacobs H T, Posakony J W, Grula J W, Roberts J W, Britten R J, Davidson E H (1983) Transcripts of three mitochondrial genes in the RNA of sea urchin eggs and embryos. Dev Biol 97: 500–505

Cameron R A, Hinegardner R T (1974) Initiation of metamorphosis in laboratory cultured sea urchins. Biol Bull 146: 335–346

Cameron R A, Hinegardner R T (1978) Early events in sea urchin metamorphosis, description and analysis. J Morphol 157: 21–32

Campisi J, Scandella C J (1978) Fertilization induced changes in membrane fluidity of sea urchin eggs. Science 199: 1336–1337

Campisi J, Scandella C J (1980a) Bulk membrane fluidity increases after fertilization or partial activation of sea urchin eggs. J Biol Chem 255: 5411–5419

Campisi J, Scandella C J (1980b) Calcium-induced decrease in membrane fluidity of sea urchin egg cortex after fertilization. Nature (London) 286: 185–186

Candelas G C, Iverson R M (1966) Evidence for translational level control of protein synthesis in the development of sea urchin eggs. Biochem Biophys Res Commun 34: 867–871

Cantatore P, Nicotra A, Loria P, Saccone C (1974) RNA synthesis in isolated mitochondria from sea urchin embryos Cell Differ 3: 45–53

Card C O, Morris G F, Brown D T, Marzluff W F (1982) Sea urchin small nuclear RNA genes are organized in distinct tandemly repeating units. Nucleic Acid Res 10: 7677–7688

Cardasis C A, Schuel H, Herman L (1978) Ultrastructural localization of calcium in unfertilized sea-urchin eggs. J Cell Sci 77: 101–115

Cariello L, Brown J C, Storrie B (1980) Properties of isolated fertilization envelopes from *Arbacia peructulata*. Biol Bull 159: 467 (Abstr)

Carpenter C D, Bruskin A M, Spain L M, Eldon E D, Klein W H (1982) The 3' untranslated regions of two related mRNAs contain an element highly repeated in the sea urchin genome. Nucleic Acid Res 10: 7829–7842

Carroll A G, Ozaki H (1979) Changes in the histones of the sea urchin Strongylocentrotus purpuratus at fertilization. Exp Cell Res 119: 307–316

Curroll A G, Eckberg W R, Ozaki H (1975) A comparison of protein synthetic patterns in normal and animalized sea urchin embryos. Exp Cell Res 90: 328–332

Carroll E J Jr, Baginski R M (1977) Isolation, solubilization and characterization of sea urchin fertilization envelope. J Cell Biol 75: 169 (Abstr)

Carroll E J Jr, Baginski R M (1978) Sea urchin fertilization envelope: isolation, extraction, and characterization of a major protein fraction from *Strongylocentrotus purpuratus* embryos. Biochemistry 17: 2605–2612

Carroll E J Jr, Endress A G (1982) Sea urchin fertilization envelope: Uncoupling of cortical granule exocytosis from envelope assembly and isolation of an envelope intermediate from *Strongylocentrotus purpuratus* embryos. Dev. Biol 94: 252–258

Carroll E J Jr, Epel D (1975a) Elevation and hardening of the fertilization membrane in sea urchin eggs. Role of fertilization product. Exp Cell Res 90: 429–432

Carroll E J Jr, Epel D (1975b) Isolation and biological activity of the proteases released by sea urchin eggs following fertilization. Dev Biol 44: 22–32

Carrol E J Jr, Epel D (1981) Reevaluation of cell surface protein release at fertilization and its role in regulation of sea urchin egg protein synthesis. Dev Biol 87: 374–378

Carroll E J Jr, Levitan H (1978a) Fertilization in sea urchin, *Strongylocentrotus purpuratus* is blocked by fluorescein dyes. Dev Biol 63: 432–440

Carroll E J Jr, Levitan H (1978b) Fertilization is inhibited in five diverse animal phyla by erythrosin B. Dev Biol 64: 329–331

Carroll E J Jr, Endress A G, du Blaine P J, Kitasako J T (1979) Isolation and characterization of an intermediate stage in the assembly of sea urchin fertilization envelope. J Cell Biol 83: 216 (Abstr)

Carroll E J Jr, Byrd E W, Epel D (1977) A novel procedure for obtaining denuded sea urchin eggs and observations on the role of the vitelline layer in sperm reception and egg activation. Exp Cell Res 108: 365–374

Carroll E J Jr, Johnson J D, Byrd E W (1975) A novel procedure for obtaining denuded sea urchin eggs and observation on the role of the vitelline layer in sperm reception and egg activation. J Cell Biol 67: 57 (Abstr)

Carron C P, Longo F J (1980) Cytoplasmic alkalinization and microvillar elongation in the sea urchin egg. J Cell Biol 87: 141 (Abstr)

Carron C P, Longo F J (1980b) Relation of intracellular pH and pronuclear development in the sea urchin, *Arbacia punctulata*. A fine structural analysis. Dev Biol 79: 478–487

Carron C P, Longo F J (1982) Relation of cytoplasmic alkalinization to microvillar elongation and microfilament formation in the sea urchin egg. Dev Biol 89: 128–137

Carron C P, Longo F J (1983) Filipin/sterol complexes in fertilized and unfertilized sea urchin egg membranes. Dev Biol 99: 482–488

Carson D D, Lennarz W J (1981) Relationship of dolichol synthesis to glycoproteic synthesis during embryonic development. J Biol Chem 256: 4679–4686

Case S T, Baker R F (1975a) Detection of long eukaryote-specific pyrimidine runs in repetitive DNA sequences and their relation to single-stranded regions in DNA isolated from sea urchin embryos. J Mol Biol 98: 69–92

Case S T, Baker R F (1975b) Position of regularly spaced single-stranded regions relative to 5-bromo-deoxyuridine-sensitive sites in sea urchin morula DNA. Nature (London) 253: 64–66

Case S T, Mongeon R L, Baker R F (1974) Single stranded regions in DNA isolated from different developmental stages of the sea urchin. Biochim Biophys Acta 349: 1–12

Case S T, Talkington C A, Baker R F (1975) The effect of 5-bromodeoxyridine on the synthesis of nuclear and cytoplasmic DNA-binding proteins isolated from sea urchin embryos. Cell Differ 4: 55–62

Castaneda M (1969) The activity of ribosomes of sea urchin eggs in response to fertilization. Biochim Biophys Acta 79: 381–388

Ceccarini C, Maggio R (1969) A study of aminoacyl tRNA synthetases by methylated albumin kieselguhr column chromatography in Paracentrotus lividus. Biochim Biophys Acta 190: 556–559

Ceccarini C, Maggio R, Barbata G (1967) Aminoacyl-SRNA synthetases as possible regulators of protein synthesis in the embryo of the sea urchin Paracentrotus lividus. Proc Natl Acad Sci USA 58: 2235–2239

Cestelli A M, Albeggiani G, Allotta S, Vittorelli M L (1975) Isolation of the plasma membrane from sea urchin embryos. Cell Differ 4: 305–313

Chaffee R R, Mazia D (1963) Echinochrome synthesis in hybrid sea urchin embryos. Devel Biol 7 502–512

Chamberlain J (1968) Extranuclear RNA synthesis in sea urchin embryos. J Cell Biol 38: 23 (Abstr)

Chamberlain, J P (1970) RNA synthesis in anucleate egg fragments and normal embryos of the sea urchin Arbacia punctulata. Biochim Biophys Acta 213: 183–193

Chamberlain J P (1977) Protein synthesis by separated blastomeres of sixteen-cell sea urchin embryos. J Cell Biol 75: 33 (Abstr)

Chamberlain J P, Metz C B (1972) Mitochondrial RNA synthesis in sea urchin embryos. J Mol Biol 64: 593–607

Chambers E L (1974) Effects of ionophores on marine eggs and cation requirements for activation. Biol Bull 147: 471 (Abstr)

Chambers E L (1975a) Na+ is required for nuclear and cytoplasmic action of sea urchin eggs by sperm and divalent ionophores. J Cell Biol 67: 60 (Abstr)

Chambers E L (1975b) Potassium exchange in unfertilized sea urchin (Arbacia punctulata eggs). Biol Bull 149: 422–423 (Abstr)

Chambers E L and De Armendi J (1979) Membrane potential, action potential and activation potential of eggs of the sea urchin, Lytechinus variegatus. Exp Cell Res 122: 203–218

Chambers E L, Hinkley R E (1979a) The divalent ionophore A23187 can induce non-propagated cortical reactions in eggs of the sea urchin, Lytechinus variegatus. J Cell Biol 83: 210 (Abstr)

Chambers E L, Hinkley R E (1979b) Non propagated cortical reaction induced by the divalent ionophore A23187 in egg of the sea urchin, Lytechinus variegatus. Exp Cell Res 124: 441–446

Chambers E L, Whiteley A H (1966) Phosphate transport in fertilized sea urchin eggs. I Kinetics aspects. J Cell Physiol 68: 289–308

Chambers E L, Azarmia R, McGowan W E (1970) The effect of temperature on the efflux of [45]Ca from the eggs of Arbacia punctulata. Biol Bull 139: 417–418

Chambers E L, Pressman B C, Rose B (1974) The activation of sea urchin eggs by the divalent ionophores A23187 and X-537A. Biochem Biophys Res Commun 60: 126–132

Chambers S A M, Vaughn J P, Shaw B R (1983) Shortest nucleosomal repeat lenghts during sea urchin development are found in two-cell embryos. Biochemistry 22: 5626–5630

Chandler D E, Heuser J (1979) Membrane fasion during secretion. Cortical granule exocytosis in sea urchin eggs as studied by quick-freezing and freeze-fracture. J Cell Biol 83; 91–108

Chang D C, Afzelius B A (1973) Electron microscopic study on membrane junctions of Arbacia punctulata blastomeres. Biol Bull 145: 428 (Abstr)

Chaudari N, Craig S P (1979a) The evolution of the long and short repetitive DNA sequences in sea urchins. Biochim Biophys Acta 563: 433–452

Chaudari N, Craig S P (1979b) Internal organization of long repetitive DNA sequences in sea urchin genome. Proc Nat Acad Sci USA 76: 6101–6105

Chargaff E, Davidson I N (eds) (1955) In: The nucleic acids, vol I. Academic Press, London New York p 357

Chatlynne L G (1969) A histochemical study of oogenesis in the sea urchin *Strongylocentrotus purpuratus.* Biol Bull 136: 167–184

Chen P S, Baltzer F (1962) Experiments concerning the incorporation of labelled adenine into ribonucleic acid in normal sea urchin embryos and in the hybrid Paracentrotus ♀ x Arbacia ♂. Experientia 18 522–524

Chia F S, Atwood D, Crawford B (1975) Comparative morphology of echinoderm sperm and possible phylogenetic implications. Am Zool 15: 553–565

Childs G, Lovy S, Kedes L H (1979a) Rapid purification of biologically active individual histone messenger RNAs by hybridization to cloned DNA linked to cellulose. Biochemistry 18: 208–213

Childs G, Maxon R, Kedes L H (1979b) Histone gene expression during sea urchin embryogenesis. Isolation and characterization of early and late messenger RNAs of *Strongylocentrotus purpuratus* by gene-specific hybridization and template activity. Dev Biol 73: 153–173

Childs G, Maxon R, Cohn H, Kedes L (1981) Orphons: Dispersed genetic elements derived from tandem repetitive genes of eucaryotes. Cell 23: 651–663

Childs G, Nocente C, Mc Grath, Lieber T, Holt C, Knowles J A (1982) Sea urchin (Lytechinus pictus) late-stage histone H3 and H4 genes: characterization and mapping of a clustered but nontandemly linked multigene family. Cell 31: 383–393

Chilton B S, Lanfer M R, Nicosia S V (1979) Some biochemical properties of an RNA-instructed DNA polymerase in developing sea urchin (*Lytechinus pictus*) embryos. Biol Bull 157: 362 (Abstr)

Christen R, Schackman R W, Shapiro B M (1980) Regulation of viability and the acrosome reaction of *Strongylocentrotus purpuratus.* J Cell Biol 87: 140 (Abstr)

Christen R, Schackmann R W, Shapiro B M (1983a) Interactions between sperm and sea urchin egg jelly. Dev Biol 98: 1–14

Christen R, Schackmann R W, Shapiro B M (1983b) Metabolism of the sea urchin sperm. Interrelationships between intracellular pH, ATPase activity, and mitochondrial respiration. J Biol Chem 258: 5392–5399

Christen R, Schackmann R W, Dahlquist F W, Shapiro B M (1983c) ^{31}P-NMR analysis of sea urchin sperm activation. Reversible formation of high energy phosphate compounds by changes in the intracellular pH. Exp Cell Res 149: 289–294

Citkowitz E (1971) The hyaline layer: its isolation and role in echinoderm development. Dev Biol 24: 348–362

Citkowitz E (1972) Analysis of the isolated hyaline layer of sea urchin embryos. Dev Biol 27: 494–503

Clegg K B, Denny P C (1974) Synthesis of rabbit globin in a cell-free protein synthesis system utilizing sea urchin egg and zygote ribosomes. Dev Biol 37: 263–272

Cleland K W, Rotschild Lord (1952a) The metabolism of the sea urchin egg. Anaerobic breakdown of carbohydrate. J Exp Biol 29: 285–294

Cleland K W, Rotschild Lord (1952b) The metabolism of the sea urchin egg. Oxidation of carbohydrate. J. Exp Biol 29: 416–428

Coburn M, Schuel H, Troll W (1981) A hydrogen peroxide block to polyspermy in the sea urchin *Arbacia punctulata.* Dev Biol 84: 235–238

Coffe G, Rola F H, Soyer M O, Pudles J (1982) Parthenogenetic activation of sea urchin egg induces a cyclical variation of the cytoplasmic resistance to hexylene glycol-Triton X 100 treatment. Exp Cell Res 137: 63–72

Coffe G, Foucault G, Raymond M N, Pudles J (1983) Tubulin dynamics during the cytoplasmic cohesive cycle in artificially activated sea urchin eggs. Exp Cell Res 149: 409–418

Cognetti G (1982) Nutrition of embryos. In: Jangoux M, Laurence J M (eds) Echinoderm nutrition''. Balkema, Rotterdam, pp 469–476

Cognetti G, Shaw B R (1981) Structural differences in the chromatin from compartimentalized cells of the sea urchin embryo: differential nuclease accessibility of micromere chromatin. Nucleic Acid Res 9: 5609

Cognetti G, Settineri D, Spinelli G (1972) Developmental changes of chromatin non-histone proteins in sea urchins. Exp Cell Res 71: 465–468

Cognetti G, Spinelli G, Vivoli A (1974) Synthesis of histones during sea urchin oogenesis. Biochim Biophys Acta 349: 447–455

Cognetti G, Platz R D, Meistrich M L, Di Liegro I (1977a) Studies on protein synthesis during sea urchin oogenesis. I. Synthesis of histone F2 b. Cell Differ 5: 283–291

Cognetti G, Di Liegro I, Cavarretta F (1977b) Studies of protein synthesis during sea urchin oogenesis. II. Synthesis of tubulin. Cell Differ 6: 159–165

Cohen L H, Newrock K M, Zweidler A (1975) Stage-specific switches in histone synthesis during embryogenesis of the sea urchin. Science 190: 994–997

Cohn R H, Kedes L M (1979a) Nonallelic histone gene clusters of individual sea urchins (*Lytechinus pictus*): Polarity and gene organization. Cell 18: 843–853

Cohn R H, Kedes L H (1979b) Nonallelic histone gene clusters of individual sea urchins (*Lytechinus pictus*): Mapping of homologies in coding and spacer DNA. Cell 18: 855–864

Cohn R H, Lowry J C, Chang A C Y, Cohen S M, Kedes L H (1975) Organization of sea urchin histone genes cloned in *Escherichia coli*. J Cell Biol 67: 76 (Abstr)

Cohn R H, Lowry J C, Kedes L H (1976) Histone genes of the sea urchin (*S. purpuratus*) cloned in *E. coli*: order, polarity and strandedness of the five histone-coding and spacer regions. Cell 9°, 147–161

Collins F, Epel D (1977) The role of calcium ions in the acrosome reaction of sea urchin sperm. Exp Cell Res 106; 211–222

Colombera D (1974) Chromosome evolution in the phylum echinodermata. Zool Syst Evolutionsforsch 12; 299–308

Comb D G (1965) Methylation of nucleic acids during sea urchin embryo development. J Mol Biol 11: 851–855

Comb D G, Sarkar N, De Vallet J, Pinzino C J (1965a) Properties of transfer-like RNA associated with ribosomes. J Mol Biol 12: 509–513

Comb D G, Katz S, Branda R, Pinzino G J (1965b) Characterization of RNA species synthesized during early development of sea urchins. J Mol Biol 14: 195–213

Conway A F, Metz C B (1976) Phospholipase activity of sea urchin sperm: its possible involvement in membrane fusion. J Exp Zool 198: 39–48

Cordle C T, Metz C B (1973) Isolation of Arbacia sperm antigens important in fertilization. Biol Bull 145: 430 (Abstr)

Cornudella L, Rocha E (1979) Nucleosoma organization during germ cell development in the sea cucumber *Holoturia tubulosa*. Biochemistry 18: 3724–3732

Cosson M P, Gibbons I R (1977) Modification of the flagellar waweform of reactivated sea urchin sperm by treatment with mono and di-maleimide derivateves. J Cell Biol 75: 278a

Cosson M P, Tang W-J Y, Gibbons I R (1983) Modification of flagellar wave form and adenosine triphosphatase activity in reactivated sea urchin sperm treated with N-ethylmaleinide. J Cell Sci 60: 231–249

Costantini F D, Scheller R H, Britten R J, Davidson E H (1978) Repetitive sequence transcripts in the mature sea urchin oocyte. Cell 15: 173–187

Costantini F D, Britten R J, Davidson E H (1980) Message sequences short repetitive sequences are interspersed in sea urchin egg poly (A)$^+$ RNAs. Nature (London) 287: 111–117

Costello D P (1973) A new theory on the mechanics of ciliary and flagellar motility. II. Theoretical considerations. Biol Bull 145: 299–309

Crabb W D, Firshein W, Infante A A (1980) A potential DNA replication complex isolated from sea urchin embryos by renografin gradient centrifugation. Biochim Biophys Acta 609: 456–463

Craig S P (1970) Synthesis of RNA in non-nucleate fragments of sea urchin eggs. J Mol Biol 47: 615–618

Craig S P, Innis M A (1978) Mitochondrial regulation in sea urchins. III Rapid degradation of mitochondrial RNA in association with a failure to form mitochondrial polyribosomes in eggs activated with ionophore A 23187. Exp Cell Res 117: 145–163

Craig S P, Piatigorsky J (1971) Protein synthesis and development in the absence of cytoplasmic RNA synthesis in nonnucleate egg fragments and embryos of sea urchins: Effect of ethidium bromide. Dev Biol 24: 214–232

Craig S P, Chaudari N, Steinert M (1979) Characterization of long and short repetitive sequences in sea urchin genome. Biochim Biophys Acta 565: 33–50

Crain W R Jr, Bushman F D (1983) Transcripts of paternal and maternal actin gene alleles are present in interspecific sea urchin embryo hybrids. Dev Biol 100: 190–196

Crane R K, Keltch A K (1949) Dinitrocresol and phosphate stimulation of the oxygen consumption of a cell-free oxidative system obtained from sea urchin eggs. J Gen Physiol 39: 503–509

Crane C M, Villee C A (1971) The synthesis of nuclear histones in early embryogenesis. J Biol Chem 246: 719–723

Crkvenjakov R, Bajkovic N, Glisin V (1970) The effect of 5-azacytidine on development, nucleic acid and protein metabolism in sea urchin embryos. Biochem Biophys Res Commun 39: 655–660

Cross N L (1983) Isolation and electrophoretic characterization of the plasma membrane of sea-urchin sperm. J Cell Sci 59: 13–26

Csernansky J G, Rosman G A, Grossman A, Zimmerman M, Troll W (1977) Egg proteases of Arbacia punctulata: evidence for a new elastase-like enzyme. Biol Bull 153: 421–422a

Cummins C, Hunt T (1977) Characterization of messenger ribonucleoprotein particles from the sea urchin Lytechinus pictus. Biol Bull 153: 422 (Abstr)

Cuthbert A, Cuthbert A W (1978) Fertilization acid production in Psammechinus eggs under pH clamp conditions, and the effect of some pyrazine derivatives. Exp Cell Res 114: 409–416

Czihak G (1974) The role of astral rays in early cleavage of sea urchin eggs. Exp Cell Res 83: 424–426

Czihak G (1975) The sea urchin embryo. Springer, Berlin Heidelberg New York

Czihak C (1977) Kinetics of RNA synthesis in the 16-cell stage of the sea urchin Paracentrotus lividus. Wilhelm Roux' Arch Dev Biol 182: 59–68

Czihak G (1978) Effect of 5-bromodeoxyuridine on differentiation. II the effect of early BUdR treatment on gastrulation of sea urchin embryos. Differentiation 11: 103–107

Dale B, De Santis A (1981a) The effect of cytochalasin B and D on the fertilization of sea urchins. Dev Biol 83: 232–237

Dale B, De Santis A (1981b) Maturation and fertilization of the sea urchin oocyte: An electrophysiological study. Dev Biol 85: 474–484

Dale B, Monroy A (1981) How is polyspermy prevented? J Gam Res 4: 151–169

Dale B, De Felice L J, Taglietti V (1978) Membrane noise and conductance increase during single spermatozoon-egg interaction. Nature (London) 275: 217–219

Dale B, De Santis A, Ortolani G, Rasotto M, Santella L (1982) Electrical coupling of blastomeres in early embryos of ascidians and sea urchins. Exp Cell Res 140: 457–461

Dan J C (1953) Studies on the acrosome. I. Reaction to egg water and other stimuli. Biol Bull 103: 54–66

Dan J C (1954) Studies on the acrosome. II. Acrosome reaction in starfish spermatozoa. III. Effect of calcium deficiency. Biol Bull 107: 203–218

Dan J C (1956) The acrosome reaction. Int Rev Cytol 5: 365–393

Dan J C (1960) Studies on the acrosome. VI Fine structure of the starfish acrosome. Exp Cell Res 19: 13–28

Dan J C (1967) Acrosome reaction and lysins. In: Metz CB, MonroyA(eds) Fertilization, voll I. Academic Press, London New York, pp 237–294

Dan J C (1970) Morphogenetic aspects of acrosome formation and reaction. Adv Morphogen 8: 1–37

Dan K (1952) Cyto-embryological studies of sea urchins. II. Blastula stage. Biol Bull 102: 74–89

Dan K (1960) Cyto-embryology of echinoderms and amphibia. Int Rev Cytol 9: 321–367

Dan K (1978) Unequal division: Its cause and significance. In: Cell Reproduction. ICN-UCCA Symp Mol Cell Biol, vol I. Academic Press, London New York, pp 557–561

Dan K, Ikeda M (1971) On the system controlling the time of micromere formation in sea urchin embryos. Dev Growth Differ 13: 285–301

Dan K, Endo S, Uemura I (1983) Studies on unequal cleavage in sea urchins II. Surface differentiation and the direction of nuclear migration. Dev Growth Differ 25: 227–237

Danilchick M V, Hille M B (1981) Sea urchin egg and embryo ribosomes: Differences in translational activity in a cell-free system. Dev Biol 84: 291–298

Dasgupta J D, Garbers D L (1983) Tyrosine protein kinase activity during embryogenesis. J Biol Chem 258: 6174–6178

Davidson E H, Britten R J (1979) Regulation of gene expression: possible role of repetitive sequences. Sciences 204: 1052–1059

Davidson E H, Hough B R, Klein W H, Britten, R J (1975) Structural genes adjacent to interspersed repetitive DNA sequences. Cell 4: 217–238

Davidson E H, Thomas T L, Scheller R H, Britten R J (1982) In: Dows G, Elawll R B (eds) genome evolution. Academic Press, London New York, p 177

Decker G L, Lennarz W J (1979) Sperm binding and fertilization envelope formation in a cell surface complex isolated from sea urchin eggs. J Cell Biol 81: 92–103

Decker G L, Joseph D B, Lennarz W J (1976) A study of factors involved in induction of the acrosomal reaction in sperm of the sea urchin, *Arbacia punctulata*. Dev Biol 53: 115–125

Decker S J, Kinsey W H (1983) Characterization of cortical secretory vesicles from the sea urchin egg. Dev Biol 96: 37–45

De Felice L T, Dale B (1979) Voltage response to fertilization and polyspermy in sea urchin eggs and oocytes. Dev Biol 72: 327–341

De Leo G, Rinaldi A M, Salcher-Cillari I, Mutolo V (1979) Fine structure and distribution of cell organelles in nucleated and non-nucleated halves of *Paracentrotus lividus* eggs. Acta Embryol Exp 2: 141–160

De Leon D V, Cox K H, Angerer L M, Angerer R C (1983) Most early-variant histone mRNA is contained in the pronucleus of sea urchin eggs. Dev Biol 100: 197–206

Delgado N M, Reyes R, Huaciya L, Carranco A, Merchant H, Rosado A (1983) Decondensation of human sperm nuclei by glycosamineglycan-sulfate from sea urchin egg. J Exp Zool 224: 457–460

Delobel N (1971) Determination de nombre chromosomique chez une asteride: Echinaster sepositus. Caryologia 24: 247–250

Denis S (1968) Changes in the level of triphosphopyridine nucleotides during development of sea urchin eggs (normal and letal hybrids) Biochim Biophys Acta 157: 212–214

Denis H, Brachet J (1969a) Gene expression in interspecific hybrids. I DNA synthesis in the lethal cross *Arbacia lixula* ♂ X *Paracentrotus lividus* ♀ Proc Natl Acad Sci 62: 194–201

Denis H, Brachet J (1969b) Gene expression in interspecific hybrids. II RNA synthesis in the lethal cross *Arbacia lixula* ♂ X *Paracentrotus lividus* ♀ Proc Natl Acad Sci 62: 438–445

Denny P C, Tyler A (1964) Activation of protein biosynthesis in non-nucleate fragments of sea urchin eggs. Biochem Biophys Res Commun 14: 245–249

De Petrocellis B, Parisi E (1972) Changes in alkaline deoxyribonuclease activity in sea urchin during embryonic development. Exp Cell Res 73: 496–500

De Petrocellis B, Parisi E (1973a) Deoxyribonuclease in sea urchin embryos. Comparison of the activity present in developing embryos in nuclei and in mitochondria. Exp Cell Res 79: 53–62

De Petrocellis B, Parisi E (1973b) Effect of actinomycin and puromycin on the deoxyribonuclease activity in *P. lividus* embryos at various stages of development. Exp Cell Res 82: 351–356

De Petrocellis B, Rossi M (1976) Enzymes of DNA biosynthesis in developing sea urchins. Changes in ribonucleotide reductase, thymidine, and thymidilate kinase activities. Dev Biol 18: 250–257

De Petrocellis B, Vittorelli M L (1975) Role of cell interactions in development and differentiation of the sea urchin *Paracentrotus lividus*. Changes in the activity of some enzymes of DNA biosynthesis after cell dissociation. Exp Cell Res 94: 392–400

De Petrocellis B, Parisi E, Filosa S, Capasso A (1976) Separation and partial characterization of DNA polymerases in sea urchin *Paracentrotus lividus* eggs. Biochem Biophys Res Commun 68: 954–960

De Petrocellis B, De Petrocellis L, Lancier M, Geraci G, (1980) Species specificity and individual variability of sea urchin sperm H2B histones. Cell Differ 9: 195–202

Deretic V, Glisin V (1982) On the evolution of sea urchin "early" histone genes. Bull Mol Biol Med 7: 65–76

De Rosier D J, Censullo R (1981) Structure of F-actin needles from extracts of sea urchin oocytes. J Mol Biol 146: 77–99

Detering N K, Decker G L, Schmell E D, Lennarz W J (1977) Isolation and characterization of plasma membrane-associated cortical granules from sea urchin eggs. J Cell Biol 75: 899–914

Detrich H W III, Wilson L (1983) Purification, characterization, and assembly properties of tubulin from unfertilized eggs of the sea urchin *Strongylocentrotus purpuratus*. Biochemistry 22: 2453–2462

Devlin R (1976) Mitochondrial poly (A) RNA synthesis during early sea urchin development. Dev Biol 50: 443–456

Devlin R, Swanson R F (1975) Mitochondrial poly A RNA synthesis during sea urchin development. J Cell Biol 67: 95 (Abstr)

De Vincentiis M, Hörstadius S, Runnström J (1966) Studies on controlled and released respiration in animal and vegetal halves of the embryo of the sea urchin *Paracentrotus lividus*. Exp Cell Res 41: 535–544

Dickinson D C, Baker R F (1978) Evidence for translocation of DNA sequences during sea urchin embryogenesis. Proc Natl Acad Sci USA 75: 5627–5630

Dickinson D G, Baker R F (1979) 5-bromodeoxyridine inhibits sequence changes within inverted repeat DNA during embryogenesis. Science 205: 816–818

Di Liegro I, Cestelli A, Ciaccio M, Cognetti G (1978) Block of histone synthesis in isolated sea urchin cells actively synthesizing DNA. Dev Biol 67: 266–273

Di Mauro E, Finotti R, Pomponi M (1977) Transcription in sea urchin I. Initiation sites on DNA and chromatin. Exp Cell Res 105: 207–216

Di Mauro E, Ballario P, Pedone F (1979) Transcription in sea urchin. The role of non-histone chromosomal proteins in preventing unproductive binding of RNA polymerase to template. Cell Differ 8: 291–304

Di Mauro E, Pedone F, Ballario P (1980) Analysis of the sea-urchin genome by homologous RNA polymerase II binding. Eur J Biochem 105: 235–243

Dolecki G J, Duncan R F, Humphreys T (1977) Complete turnover of poly (A) on maternal mRNA of sea urchin embryos. Cell 11: 339–344

Doree M, Guerrier P (1974) A kinetic analysis of the changes in membrane permeability induced by fertilization in the egg of the sea urchin *Sphaerechinus granularis*. Dev Biol 41: 124–136

Driesch H (1891) Entwicklungsmechanische Studien I–II. Z Wiss Zool 53: 160–178

Driesch H (1892) Entwicklungsmechanische Studien III–IV. Z Wiss Zool 55: 1–62

Driesch H (1900) Die isolierten Blastomeren des Echinidenkeimes. Wilhelm Roux' Arch Entwicklungsmech Org 10: 361–410

Driesch H (1903) Drei Aphorismen zur Entwicklungsphysiologie jüngster Stadien. Wilhelm Roux' Arch Entwicklungsmech Org 17: 41–63

Dubé F, Guerrier P (1983) Ca^{2+} influx and stimulation of protein synthesis in sea urchin eggs. Exp Cell Res 147: 209–215

Dubroff L M (1977) Oligouridylate stretches in heterogeneous nuclear RNA. Proc Natl Acad Sci USA 74: 2217–2221

Dubroff L M (1980a) Oligomeric sequences in the cytoplasmic RNA of sea urchin embryos. Biochim Biophys Acta 607: 115–121

Dubroff L M (1980b) Developmental changes in the molecular weight of heterogeneous nuclear RNA. Biochim Biophys Acta 608: 378–386

Dubroff L M, Nemer M (1975) Molecular classes of heterogeneous nuclear RNA in sea urchin embryos. J Mol Biol 95: 455–476

Dubroff L M, Nemer M (1976) Developmental shifts in the synthesis of heterogeneous nuclear RNA classes is the sea urchin embryo. Nature (London) 260: 120–124

Dunbar B S, Johnson J, Epel D (1974) [125]I iodination and autoradiography of the plasma membrane of *Arbacia punctulata* eggs. Biol Bull 147: 474 (Abstr)

Duncan R, Dower W (1973) Normal synthesis transport and decay of messenger RNA in the absence of translation of the RNA. J Cell Biol 59: No 2 part 2, 85 (Abstr)

Duncan R, Humphreys T (1977) Changes in poly (A) and poly (U) sequences during sea urchin embryogenesis. J Cell Biol 75: 352 (Abstr)

Duncan R, Humphreys T (1979) All the poly (A) (+) mRNA sequence complexity also occurs in poly (A) (−) mRNA in sea urchin embryos. Biol Bull 157: 366 (Abstr)

Duncan R, Humphreys T (1981a) Most sea urchin maternal mRNA sequences in very abundance class appear in both polyadenylated and non-polyadenylated molecules. Dev Biol 88: 201–210

Duncan R, Humphreys T (1981b) Multiple digo (A) tracts associated with inactive sea urchin maternal mRNA sequences. Dev Biol 88: 211–219

Duncan R, Humphreys T (1983) Oligo (U) sequences present in sea urchin maternal RNA decrease following fertilization. Dev Biol 96: 258–262

Duncan R, Dower W, Humphreys T (1975) Normal synthesis, transport and decay of mRNA in the absence of its translation. Nature (London) 253: 751–753

Dunham P, Nelson L, Vosshall L, Weissman G (1982) Effect of enzymatic and non-enzymatic proteins on *Arbacia* spermatozoa: Reactivation of aged sperm and the induction of polyspermy. Biol Bull 163: 420–430

Durica D S, Schloss J A, Crain W R (1979) Studies on actin mRNA-complementary genomic DNA sequences in the sea urchin, *S. purpuratus*. Biol Bull 157: 367 (Abstr)

Durica D S, Schloss J A, Crain W R Jr (1980) Organization of actin gene sequences in the sea urchin: Molecular cloning of an intron-containing DNA sequence coding for a cytoplasmic actin. Proc Natl Acad Sci USA 77: 5683–5687

Dworkin M B, Infante A A (1976) Relationship between the mRNA of polysomes and free RNP particles in the early sea urchin embryo. Dev. Biol 53: 73–90

Dworkin M B, Infante A A (1978) RNA synthesis in unfertilized sea urchin eggs. Dev Biol 62: 247–257

Dworkin M B, Rudensey L M, Infante A A (1977) Cytoplasmic non polysomal ribonucleoprotein particles in sea urchin embryos and their relationship to protein synthesis. Proc Natl Acad Sci USA 74: 2231–2235

Easton D, Chalkey R (1972) High-resolution electrophoretic analysis of the histones from embryo and sperm of *Arbacia punctulata*. Exp Cell Res 71: 503–508

Eckberg W R and Ozaki H (1975) RNA complementary to unique DNA sequences in normal and animalized sea urchin embryos. Exp Cell Res 92: 403–411

Eckberg W R, Perotti M E (1983) Inhibition of gamete membrane fusion in the sea urchin by quercetin. Biol Bull 164: 62–70

Eddy B M, Shapiro B M (1976) Changes in the topography of the sea urchin egg after fertilization. J Cell Biol 71: 35–48

Eden F C, Graham D E, Davidson B H, Britten R J (1977) Exploration of long and short repetitive sequence relationships in the sea urchin genome. Nucleic Acid Res 5: 1553–1568

Egrie J C, Wilt F H (1979) Changes in poly (adenylic acid) polymerase activity during sea urchin embryogenesis. Biochemistry 18: 269–274

Elhai J, Scandella C J (1983) Arachidonic acid and other fatty acids inhibit secretion from sea urchin eggs. Exp Cell Res 148: 63–71

Ellis C H Jr (1966) The genetic control of sea urchin development: A chromatographic study of protein synthesis in the *Arbacia punctulata* embryo. J Exp Zool 163: 1–22

Ellis C H Jr, Wintex R J (1967) Protein synthesis and skeletal spicule formation in the sea urchin larva. Am Zool 7: 750 (Abstr)

Emerson C P Jr, Humphreys T (1970) Regulation of DNA-like RNA and apparent activation of ribosomal RNA synthesis in sea urchin embryos: Quantitative measurements of newly synthesized RNA. Dev Biol 23: 86–112

Emerson C P Jr, Humphreys T (1971) Ribosomal RNA synthesis and the multiple, atypical nucleoli in cleaving embryos. Science 171: 898–901

Emlet R B (1983) Locomotion, drag and the rigid skeleton of larval echinoderms. Biol Bull 164: 433–445

Endo Y (1961) Changos in the cortical layer of sea urchin eggs at fertilization as studied with the electron microscope. Exp Cell Res 25: 383–397

Enesco H B, Man K H (1974) Cytoplasmic DNA in sea urchin oogenesis studied by ^{3}H-actinomycin-D-binding and radioautography. Biol Bull 147: 586–593

Enger M D, Hanners J L (1978) Informosomal and polysomal messenger RNA. Differential kinetics of polyadenylation and nucleocytoplasmic transport in Chinese hamster ovary cells. Biochim Biophys Acta 521: 606–618

Epel D (1964a) A primary metabolic change of fertilization: interconversion of pyridine nucleotides. Biochem Biophys Res Commun 17: 62–68

Epel D (1964b) Simultaneous measurement of TPNH formation and respiration following fertilization of the sea urchin egg. Biochem Biophys Res. Commun 17: 69–73

Epel D (1967) Protein synthesis in sea urchin eggs: A "late" response to fertilization. Proc Natl Acad Sci USA 57: 899–906

Epel D (1972) Activation of an Na$^+$-dependent amino acid transport system upon fertilization of sea urchin eggs. Exp Cell Res 73: 74–89

Epel D (1975) The program and mechanisms of fertilization in the echinoderm egg. Am Zool 15: 507–522

Epel D (1978) Mechanisms of activation of sperm and egg during fertilization of sea urchin gametes. Curr Top Dev Biol 12: 186–246

Epel D, Steinhardt R, Humphreys T, Mazia D (1974) An analysis of the partial metabolic derepression of sea urchin eggs by ammonia: the existence of independent pathways. Dev Biol 40: 245–255

Epel D, Patton C, Wallace R W, Cheung W Y (1981) Calmodulin activates NAD kinase of sea urchin eggs: an early event of fertilization. Cell 23: 543–549

Ernst S G, Britten R J, Davidson B H (1979) Distinct single copy sequence sets in sea urchin nucleas RNAs. Proc Natl Acad Sci USA 76: 2209–2212

Ernst S G, Hough-Evans B R, Britten R J, Davidson E H (1980) Limited complexity of the RNA in micromeres of sixteen-cell sea urchin embryos. Dev. Biol 79: 119–127

Etkin L D, Maxon R E Jr (1980) The synthesis of authentic sea urchin transcriptional and translational products by sea urchin histone genes injected into *Xenopus laevis* oocytes. Dev Biol 75: 13–25

Etkin L D, Roberts M (1983) Transmission of integrated sea urchin histone genes by nuclear transplantation in *Xenopus laevis*. Science 221: 67–69

Evans I M, Bosman H B (1977) Glycosyltransferase activity in developing sea-urchin embryos. J Cell Sci 25: 355–366

Evans I M, Gross P R (1978) 5-bromodeoxyuridine does not affect development of the sea urchin, *Arbacia punctulata*. Exp Cell Res 114: 85–93

Evans I B, Ozaki H (1973) Nuclear histones of unfertilized sea urchin eggs. Exp Cell Res 79: 228–231

Evans T, Rosenthal E T, Youngblom J, Distel D, Hunt T (1983) Cylin: a protein specified by maternal mRNA in sea urchin eggs that is destroyed at each cleavage division. Cell 33: 389–396

Evola-Maltese C (1957) Histochemical localization of alkaline phosphatase in the sea urchin embryo. Acta Embryol Morphol Exp 1: 99–103

Fansler B, Loeb L. A (1969) Sea urchin nuclear DNA polymerase. II. Changing localization during early development. Exp Cell Res 57: 305–310

Fansler B, Loeb L A (1972) Sea urchin nuclear DNA polymerase. IV. Reversible association of DNA polymerase with nuclei during the cell cycle. Exp Cell Res 75: 433–441

Farquhar M N, McCarthy B J (1973) Evolutionary stability of the histone genes of sea urchins. Biochemistry 12: 4113–4122

Farrell K W, Wilson L (1977) Microtubule reassembly in vitro with tubulin from outer doublet microtubules. J Cell Biol 75: 252 (Abstr)

Farrell K W, Wilson I (1978) Microtubule reassembly in vitro of *Strongylocentrotus purpuratus* sperm tail outer doublet tubulin. J Mol Biol 121: 393–410

Farrell K W, Kassis J A, Wilson L (1979a) Outer doublet tubulin reassembly Evidence for opposite end assembly-disassembly at steady state and a disassembly end equilibrium. Biochemistry 18: 2642–2647

Farrell K W, Morse A, Wilson L (1979b) Characterization of the in vitro reassembly of tubulin derived from stable *Strongylocentrotus purpuratus* outer doublet microtubules. Biochemistry 18: 905–910

Faust M, Millward S, Fromson D (1975) Methylated nucleosides in sea urchin poly $(A)^+$ and poly $(A)^-$ RNA. J Cell Biol 67: 114 (Abstr)

Faust M, Millward S, Duchastel A, Fromson D (1976) Methylated constituents of poly $(A)^-$ and poly $(A)^+$ polyribosomal RNA of sea urchin embryos. Cell 9: 597–604

Feit H, Shusarek L, Shelanski M L (1971) Heterogeneity of tubulin subunits. Proc Natl Acad Sci USA 68: 2028–2031

Felicetti L, Metafora S, Gambino R (1972) Characterization and activity of the elongation factor T1 and T2 in the unfertilized egg and in the early development of sea urchin. Cell Diff 1: 265–277

Ficq A (1964) Effects de l'actinomycin D et de la puromycine sur le métabolism de l'oocyte on croissance. Exp Cell Res 34: 581–594

Ficq A, Brachet J (1963) Metabolisme des acides nucléiques et des proteines chez les embryous normaux et des hybrides létaux entre échinodermes Exptl Cell Res 32: 90–108

Fink R D, McClay D R (1980) Sea urchin mesenchyme cells temporal changes in hyaline-binding affinity. Am Zool 20: 942 (Abstr)

Finkel T, Levitan H, Carroll E J (1981) Fertilization in the sea urchin *Arbacia punctulata* inhibited by fluorescein dyes: evidence for a plasma membrane mechanism. Gam Res 4: 219–229

Fisher G W, Rebhun L I (1983) Sea urchin egg cortical granule exocytosis is followed by a burst of membrane retrieval via uptake into coated vesicles. Dev Biol 99: 456–472

Fitzmaurice L C, Baker R F (1973) Sequence differences in chromatin DNA and nuclear membrane associated DNA in the sea urchin embryo. Biochem Biophys Res Commun 55: 328–332

Fitzmaurice L C, Baker R F (1974a) Effect of 5-bromodeogyuridine on incorporation of RNA precursors in sea urchin embryos. J Cell Physiol 83: 259–261

Fitzmaurice L C, Baker R F (1974b) Deoxyadenylate-rich sequences in sea urchin DNA during early development. Cell Differ 3: 117–126

Flemming W (1881) Beiträge zur Kenntnis der Zolle und ihrer Lebenserscheinungen 3. Arch Mikrosk Anat 30: 1–86

Flytzanis C N, Brandhorst B P, Britten R J, Davidson E H (1982) Developmental patterns of cytoplasmic transcript prevalence in sea urchin embryos. Dev Biol 91: 27–35

Foerder C A, Shapiro B M (1977) Release of ovoperoxydase from sea urchin eggs hardens the fertilization membrane with tyrosine crosslinks. Proc Natl Acad Sci USA 74: 4214–4218

Foerder C A, Eddy E M, Klebanoff S J, Shapiro B M (1977) Chemiluminescence accompanics the ovoperoxydase catalyzed cross link of the sea urchin fertilization membrane. J Cell Biol 75: 409 (Abstr)

Foerder C A, Klebanoff S J, Shapiro B M (1978) Hydrogen peroxide production, chemiluminescence, and the respiratory burst of fertilization: interrelated events in early sea urchin development. Proc Natl Acad Sci USA 75: 3183–3187

Fodor B J B, Ako H, Walsh K W (1975) Isolation of a protease from sea urchin eggs before and after fertilization. Biochemistry 14: 4923–4927

Fox D L, Hopkins T S (1966) In: Boolotian RA (ed) Physiology of echnodermata. Whiley Interscience, New York, London, Sidney, pp 277–300

Frederiksen S, Hellung-Larsen P (1974) Synthesis of small molecular weight RNA components during the early stages of sea urchin embryo development. Exp Cell Res 89: 217–227

Frederiksen S, Hellung-Larsen P, Enberg J (1973) Small molecular weight RNA components in sea urchin embryos. Exp Cell Res 78: 287–294

Fromson D, Duchastel A (1973) Effect of cordycepin on sea urchin embryo polyribosomes. J Cell Biol 59: No 2, part 2, 105 (Abstr)

Fromson D, Duchastel A (1974) Poly (A)-containing polyribosomal RNA in sea urchin embryos: Changes in proportion during development. J Cell Biol 63: 105 (Abstr)

Fromson D, Duchastel A (1975) Poly (A)-containing polyribosomal RNA in sea urchin embryos: changes in proportion during development. Biochim Biophys Acta 378: 394–404

Fromson D, Verma D P S (1976) Translation of non poly adenylated messenger RNA of sea urchin embryos. Proc Natl Acad Sci USA 73: 148–151

Fromson D, Duchastel A, Tufaro F, Brandhorst B (1977) Polyacrilamide gel analysis of in vitro synthesized polypeptides coded for by poly (A)$^+$ and poly (A)$^-$ messenger RNA. J Cell Biol 75: 359 (Abstr)

Fronk E, Gibbons I R, Ogawa K (1975) Multiple forms of dynein associated with flagellar axonemes from sea urchine sperm. J Cell Biol 67: 125 (Abstr)

Fry B J, Gross P R (1970a) Patterns and rates of protein synthesis in sea urchin embryos. I Uptake and incorporation of amino acids during the first cleavage. Dev. Biol 21: 105–124

Fry B J, Gross P R (1970b) Patterns and rates of protein synthesis in sea urchin embryos. II The calculation of absolute rates. Dev. Biol 21: 125–146

Frydenberg O, Zeuthen E (1960) Oxygen uptake and carbon dioxide output related to the mitotic rhythm in the cleaving eggs of *Dendraster excentricus* and *Urechis caupo*. C R Trav Lab Carlsberg 31: G 23

Fujino Y, Yasumasu I (1975) Activation of adenyl cyclase with hexose monophosphates. Biochem Biophys Res Commun 65: 1067–1072

Fujino Y, Yasumasu I (1978) Adenosine 3′,5′-cyclic monophosphate-binding protein in sea urchin eggs and embryos. Gam. Res 1: 137–143

Fujino Y, Yasumasu I (1981) cMPA-dependent protein kinase in sea urchin embryos. Gam Res 4: 395–406

Fujino Y, Yasumasu I (1982) Change in the intracellular distribution of cAMP-binding capacity in sea urchin embryos during early development. Gam Res 6: 19–27

Fujiwara A, Yasumasu I (1974) Morphogenetic substances found in the embryos of sea urchin, with special reference to the antivegetalizing substance. Dev Growth Differ 16: 93–103

Fujiwara A, Hino A, Yasumasu I (1980) Inhibition of respiration in sea urchin spermatozoa following interaction with fixed unfertilized eggs. III Inhibition of sperm respiration by heat-stable substance removed from glutaraldehyde-fixed eggs. Dev. Growth Differ 22: 763–771

Fujiwara A, Hino A, Hiruma T, Yasumasu I (1982a) Inhibition of respiration in sea urchin spermatozoa following interaction with fixed unfertilized eggs. VI Probable difference between the species in the mechanism for the fixed-egg-induced inhibition of sperm respiration. Dev Growth Differ 22: 145–154

Fujiwara A, Yokokawa M, Hino A, Yasumasu I (1982b) Inhibition by palmitoyl CoA of dynein ATPase from sea urchin spermatozoa. J Biochem (Tokyo) 92: 441–447

Fujimoto N, Yasumasu I (1979) Change in the fructose 1,6-biphosphatase activity in sea urchin eggs following fertilization. J Biochem (Tokyo) 86: 719–724

Gabel C A, Eddy E M, Shapiro B M (1979) After fertilization, sperm surface components remain as a patch in sea urchin and mouse embryos. Cell 18: 207–215

Gabers D L (1981) The elevation of cyclic AMP concentration in flagelalless sea urchin sperm heads. J Biol Chem 256: 620–648

Gache C, Vacquier V D (1983) Transport of methionine in sea-urchin sperm by a neutral amino-acid carrier. Eur J Biochem 133: 341–347

Gache C, Niman H L, Vacquier V D (1983) Monoclonal antibodies to the sea urchin egg vitelline layer inhibit fertilization by blocking sperm adhesion. Exp Cell Res 147: 75–84

Gadaleta M N, Nicotra A, Del Prete M G, Saccone C (1977) DNA polymerase activity in isolated mitochondria of *Paracentrotus lividus* at various stages of development. Cell Differ 6: 85–94

Gafurov N N, Rasskazov V A (1974) ATP-dependent DNAase in differentiating cells of sea urchin embryos. Dokl Biochem 219: 627–629

Galau G A, Britten R J, Davidson E H (1974) A measurement of the sequence complexity of polysomal mRNA in sea urchin embryos. Cell 2: 9–20

Galau G A, Klein W H, Davis M M, Wold B J, Britten R J, Davidson E H (1976) Structural gene sets active in embryos and adult tissues of the sea urchin. Cell 7: 487–506

Galau G A, Lipson B D, Britten R J, Davidson E H (1977) Synthesis and turnover of polysomal mRNA in sea urchin embryos. Cell 10: 415–432

Galli G, Hofstetter H, Stunnemberg H G, Birnstiel M L (1983) Biochemical complementation with RNA in the *Xenopus* oocyte: A small RNA is required for the generation of 3'histone mRNA termini. Cell 34: 823–828

Gama-Sosa M A, Midgett R M, Slagel V A, Githens S, Kuo K C, Gehrke C W, Ehrlich M (1983) Tissue-specific differences in DNA methylation in various mammals. Biochim Biophys Acta 740: 212–219

Gambino R, Metafora S, Felicetti L, Raisman J (1973) Properties of the ribosomal salt wash from unfertilized and fertilized sea urchin eggs and its effect on natural mRNA translation. Biochim Biophys Acta 312: 377–391

Garbers L D, Tubb D J, Kopf G S (1980) Regulation of sea urchin sperm cyclic AMP dependent protein kinases by an egg associated factor. Biol Reprod 22: 526–532

Garling D, Hunt T (1977) Phosphorylation of *Arbacia punctulata* sperm histones by cytoplasm from eggs andeearly embryos. Biol Bull 153: 436 (Abstr)

Geraci G, Lancieri M, Marchi P, Noviello L (1979) The sea urchin (*Sphaerechinus granularis*) codes different H2B histones to assemble sperm and embryo chromatin. Cell Differ 8: 187–194

German J (1964) The chromosomal complement of blastomeres in *Arbacia punctulata*. Biol. Bull 127: 370–371 (Abstr)

German J (1966) The chromosomal complement of blastomeres in *Arbacia punctulata*. Chromosoma 20: 195–201

Geuskens M (1968) Etude ultrastructurale des embryons normaux et des hybrides létaux entre echinodermes Exptl Cell Res 49: 477 (Abstr)

Gezelius G (1974a) Effect of sulphate deficiency on the morphogenesis of sea urchin larvae. Zoon 2: 37–47

Gezelius G (1974b) Incorporation of sulphate ions during early sea urchin development. Zoon 2: 105–116

Gezelius G (1976) Further aspects on the effect of sulphate deficiency on the RNA synthesis in sea urchin embryos. Zoon 4: 43–46

Ghiretti F, Ghiretti-Magaldi A, Rothschild H A, Tosi L (1958) A study of the cytochromes of marine invertebrates. Acta Physiol Latinoam 8: 239–247

Gibbins J R, Tilney L G, Porter K R (1969) Microtubules in the formation and development of the primary mesenchyme in *Arbacia punctulata*. I The distribution of microtubules. J Cell Biol 41: 201–206

Gibbons B G, Fronk E (1979) A latent adenosine triphosphatase form of dynein 1 from sea urchin sperm flagella. J Biol. Chem 254: 187–196

Gibbons B G, Gibbons I R (1979) Relationship between the latent adenosinetriphosphatase state of dynein 1 and its ability to recombine functionally with KCl extracted sea urchin sperm flagella. J Biol Chem 254: 197–201

Gibbons B H (1980) Intermittent swimming in live sea urchin sperm. J Cell Biol 84: 1–12

Gibbons B H (1982) Effects of organic solvents on flagellar asymmetry and quiescence in sea urchin sperm. J Cell Sci 54: 115–136

Gibbons B H, Gibbons I R (1972a) Effect of partial extraction of dynein arms on the movement of Triton-extracted sea urchin sperm. J Cell Biol 55: 84 (Abstr)

Gibbons B H, Gibbons I R (1972b) Flagellar movement and adenosine triphosphatase activity in sea urchin sperm extracted with Triton X 100. J Cell Biol 54: 75–97

Gibbons B H, Gibbons I R (1973) The effect of partial extraction of dynein arms on the movement of reactivated sea urchin sperm. J Cell Sci 13: 337–357

Gibbons B H, Gibbons I R (1976) Functional recombination of dynein 1 with demembranated sea urchin sperm partially extracted with KCl. Biochem Biophys Res. Commun 73: 1–6

Gibbons B H, Gibbons I R (1977) Transient waveforms during intermittent swimming in live sea urchin sperms. J Cell Biol 75: 276 (Abstr)

Gibbons B H, Gibbons I R (1980) Calcium-induced quiescence in reactivated sea urchin sperm. J Cell Biol 84: 13–27

Gibbons B H, Gibbons I R (1981) Organic solvents modify the calcium control of flagellar movement in sea urchin sperm. Nature (London) 292: 85–86

Gibbons B H, Ogawa K, Gibbons I R (1975) Properties of reactivated sea urchin sperm treated with anti-dynein serum. J Cell Biol 67: 134 (Abstr)

Gibbons B H, Ogawa K, Gibbons I R (1976) The effect of antidynein 1 serum on the movement of reactivated sea urchin sperm. J Cell Biol 71: 823–831

Gibbons I R (1972) ATP-induced sliding of tubules in trypsin treated flagella of sea urchin sperm. J Cell Biol 55: 85 (Abstr)

Gibbons I R, Fronk E (1972) Some properties of bound and soluble dynein from sea urchin sperm flagella. J Cell Biol 54: 365–381

Gibbons I R, Fronk B H, Ogawa K (1976) Multiple forms of dynein in sea urchin sperm flagella. In: Cell mobility. (Cold Spring Harbor Laboratory), Cold Spring Harbor, New York, pp 915–932

Gillies R J, Rosemberg M, Deamer D W (1980) Inorganic carbonate release and pH changes during the activation sequence in sea urchin eggs. J Cell Biol 87: 136 (Abstr)

Gilula N B (1973) Septate junction development in sea urchin embryos. J Cell Biol 55: 172 (Abstr)

Gineitis A R, Stankeviciute J V, Voraboiev V (1976a) Chromatin proteins from normal, vegetalized, and animalized sea urchin embryos. Dev. Biol 52: 181–193

Gineitis A A, Nivinskae H H, Voraboiev V I (1976b) Nonhistone chromatin proteins from the sea urchin Strongylocentrotus droebachiensis sperm and embryo. Exp Cell Res 98: 248–252

Giudice G (1962a) Restitution of whole larvae from disaggregated cells of sea urchin embryos. Dev Biol 5: 402–411

Giudice G (1962b) Amino acid incorporation into the proteins of isolated cells and total homogenates of sea urchin embryos. Arch Biochem Biophys 99: 447–450

Giudice G (1963) Aggregation of cells isolated from vegetalized and animalized sea urchin embryos. Experientia 19: 83–86

Giudice G (1965) The mechanism of aggregation of embryonic sea urchin cells; a biochemical approach. Dev Biol 12: 233–247

Giudice G (1973) Developmental biology of the sea urchin embryo. Academic Press, London, New York, 469 p

Giudice G, Hörstadius S (1965) Effect of actinomycin D on the segregation of animal and vegetal potentialities in the sea urchin egg. Exp Cell Res 39: 117–120

Giudice G, Monroy A (1958) Incorporation of S^{35}-methionine in the proteins of the mitochondria of developing and parthenogenetically activated sea urchin eggs. Acta Embryol Morphol Exp 3: 58–65

Giudice G, Mutolo V (1967) Synthesis of ribosomal RNA during sea urchin development. Biochim Biophys Acta 138: 276–285

Giudice G, Mutolo V (1969) Synthesis of ribosomal RNA during sea urchin development. II. Electrophoretic analysis of nuclear and cytoplasmic RNA's. Biochim Biophys Acta 179: 345–347

Giudice G, Mutolo V (1970) Reggregation of dissociated cells of sea urchin embryos. In: M Abercrombie, J Brachet, T J King (eds.) Advances in Morphogenesis, Vol VIII. Academic Press, London New York, pp 115–158

Giudice G, Vittorelli M L, Monroy A (1962) Investigations on protein metabolism during the early development of the sea urchin. Acta Embryo Morphol Exp 5: 113–122

Giudice G, Mutolo V, Donatuti G (1968) Gene expression in sea urchin development. Wilhelm Roux' Arch Entwicklungsmech Org 161: 118–128

Giudice G, Mutolo V, Donatuti G, Bosco M (1969) Reggregation of mixture of cells from different developmental stages of sea urchin embryos. Exp Cell Res 54: 279–281

Giudice G, Sconzo G, Ramirez F, Albanese I (1967a) Giant RNA is also found in the cytoplasm in sea urchin embryos. Biochim Biophys Acta 262: 401–403

Giudice G, Pirrone A M, Roccheri M, Trapani M (1973) Maturational cleavage of nucleolar ribosomal RNA precursor can be catalyzed by non-specific endonuclease. Biochim Biophys Acta 319: 72–80

Giudice G, Sconzo G, Bono N, Albanese I (1972b) Studies on sea urchin oocytes. I. Purification and cell fractionation. Exp Cell Res 72: 90–94

Giudice G, Sconzo G, Albanese I, Ortolani G, Cammarata M (1974) Cytoplasmic giant RNA in sea urchin embryos. I. Proof that it is not derived from artifactual nuclear leakage. Cell Differ 3: 287–295

Giudice G, Roccheri M C, Di Bernardo M G (1980) Synthesis of "heat shock" proteins in sea urchin embryos. Cell Biol Int Rep 4: 69–74

Glabe C G, Lennarz W (1979) Species-specific sperm adhesion in sea urchins. A quantitative investigation of bindin-mediated egg agglutination. J Cell Biol 83: 595–604

Glabe C G, Vacquier V D (1977a) Isolation and characterization of the vitellino layer of sea urchin eggs. J Cell Biol 75: 410–421

Glabe C G, Vacquier V D (1977b) The surfaces of sea urchin eggs possess a species-specific receptor for bindin. J Cell Biol 75: 56 (Abstr)

Glabe C G, Vacquier V D (1978) Egg surface glycoprotein receptor for sea urchin sperm bindin. Proc Natl Acad Sci USA 75: 881–885

Glabe C G, Buchalter M, Lennarz W J (1981) Studies on the interactions of sperm with the surface of the sea urchin egg. Dev Biol 84: 397–406

Glabe C G, Grabel L B, Vacquier V D, Rosen S D (1982) Carbohydrate specificity of sea urchin sperm binding: a cell surface lectin mediating sperm-egg adhesion. J Cell Biol 94: 123–128

Glisin V R, Glisin M V, Doty P (1966) The nature of messenger RNA in the early stages of sea urchin development. Proc Natl Acad Sci USA 56: 285–289

Goldberg R B, Galau G A, Britten R J, Davidson E H (1973) Nonrepetitive DNA sequence representation in sea urchin embryo messenger RNA. Proc Natl Acad Sci USA 70: 3516–3520

Goldinger J M, Barron E S G (1946) The pyruvate metabolism of sea urchin eggs during the process of cell division. J Gen Physiol 30: 73–82

Goldstein S F (1979) Starting transients in sea urchin sperm flagella. J Cell Biol 80: 61–68

Goldstein S F (1981) Motility of basal fragments of sea urchin sperm flagella. J Cell Sci 50: 65–78

Golob R, Chetsanga C J, Doty P (1974) The onset of collagen synthesis in sea urchin embryos. Biochim Biophys Acta 349: 135–141

Gordon K, Infante A A (1983) Utilization of maternal and embryonic histone RNA in early sea urchin development. Dev Biol 95: 414–420

Goudsmith E M (1972) Glycogen content of eggs and embryos of *Arbacia punctulata*. Dev Biol 27: 329–336

Gould D, Benson S C (1978) Selective inhibition of collagen synthesis in sea urchin embryos by a low concentration of actinomycin D. Exp Cell Res 112: 73–78

Gourlie B B, Infante A A (1975) Pool sizes of the deoxynucleotide triphosphates in the sea urchin egg and developing embryo. Biochem Biophys Res Commun 64: 1206–1214

Goustin A S (1981) Two temporal phases for the control of histone gene activity in cleaving sea urchin embryos (*S. purpuratus*). Dev Biol 87: 163–175

Goustin A S, Wilt F H (1981) Protein synthesis, polyribosomes, and peptide elongation in early development of *Strongylocentrotus purpuratus*. Dev Biol 82: 32–40

Goustin A S, Wilt F H (1982) Direct measurement of histone peptide elongation rate in cleaving sea urchin embryos. Biochim Biophys Acta 699: 22–27

Graham D B, Neufeld B R, Davidson B H, Britten R J (1974) Interspersion of repetitive and non-repetitive DNA sequences in the sea urchin genome. Cell 1: 129–137

Grainger J L, Barrett D (1973) Radio-labelling proteins of the sea urchin eggs surface. Biol Bull 145: 437 (Abstr)

Grainger J L, Hinegardner R T (1974) A comparison of 5-bromodeoxyuridine and thymidine incorporation into fertilized sea urchin eggs. Exp Cell Res 84: 395–398

Grainger J L, Winkler M M, Shen S S, Steinhardt R A (1979) Intracellular pH controls protein synthesis rate in sea urchin egg and early embryo. Dev Biol 68: 396–406

Grainger R M, Wilt F H (1976) Incorporation of ^{13}C, ^{15}N-labeled nucleosides and measurement of RNA synthesis and turnover in sea urchin embryos. J Mol Biol 104: 589–601

Greco M, Baykovic, Moskov N, Saccone C, Glicin V (1977) Mitochondrial RNA synthesis during the cleavage stages of sea urchin embryos. Bull Mol Biol Med 2: 145–155

Green C R (1981) Fixation-induced intramembrane particle movement demonstrated in freeze-fracture replicas of a new type of uptake junction in echinoderm epithelia. J Ultrastruct Res 75: 11–22

Green C R, Bergquist P R, Bullivant S (1979) An anastomosing septate junction in endotelial cells of the Phylum Echindermata. J Ultrastruct Res 67: 72–80

Green J D, Summers R G (1979) Are fertilization envelope elevation and cortical contraction related to secretion of macromolecules into the perivitelline space of the sea urchin egg? J Cell Biol 83: 211 (Abstr)

Green J D, Summers R G (1980) Ultrastructural demonstration of trypsin-like protease in acrosomes of sea urchin sperm. Science 209: 398–400

Green J D, Summers R G (1982) Effects of protease inhibitors on sperm related events in sea urchin fertilization. Dev Biol 93: 139–144

Griffith J K, Humphreys T D (1979) Ribosomal nucleic acid synthesis and processing in embryos of the Hawaian sea urchin *Tripneustes gratilla*. Biochemistry 18: 2178–2185

Griffith J K, Griffith B B, Humphreys T (1981) Regulation of ribosomal RNA synthesis in sea urchin embryos and oocyte. Dev Biol 87: 220–228

Grippo P, Iaccarino M, Parisi E, Scarano E (1968) Methylation of DNA in developing sea urchin embryos. J Mol Biol 36: 195–208

Grippo P, Parisi E, Carestia C, Scarano E (1970) A novel origin of some deoxyribonucleic acid thymine and its monrandom distribution. Biochemistry 9: 2605–2609

Gross K, Ruderman J, Jacobs-Lorena M, Baglioni C, Gross P R (1973a) Cell-free synthesis of histones directed by messenger RNA from sea urchin embryos. Nature (London) New Biol 241: 273–274

Gross K, Jacobs-Lorena M, Baglioni C, Gross P R (1973b) Cell-free translation of maternal messenger RNA from sea urchin eggs. Proc Natl Acad Sci USA 70: 2614–2618

Gross K, Probst E, Schaffner W, Birnstiel M (1976a) Molecular analysis of the histone gene cluster of *Psammechinus miliaris*: I. Fractionation and identification of five individual histone mRNA. Cell 8: 455–470

Gross K, Schaffner W, Telford J, Birnstiel M (1976b) Molecular analysis of the histone gene cluster of *Psammechinus miliaris*: III Polarity and asimmetry of the histone coding sequences Cell 8: 479–484

Gross P R, Cousineau G H (1963a) Effect of actinomycin D on macromolecule synthesis and early development in sea urchin eggs. Biochem Biophys Res Commun 10: 321–326

Gross P R, Cousineau G H (1963b) Synthesis of spindle associated proteins in early cleavage. J Cell Biol 19: 260–265

Gross P R, Cousineau G H (1964) Macromolecule synthesis and the influence of actinomycin on early development. Exp Cell Res 33: 368–395

Gross P R, Fry B J (1966) Continuity of protein synthesis through cleavage metaphase. Science 153: 749–751

Gross P R, Spindel W, Cousineau G H (1963) Decoupling of protein and RNA synthesis during deuterium parthenogenesis in sea urchin eggs. Biochem Biophys Res Commun 13: 405–410

Gross P R, Kraemer K, Malkin L I (1965) Base composition of RNA synthesized during cleavage of the sea urchin embryo. Biochem Biophys Res Commun 18: 569–575

Grosscheld R, Birnstiel M L (1980) Identification of regulatory sequences in the prelude sequences of an H2A histone gene by the study of specific deletion mutants in vivo. Proc Natl Acad Sci USA 77: 1432–1436

Grosscheld R, Wasyly B, Chanbon P, Birnstiel M L (1981) Point mutation in the TATA box curtails expression of sea urchin H2A histone gene in vivo. Nature (London) 294: 178–180

Grosscheld R, Mächler M, Rohrer V, Birnstiel M L (1983) A functional component of the sea urchin H2A gene modulator contains an extended sequence homology to a viral enhancer. Nucleic Acid Res 11: 8123–8136

Grossman A, Cagan L, Levy M, Troll W, Weck S, Weissman G (1971) Is the redistribution of tosylarginine methyl ester hydrolase activity that follows fertilization of *Arbacia punctulata* eggs intimately related to embryogenesis? Biol Bull 141: 387 (Abstr)

Grossman A, Levy M, Troll W, Weissmann G (1973a) Redistribution of tosylarginine methylester hydrolase activity after fertilization of sea urchin (*Arbacia punctulata*) eggs. Nature (London) New Biol 243: 277–278

Grossman A, Inoue S, Fishman L (1973b) Release of particulate structures and TAME hydrolase activity from sea urchin (*Arbacia punctulata*) eggs after fertilization. Nature (London) New Biol 243: 279–281

Grunstein M (1978) Hatching in the sea urchin *Lytechinus pictus* is accompanied by a shift in histone H4 gene activity. Proc Natl Acad Sci USA 75: 4135–4139

Grunstein M, Grunstein J B (1977) The histone H4 gene of *Strongylocentrotus purpuratus*: DNA and mRNA sequences at the 5' end. Cold Spring Harbor Symp Quant Biol 42: 1083–1092

Grundstein M, Schedl P (1976) Isolation and sequence analysis of sea urchins (*Lytechinus pictus*) histone H4 mRNA. J Mol Biol 104: 323–349

Grunstein M, Schedl P, Kedes L (1976) Sequences analysis and evolution of sea urchin *Lytechinus pictus* and *Strongylocentrotus purpuratus*) histone H mRNAs. J Mol Biol 104: 351–369

Grunstein M, Diamond K E, Knoppel E, Grunstein J (1981) Comparison of the early histone H4 gene sequence of *Strongylocentrotus purpuratus* with maternal, early, late histone H4 mRNA sequences. Biochemistry 20: 1216–1223

Gundersen G C, Gabel C A, Shapiro B M (1982) An intermediate state of fertilization involved in internalization of sperm components. Dev Biol 93: 59–72

Gustafson T, Hasselberg I (1951) Studies of enzymes in the developing sea urchin. Exp Cell Res 2: 642–672

Gustafson T, Toneby M (1970) On the role of serotonin and acetylcholine in sea urchin morphogenesis. Exp Cell Res 62: 102–117

Gustafson T, Toneby M (1971) How genes control morphogenesis the role of serotonin and acetylcholine in morphogenesis. Am Sci 59: 452–462

Gustafson T, Wolpert L (1961a) Studies on the cellular basis of morphogenesis in the sea urchin embryo. Exp Cell Res 24: 64–79

Gustafson T, Wolpert L (1961b) Studies on the cellular basis of morphogenesis in the sea urchin embryo. Exp Cell Res 22: 432–449

Gustafson T, Wolpert L (1961c) Cellular mechanisms in the morphogenesis of the sea urchin larva. Exp Cell Res 22: 509–520

Gustafson T, Wolpert L (1962) Cellular mechanisms in the morphogenesis of the sea urchin larva. Exp Cell Res 27: 260–279

Gustafson T, Wolpert L (1963a) Studies on the cellular basis of morphogenesis in the sea urchin embryo. Formation of the coelom, the mouth, and the primary pore-canal. Exp Cell Res 39: 561–382

Gustafson T, Wolpert L (1963b) The cellular basis of morphogenesis and sea urchin development. Int Rev Cytol 15: 139–214

Gustafson T, Wolpert L (1967) Cellular movement and contact in sea urchin morphogenesis. Biol Rev 42: 442–498

Gustafson T, Lundgren B, Trenfeldt R (1972a) Serotonin and contractile activity in the echinopluteus. A study of the cellular basis of larval behaviour. Exp Cell Res 72: 115–139

Gustafson T, Ryberg E, Trenfeldt R (1972b) Acetylcholine and contractile activity in the echinopluteus. A study of the cellular basis of larval behaviour. Acta Embryol Morphol Exp 2: 199–223

Habara A, Nagano H, Mano Y (1979) Identification of γ-like DNA polymerase from sea urchin embryos. Biochim Biophys Acta 561: 17–28

Habara A, Nagano H, Mano Y (1980) Characterization of DNA polymerase in mature sperm of the sea urchin. Biochim Biophys Acta 608: 287–294

Hagedorn A L (1909) On the provely motherly character of the hybrids produced from the eggs of Strongylocentrotus Arch f Entwick 27 1–20

Haggerty J G, Jackson R C (1983) Release of granule contents from sea urchin egg cortices. New assay procedures and inhibition by sulphydryl-modifying reagents. J Biol Chem 258: 1819–1850

Hagström B E (1959) Experiments on hybridization of sea urchins Arkiv Zool 12 127–135

Hall H G (1978) Hardening of the sea urchin fertilization envelope by peroxidase-catalyzed phenolic coupling of tyrosines. Cell 15: 343–356

Hall H G, Vacquier V (1982) The apical lamina of the sea urchin embryo: Major glycoproteins associated with the hyaline layer. Dev Biol 89: 168–178

Hallenstvet M, Ryberg E, Liaaen-Jensen S (1978) Animal carotenoids. XIV Carotenoids of *Psammechinus miliaris.* (sea urchin). Comp Biochem Physiol 60B: 173–175

Hamaguchi Y, Hiramoto Y (1981) Activation of sea urchin eggs by microinjection of calcium buffers. Exp Cell Res 134: 171–179

Hamaguchi Y, Kuriyama R (1982) Aster formation in sand dollar eggs by microinjection of calcium buffers and centridar complexes isolated from starfish sperm. Exp Cell Res 141: 440–454

Hansbrough J R, Garbers D L (1981a) Purification and characteristics of a peptide (speract) associated with eggs that stimulates spermatozoa. Adv Enzyme Regul 19: 351–376

Hansbrough J R, Garbers D L (1981b) Sodium-dependent activation of sea urchin spermatozoa by speract and monensin. J Biol Chem 256: 2235–2241

Harding C V, Harding D, Perlmann P (1954) Antigens in sea urchin hybrid embryos. Exptl Cell Res 6 202–210

Harding C V, Harding D, Bamberg J W (1955) On cross fertilization and the analysis of hybrid embryonic development in echinoderms. Exp Cell Res Suppl 3: 181–187

Harding C V, Harding D (1952a) The hybridization of *Echinocardium cordatum* and *Psammechinus miliaris* Arkiv f Zool 4: 403–411

Harding C V, Harding D (1952b) Cross fertilization with the sperm of *Arbacia lixula.* Exp Cell Res 3: 475–484

Harkey M A, Whiteley A H (1980) Isolation culture, and differentiation of echinoid primary mesenchyme cells. Wilhelm Roux's Arch Dev Biol 189: 111–122

Harkey M A, Whiteley A H (1982a) Cell-specific regulation of protein synthesis in the sea urchin gastrula: A two-dimensional electrophoretic study. Dev Biol 93: 453–462

Harkey M A, Whiteley A H (1982b) The translational program during the differentiation of isolated primary mesenchyme cells. Cell Differ 11: 325–329

Harkey M A, Whiteley A H (1983) The program of protein synthesis during the development of the micromere-primary mesenchyme cell line in sea urchin embryo. Dev Biol 100: 12–28

Harpold M M, Craig S P (1975) A possible mechanism for regulation of mitochondrial RNA polymerase during development of the sea urchin embryo. J Cell Biol 67: 157 (Abstr)

Harpold M M, Craig S P (1977) The evolution of repetitive DNA sequences in sea urchins. Nucleic Acid Res 4: 4425–4437

Harpold M M, Craig S P (1978) The evolution of non-repetitive DNA in sea urchins. Differentiation 10: 7–11

Harrington F E, Easton D F (1980) Studies of yolk glycoproteins in sea urchin oogenesis. J Cell Biol 87: 132 (Abstr)

Harrington F E, Easton D P (1982) A putative precursor of the major yolk protein of the sea urchin. Dev Biol 94: 505–508

Harris R S, Schuel H, Troll W (1977) Inhibition of fertilization membrane elevation in sea urchin eggs by procain and tetracaine. Biol Bull 151: 412 (Abstr)

Harris P (1968) Cortical fibers in fertilized eggs of the sea urchin *Strongylocentrotus purpuratus*. Exp Cell Res 52: 677–681

Harris P (1979) A spiral cortical fiber system in fertilized sea urchin eggs. Dev Biol 68: 525–532

Harris P (1983) Caffeine-induced monaster cycling in fertilized eggs of the sea urchin *Strongylocentrotus purpuratus*. Dev Biol 96: 277–284

Harris P, Osborne M, Weber K (1980a) A spiral array of microtubules in the fertilized sea urchin egg cortex examined by indirect immunofluorescence and electron microscopy. Exp Cell Res 126: 227–236

Harris P, Osborn M, Weber K (1980b) Distribution of tubulin-containing structures in the egg of the sea urchin *Strongylocentrotus purpuratus* from fertilization through first cleavage. J Cell Biol 84: 668–679

Harrison M F, Wilt F H (1982) The program of H1 histone synthesis in *Strongylocentrotus purpuratus* embryos and the control of its timing. J Exp Zool 223: 245–256

Hartman J F, Ziegler M M, Comb D G (1971) Sea urchin embryogenesis. I RNA synthesis by cytoplasmic and nuclear genes during development. Dev Biol 25: 209–231

Harvey E B (1936) Parthenogenic merogermy or cleavage without nuclei in *Arbacia punctulata*. Biol Bull 71: 101–121

Harvey E B (1942) Maternal inheritance in Echinoderm hybrids. J Exp Zool 91: 213–235

Harvey E B (1956) The American *Arbacia* and other sea urchins. Princeton University Press, Princeton, 298

Harvey E N (1911) Studies on the permeability of cells. J Exp Zool 10: 507–556

Hata H, Yano Y, Mohri T, Mohri H, Miki-Nomura T (1980) ATP-driven tubule extrusion from axenemes without outer dynein arms of sea-urchin sperm flagella. J Cell Biol 41: 331–340

Hayashi M (1976) Inhibition of axoneme and dynein ATPase from sea urchin sperm by free ATP. Biochim Biophys Acta 442: 225–230

Heifetz A, Lennarz W J (1979) Biosynthesis of N-glycosidically linked glycoproteins during gastrulation of sea urchin embryos. J Biol Chem 254: 6119–6127

Henlands L, Allfrey V C, Poccia D (1982) Translational regulation of histone synthesis in the sea urchin, *Strongylocentrotus purpuratus*. J Cell Biol 94: 219–223

Hentschel C (1982) Homocopolymer sequences in the spacer of a sea urchin histone gene repeat are sensitive to S-nuclease. Nature London 295: 714–716

Hentschel C, Birnstiel M L (1981) The organization and expression of histone gene families. Cell 25: 301–313

Hentschel C, Probst E, Birnstiel M L (1980a) Transcriptional fidelity of histone genes injected into *Xenopus* oocyte nuclei. Nature (London) 288: 100–102

Hentschel C, Irminger J C, Bucher P, Birnstiel M L (1980b) Sea urchin histone mRNA termini are located in gene regions downstream from putative regulatory sequences. Nature (London) 285: 147–151

Herbst C (1895) Experimentelle Untersuchungen über den Einfluss der veränderten chemischen Zusammensetzung des umgebenden Mediums auf die Entwicklung der Tiere. V. Wilhelm Roux' Arch Entwicklungsmech Org 2: 482–499

Herbst C (1900) Über das Auseinandergehen Wilhelm in Furchungs- und Gewebezellen in Kalkfreiem Medium. Arch Entwicklungsmech Org 9: 424–463

Hickey E D, Weber L A, Baglioni C (1976) Translation of RNA from unfertilized sea urchin eggs does not require methylation and is inhibited by 7-methylguanosine-5′-monophosphate. Nature (London) 361: 71–73

Hieter P A, Hendricks M B, Hemminki K, Weinberg E S (1979) Histone gene swith in the sea urchin embryo. Identification of late embryonic histone messenger ribonucleic acids and the control of their synthesis. Biochemistry 18: 2707–2716

Hill R J, Poccia D L, Doty P (1971) Towards a total macromolecular analysis of sea urchin embryo chromatin. J Mol Biol 61: 445–462

Hille M B (1974) Inhibitor of protein synthesis isolated from ribosomes of unfertilized eggs and embryos of sea urchins. Nature (London) 249: 556–558

Hille M B, Albers A A (1979) Efficiency of protein synthesis after fertilization in sea urchin eggs. Nature 278: 469–471

Hille M B, Hall D C, Yablonka-Reuveni Z, Darrilchick M V, Moon R T (1981) Translational control in sea urchin eggs and embryos: initiation is rate-limiting in blastula stage embryos. Dev Biol 86: 241–249

Hinegardner R T (1969) Growth and development of the laboratory cultured sea urchin. Biol Bull 137: 465–475

Hinegardner R T (1975) Morphology and genetics of sea urchin development. Am Zool 15: 679–689

Hinkley Jr R E (1979) Inhibition of sperm motility by the volatile anesthetic halothane. Exp Cell Res 121: 435–439

Hino A, Yasumasu I (1979) Change in the glycogen content of sea urchin eggs during early development. Dev Growth Differ 21: 229–236

Hino A, Tazawa E, Yasumasu I (1978) Two pathways from glycogen to glucose-6-phosphate in sea urchin eggs with special reference to the differeces between the species in the contribution ratio of each pathway. Gam Res 1: 117–128

Hino A, Fujiwara A, Yasumasu I (1980a) Inhibition of respiration in sea urchin spermatozoa following interaction with fixed unfertilized eggs. I Change in the respiratory rate of spermatozoa in the presence of fixed eggs. Dev Growth Differ 22: 421–428

Hino A, Hiruma T, Fujiwara A, Yasumasu I (1980b) Inhibition of respiration in sea urchin spermatozoa of *Hemicentrotus pulcherrimus* after their interaction with fixed unfertilized eggs. Dev Growth Differ 22: 813–820

Hirama M N, Mano Y (1976) Polysomes of the sea urchin embryo: Nature of the so-called "cyclic amino acid incorporation" and conditions to prepare cell-free systems for protein synthesis. Dev Growth Differ 18: 363–369

Hiramoto Y (1956) Cell division without mitotic apparatus in sea urchin eggs. Exp Cell Res 11: 630–636

Hiramoto Y (1965) Further studies on cell division without mitotic apparatus in sea urchin eggs. J Cell Biol 25: 161–174

Hiramoto Y (1971) Analysis of cleavage stimulus by means of micromanipulation of sea urchin eggs. Exp Cell Res 68: 291–298

Hiramoto Y (1974) Mechanical properties of the surface of the sea urchin egg at fertilization and during cleavage. Exp Cell Res 89: 320–326

Hiramoto Y, Shôji Y (1982) Location of the motive force for chromosome movement in sand dollar eggs. Cell Differ 11: 349–351

Hiramoto Y, Hamaguchi Y, Shoji Y, Shimoda S (1981a) Quantitative studies on the polarization optical properties of living cells. I Microphotometric birefrigence detection system. J Cell Biol 89: 115–120

Hiramoto Y, Hamaguchi Y, Shoji Y, Schroeder T E, Shimoda S, Nakamura S (1981b) Quantitative studies on the polarization optical propertus of living cells. II The role of microtubules in birefringence of the spindle of the sea urchin egg. J Cell Biol 89: 121–130

Hiruma T, Hino A, Fujiwara A, Yasumasu I (1982) Inhibition of respiration in sea urchin spermatozoa following interaction with fixed unfertilized eggs. V Inhibition of electron transport in a span of mitochondrial respiratory chain between cytochrome b and cytochrome c in sea urchin spermatozoa, induced by the interaction with glutaraldehyde fixed eggs. Dev Growth Differ 22: 17–24

Hisanaga S, Sakai H (1983) Cytoplasmic dynein of the sea urchin egg. II Purification, characterization and interaction with microtubules and Ca-Calmodulin. J Biochem (Tokyo) 93: 87–98

Hielle B L, Phillips J A III, Seeburg P H (1982) Relative levels of methylation in human growth hormone and chorionic somatomammotropin genes in expressing and non expressing tissues. Nucleic Acid Res 10: 3459–3474

Hnilica L S, Johnson A W (1970) Fractionation and analysis of nuclear proteins in sea urchin embryos. Exp Cell Res 63: 261–270

Ho J J L, Mazia D (1979) Calcium activated ribonuclease in *Strongylocentrotus purpuratus* egg and embryo. J Cell Biol 83: 207 (Abstr)

Hobart P M, Infante A A (1980) Persistent cytoplasmic location of a DNA polymerase B in sea urchins during development. Biochim Biophys Acta 607: 256–268

Hobart P M, Duncan R, Infante A A (1977) Association of DNA synthesis with the nuclear membrane in sea urchin embryos. Nature (London) 267: 542–544

Hogan B, Cross P R (1972) Nuclear RNA synthesis in sea urchin embryos. Exp Cell Res 72: 101–114

Holland L Z, Cross N L (1983) The pH within the jelly coat of sea urchin eggs. Dev Biol 99: 258–260

Holmes D S, Cohn R H, Kedes L H, Davidson N (1977) Position of sea urchin (*Strongylocentrotus purpuratus*) histone genes relative to restriction endonuclease sites on the chimeric plasmids pSp2 and PpSp17. Biochemistry 16: 1504–1512

Hori H, Osawa S (1979) Evolutionary change in the 5S RNA secondary structure and a phylogenic tree of 54 5S RNA species. Proc Natl Acad Sci USA 76: 381–385

Hörstadius S (1928) Über die Determination des Keimes bei Echinodermen. Acta Zool Stockholm 9: 1–191

Hörstadius S (1935) Über die Determination in Verlaufe der Eiachse bei Seeigeln. Pubbl Stn Zool Napoli 14: 251–479

Hörstadius S (1939) The mechanics of sea urchin development, studied by operative methods. Biol Rev 14: 132–179

Hörstadius S (1949) Experimental researchs on the developmental physiology of the sea urchin. Pubbl Stn Zool Napoli 21: (Suppl) 131–172

Hörstadius S (1957) On the regulation of the bilateral symmetry in plutei with exchanged meridional halves and in giant plutei. J Embryol Exp Morphol 5: 60–73

Hörstadius S (1973) Experimental embryology of echinoderms. Clarendon Press, Oxford

Hörstadius S, Josefsson L (1977) Endogenous morphogenetic substances of sea urchin influencing the larval differentiation. Acta Embryol Exp 1: 71–88

Hoshi M (1979) Exogastrulation induced by heavy water in sea urchin larvae. Cell Differ 8: 431–436

Hoshi M, Moriya T (1980) Arylsulfatase of sea urchin sperm. 2 Arylsulfatase as a lysin of sea urchins. Dev Biol 74: 343–350

Hoshi M, Nagai Y (1975) Novel sialosphingolipids from spermatozoa of the sea urchin *Anthocidaris crassispina*. Biochim Biophys Acta 388: 152–162

Hough B R, Smith M J, Britten R J, Davidson E H (1975) Sequence complexity of heterogeneous nuclear RNA in sea urchin embryos. Cell 5: 291–300

Hough-Evans B R, Wold B J, Ernst S G, Britten R J, Davidson E H (1977) Appearance and persistence of maternal RNA sequences in sea urchin development. Div Biol 60: 258–277

Houk M S, Hinegardner R T (1980) The formation and early differentiation of sea urchin gonads. Biol Bull 159: 280–294

Houk M S, Hinegardner R T (1981) Cytoplasmic inclusions specific to the sea urchin germ line. Dev Biol 86: 94–99

Howze G B, Van Holde K E (1977) Electron microscopic detection of putative replication sites in chromatin from *Arbacia* embryos. Biol Bull 153: 430 (Abstr)

Hozumi N, Tonegawa S (1976) Evidence for somatic rearrangement of immunoglobulin genes coding for variable and constant regions. Proc Natl Acad Sci USA 73: 3628–3632

Hultin T (1948a) Species cpecificity in fertilization reaction. The role of the vitelline membrane of sea urchin eggs in species specificity. Ark Zool 40A: no 12, 1–9

Hultin T (1948b) Species specificity in fertilization reaction. II Influence of certain factors on the cross-fertilization capacity of *Arbacia lixula* (L). Ark Zool 40A: no 20, 1–8

Hultin T (1950) The protein metabolism of sea urchin during early development studied by means of ^{15}N labeled ammonia. Exp Cell Res 1: 599–602

Hultin T (1952) Incorporation of N^{15}-labeled glycine and alanine into the proteins of developing sea urchin eggs. Exp Cell Res 3: 494–496

Hultin T (1953a) Incorporation of C^{14} labeled carbonate and acetate into sea urchin embryos. Ark Kemi 6: 195–200

Hultin T (1953b) Incorporation of N^{15} labeled ammonium chloride into pyrimidines and purines during the early sea urchin development. Ark Kemi 5: 267–275

Hultin T (1953c) The amino acid metabolism of sea urchin embryos studied by means of N^{15}-labeled ammonium chloride and alanine Ark Kemi 5: 543–552

196 References

Hultin T (1953 d) Incorporation of N^{15}-dl-alanine into protein fractions of sea urchin embryos. Ark Kemi 5: 559–564

Hultin T (1961) Activation of ribosomes in sea urchin eggs in response to fertilization. Exp Cell Res 25: 405–417

Hultin T (1964) Factors influencing polyribosome formation in vivo. Exp Cell Res 34: 608–611

Hultin T, Bergstrand A (1960) Incorporation of C^{14} Leucine into protein by cell-free systems from sea urchin embryos at different stages of development. Dev Biol 2: 61–75

Humphreys T (1969) Efficiency of translation of messenger-RNA before and after fertilization in sea urchins. Dev Biol 20; 435–458

Humphreys T (1971) Measurements of messenger RNA entering polysomes upon fertilization of sea urchin eggs. Dev Biol 26: 201–208

Hutchens J O, Keltch A K, Krahl M E, Clowes G H A (1942) Studies on cell metabolism and cell division. VI. Observations on the glycogen content, carbohydrate consumption, lactic acid production, and ammonia production of eggs of Arbacia punctulata. J General Physiol 25: 717–731

Hyatt E A (1967a) Polyriboadenylate synthesis by nuclei from developing sea urchin embryos I. Characterization of the ATP polymerase reaction. Biochim Biophys Acta 143: 246–253

Hyatt E A (1967) Polyriboadenylate synthesis by nuclei from developing sea urchin embryos. II. Polyriboadenylic acid priming of AT2 polymerase. Biochim Biophys Acta 142: 254–262

Hylander B L, Summers R G (1980) Hyalin is a component of cortical granules. J Cell Biol 87: 137 (Abstr)

Hylander B L, Summers R G (1981) The effect of local anesthetics and ammonia on cortical-granule-plasma membrane attachment in the sea urchin egg. Dev Biol 86: 1–11

Hylander B L, Summers R G (1982a) An ultrastructural immunocytochemical localization of hyalin in the sea urchin egg. Dev Biol 93: 368–380

Hylander B L, Summers R G (1982b) Observations on the role of the cortical reaction in surface changes at fertilization. Cell Differ 11: 267–270

Hylander B L, Summers R G, Schuel H (1979) A new technique for the isolation of sea urchin cortical granules using urethane. J Cell Biol 83: 210 (Abstr)

Hynes R O, Gross P R (1970) A method for separating cells from early sea urchin embryos. Dev Biol 21: 383–402

Hynes R O, Gross P R (1972) Informational RNA sequences in early sea urchin embryos. Biochim Biophys Acta 259: 104–111

Hynes R O, Raff R A, Gross P R (1972a) Properties of three cell types in sixteen-cell sea urchin embryos: Aggregation and microtubule protein synthesis. Dev Biol 27: 150–164

Hynes R O, Greenhouse G A, Minkoff R, Gross P R (1972b) Properties of the three cell types in sixteen-cell sea urchin embryos: RNA synthesis. Dev Biol 27: 457–478

Ichio II, Deguchi K, Kawashima S, Endo S, Ueta N (1978) Water-soluble lipoproteins from yolk granules in sea urchin eggs. J Biochem 84: 737–749

Ilan J, Ilan J (1978) Translation of maternal messenger ribonucleoprotein particles from sea urchin in a cell-free system from unfertilized eggs and product analysis. Dev Biol 66: 375–385

Ilan J, Ilan J (1981) Preferential channeling of exogenously supplied methionine into protein by sea urchin embryos. J Biol Chem 256: 2830–2834

Immers J, Runnström J (1960) Release of respiratory control by 2,4-dinitrephenol in different stages of sea urchin development. Dev Biol 2: 90–104

Imschenetzky M, Puca M, Massone R (1980) Histone analysis during the first cell cycle of development of the sea urchin. Tetrapygus niger. Differentiation 17: 111–115

Infante A A, Graves P N (1971) Stability of free ribosomes derived ribosomes and polysomes of the sea urchin. Biochim Biophys Acta 246: 100–110

Infante A A, Heilmann L J (1981) Distribution of messenger RNA in polyomes, and nonpolysomal particles: translational control of actin synthesis. Biochemistry 20: 1–8

Infante A A, Nemer M (1967) Accumulation of newly synthesized RNA templates in a unique class of polyribosomes during embryogenesis. Proc Natl Acad Sci USA 58: 681–688

Infante A A, Nemer M (1968) Heterogeneous RNP particles in the cytoplasm of sea urchin embryos. J Mol Biol 32: 543–556

Infante A A, Nauta R, Gilbert S, Hobart P, Firshein W (1973) DNA synthesis in developing sea urchins: role of a DNA-nuclear membrane complex. Nature (London) New Biol 342: 5–8

Innis M A, Craig S P (1978) Mitochondrial regulation in sea urchins. II Formation of polyribosomes within the mitochondria of 4–8 cell stage embryos of the sea urchin. Exp Cell Res 111: 223–230

Innis M A, Beers T R, Craig S P (1976) Mitochondrial regulation in sea urchins. I Mitochondrial ultrastructure transformations and changes in the ADP:ATP ratio at fertilization. Exp Cell Res 98: 47–56

Inoué H, Yoshioka T (1980) Measurement of intracellular pH in sea urchin eggs by ^{31}P NMR. J Cell Physiol 105: 461–468

Inoué H, Yoshioka T (1982) Comparison of Ca^{2+} uptake characteristic of microsomal fractions isolated from unfertilized and fertilized sea urchin eggs. Exp Cell Res 140: 283–288

Inoué S, Hardy J B (1971) Fine structure of the fertilization membranes of sea urchin embryos. Exp Cell Res 68: 259–272

Inoué S, Sato H (1967) Cell motility by labile association of molecules. The nature of mitotic spindle fibers and their role in chromosomal movement. J Gen Physiol 50: 259–269

Inoué S, Tilney L G (1982) Acrosomal reaction of Thyone sperm. I changes in the sperm head visualized by high resolution video microscopy. J Cell Biol 93: 812–819

Inoué S, Hardy J P, Cousineau G H, Bal A K (1967) Fertilization membranes structure analysis with the surface replica method. Exp Cell Res 48: 348–251

Inoué S, Buday A, Cousineau G H (1970) Developing fertilization membranes. Exp Cell Res 59: 343–346

Inoué S, Preddie E C, Buday A, Cousineau G H (1971) Use of egg membrane lysins in the preparation of sea urchin egg ghosts. Exp Cell Res 66: 164–170

Isaka A, Akino M, Hotta K, Kurakawa M (1969) Sperm isoagglutination and sialic acid content of the jelly coat of sea urchin eggs. Exp Cell Res 54: 247–249

Isaka S, Hotta K, Kurakawa M (1970) Jelly coat substances of sea urchin eggs. I. Sperm isoagglutination and sialpolysaccharide in the jelly. Exp Cell Res 59: 37–42.

Ishida K, Yasumasu I (1981) Accelerative effect of adenosine 3' 5' cyclic monophosphate and dibutyril adenosine 3', 5'-cyclic nonophosphate on the cleavage cycle of the sea urchin eggs. Acta Embryol Morphol Exp 2: 155–159

Ishida K, Yasumasu I (1982) The periodic change in adenosine 3',5'-monophosphate concentration in sea-urchin eggs. Biochim Biophys Acta 720: 266–273

Ishiguro K, Murofushi H, Sakai H (1982) Evidence that AMP-dependent protein kinase and a protein factor are involved in reactivation of Triton X-100 models of sea urchin and starfish spermatozoa. J Cell Biol 92: 777–782

Ishihara K (1968a) Chemical analysis of glycoproteins in the egg surface of the sea urchin, *Arbacia punctulata*. Biol Bull 134: 425–433

Ishihara K (1968b) An analysis of acid polysaccharides produced at fertilization of sea urchin. Exp Cell Res 51: 473–484

Ishihara K, Oguri K, Taniguchi H (1973) Isolation and characterization of fucose sulfate from jelly coat glycoprotein of sea urchin egg. Biochim Biophys Acta 320: 628–634

Isono N (1963a) Carbohydrate metabolism in sea urchin eggs. III Changes in respiratory quotient during early embryonic development. Annot Zool Spn 36: 126–132

Isono N (1963b) Carbohydrate metabolism in sea urchin eggs. V. Intracellular localization of enzymes of pentose phosphate cycle in unfertilized and fertilized eggs. J Fac Sci Univ Tokyo IV 10: 37–53

Isono N, Yanagisawa T (1966) Acid-soluble nucleotides in sea urchin egg. II Uridine diphosphate sugars. Embryologia 9: 184–195

Isono N, Yasumasu I (1968) Pathways of carbohydrate breakdown in sea urchin eggs. Exp Cell Res 50: 616–626

Isono N, Isusaka A, Nakano E (1963) Studies on glucose-6-phosphate dehydrogenase in sea urchin eggs. J Fac Sci Univ Tokyo IV 10: 55–66

Isono Y, Isono N (1975) Lipids. In: The sea urchin embryo, Biochemistry and Morphogenesis` editor Czihak G Springer Berlin pages 608–629

Ito S, Yoshioka K (1972) Real activation potential observed in sea urchin egg during fertilization. Exp Cell Res 72: 547–551

Ito S, Yoshioka K (1973) Effect of various ionic compositions upon the membrane potentials during activation of sea urchin eggs. Exp Cell Res 78: 191–200

Ito S, Revel J P, Goodenough D A (1967) Observations on the fine structure of the fertilization membrane of *Arbacia punctulata*. Biol Bull 133: 471

Ito S, Dan K, Goodenough D (1981) Ultrastructure and ^{3}H-thymidine incorporation by chromosome vesicles in sea urchin embryos. Chromosoma 83: 441–453

Iwaikawa Y (1967) Regeneration of cilia in the sea urchin embryo. Embryologya 9: 187–294

Jaffe L A (1976) Fast block to polyspermy in sea urchin eggs is electrically mediated. Nature (London) 361: 68–71

Jaffe L R (1983) Sources of calcium in egg activation: a review and hypothesis. Dev Biol 99: 265–276

Jaffe L A, Robinson K R (1978) Membrane potential of the unfertilized sea urchin egg. Dev Biol 62: 215–220

Jaffe L A, Hagiwara S, Kado R T (1978) The time course of cortical vesicle fusion in sea urchin eggs observed as membrane capacitance changes. Dev Biol 67: 213–248

Jaffe L A, Gould-Somero M, Holland L Z. (1982) Studies of the mechanism of the electrical polyspermy block using voltage clamp during cross-species fertilization. J Cell Biol 92: 616–621

Jeanmart J, Uytdenhoef P, De Sutter G, Legros F (1976) Insulin receptor sistes as membrane markers during embryonic development. Differentiation: 23–30

Jelinek W R, Schmid C W (1982) Repetitive sequences in eukaryotic DNA and their expression. Annu Rev Biochem 51: 813–844

Jenkins N, Taylor M W, Raff R A (1973) In vitro translation of oogenetic messenger RNA of sea urchin eggs and picornavirus RNA with a cell free system from sarcoma 180. Proc Natl Acad Sci USA 70: 3387–3291

Jenkins N A, Kaumeyer J F, Young E M, Raff R A (1978) A test for masked message: The template activity of messenger ribonucleoprotein particles isolated from sea urchin eggs. Dev Biol 63: 279–298

Jimenez R N, Grossman A, Rossman G, Inoue S (1978) Stimulation of DNA synthesis in vitro by a factor present in nascent *Arbacia* and *Spisula* zygotes. Biol Bull 155: 446–447

Johnson A W, Hnilica L S (1970) In vitro RNA synthesis and nuclear proteins of isolated sea urchin embryo nuclei. Biochim Biophys Acta 224: 518–530

Johnson A W, Hnilica L S (1971) Cytoplasmic and nuclear basic protein synthesis during early sea urchin development. Biochim Biophys Acta 246: 141–154

Johnson C H, Epel D (1981) Intracellular pH of sea urchin eggs measured by the dimethyloxazolinedione method. J Cell Biol 89: 284–291

Johnson C H, Chapper D L, Winkler M M, Lee H C, Epel D (1983) A volatile inhibitor immobilizes sea urchin sperm in semen by depressing the intracellular pH. Dev Biol 98: 493–501

Johnson J D, Epel D (1975) Relationship between release of surface proteins and metabolic activation of sea urchin eggs at fertilization. Proc Natl Acad Sci 72: 4474–4478

Johnson J D, Dunbar B S, Epel D (1974) Fertilization-associated changes in the plasma membrane proteins of *Arbacia punctulata* eggs. Biol Bull 147: 485 (Abstr)

Johnson J D, Epel D, Paul M (1976) Intracellular pH and activation of sea urchin eggs after fertilization. Nature (London) 262: 661—664

Jones P A, Taylor S M (1980) Cellular differentiation, cytidine analogs and DNA methylation. Cell 20: 85–93

Joseph R, Stafford W (1976) Purification of sea urchin ribosomal RNA genes with a single-strand specific nuclease. Biochim Biophys Acta 418: 167–174

Kabat-Zinn J, Singer R H (1981) Sea urchin tube feet: unique structures that allow a cytological and molecular approach to the study of actin and its gene expression. J Cll Biol 89: 109–114

Kane R E (1970) Direct isolation of the hyaline layer protein released from the cortical granules of the sea urchin egg at fertilization. J Cell Biol 45: 615–622

Kane R E (1973) Hyaline release during normal sea urchin development and its replacement after removal at fertilization. Exp Cell Res 81: 301–311

Kane R E (1974) Direct isolation of cytoplasmic actin from sea urchin eggs. J Cell Biol 63: 161 (Abstr)

Kane R E (1975a) Reversible polymerization of actin and other proteins in sea urchin egg extracte. J Cell Biol 67: 198 (Abstr)

Kane R E (1975b) Preparation and purification of polymerized actin from sea urchin egg extracts. J Cell Biol 66: 305–332

Kane R E (1976) Actin polymerization and interaction with other proteins in temperature induced gelation in sea urchin egg extracts. J Cell Biol 71: 704–714

Kane R E (1980) Induction of either contractile or structural actin-based gels in sea urchin egg cytoplasmic extract. J Cell Biol 86: 803–809

Kane R E (1982) Structural and contractile roles of actin in sea urchin egg cytoplasmic extracts. Cell Differ 11: 285–287

Kane R E, Stephens R E (1969) A comparative study of the isolation of the cortex and the role of the calcium-insoluble protein in several species of sea urchin egg. J Cell Biol 41: 133–144

Kaneko T, Terayama H (1974) Quantitative analysis of DNA in sea urchin eggs and subcellular distribution of DNA in the eggs. Anal Biochem 58: 439–448

Kaneko T, Terayama H (1975) The behavior of mitochondrial DNA and enzyme during the course of the early development of the sea urchin *Hemicentrotus pulcherrimus*. J Fac Sci Univ Tokyo Sect IV 13: 285–297

Kari B E, Rotman W L (1980) Analysis of the yolk glycoprotein of the sea urchin embryo. J Cell Biol 87: 144 (Abstr)

Karnofsky D, Simmel B (1963) Effects of growth inhibiting chemicals on the sand-dollar embryo *Echinarachinus parma*. Prog Tumor Res 3: 254

Karp G C, Solursh M (1974) Acid mucopolysaccharide metabolism, the cell surface, and primary mesenchyme cell activity in the sea urchin embryo. Dev Biol 41: 110–123

Kato K H, Ishikawa M (1982) Flagellum formation and centriolar behavior during spermatogenesis of the sea urchin. *Hemicentrotus pulcherrimus*. Acta Embryol Morphol Exp NS 3: 49–66

Kato K H, Sugiyama M (1971) On the de novo formation of the centriole in the activated sea urchin egg. Dev Growth Differ 13: 359–366

Kato K H, Sugiyama M (1978) Species-specific adhesion of spermatozoa to the surface of fixed eggs in sea urchins. Dev Growth Differ 20: 337–347

Katow H, Solursh M (1981) Ultrastructural and time-lapse studies of primary mesenchyme cell behaviour in normal and sulfate-deprived sea urchin embryos. Exp Cell Res 136: 233–245

Katow H, Solursh M (1982) In situ distribution of con A-binding sites in mesenchyme blastulae and early gastrulae of the sea urchin *Lythechinus pictus*. Exp Cell Res 139: 171–180

Katsura S, Tominaga A (1974) Peroxidatic activity or catalase in the cortical granules of sea urchin eggs. Dev Biol 40: 292–297

Katula K S (1983) High mobility groups non histone chromosomal proteins of the developing sea urchin embryo. Dev Biol 98: 15–27

Kaumeyer J F, Jenkins, Raff R A (1978) Messenger ribonucleoprotein particles in unfertilized sea urchin eggs. Dev Biol 63: 266–278

Kawabe T T, Armstrong P B, Pollock E G (1981) An extracellular fibrillar matrix in gastrulating sea urchin embryos. Dev Biol 85: 509–515

Kay E, Turner E, Weidman P, Shapiro B M (1980) Assembly of the fertilization membrane of *Strongylocentrotus purpuratus*. J Cell Biol 87: 143 (Abstr)

Kedes L H (1976) Histone messengers and histone genes. Cell 8: 321–332

Kedes L H (1979) Histone genes and histone messengers. Annu Rev Biochem 48: 837–870

Kedes L H, Birnstiel M L (1971) Reiteration and clustering of DNA sequences complementary to histone messenger RNA. Nature (London) New Biol 230: 165–169

Kedes L H, Gross P R (1969a) Synthesis and function of messenger RNA during early embryonic development. J Mol Biol 42: 559–576

Kedes L H, Gross P R (1969b) Identification in cleaving embryos of three RNA species serving as templates for the synthesis of nuclear proteins. Nature (London) 223: 1335–1339

Kedes L H, Stavy L (1969) Structural and functional identity of ribosomes from eggs and embryos of sea urchins. J Mol Biol 43: 337–340

Kedes L H, Chang A C Y, Houseman D, Cohen S N (1975a) Isolation of histone genes from unfractionated sea urchin DNA by subculture cloning in E. coli. Nature (London) 255: 533–538

Kedes L H, Cohn R H, Lowry J C, Chang A C Y, Cohen S N (1975b) The organization of sea urchin histone genes. Cell 6: 359–369

Keichline L D, Wasserman P M (1977) Developmental study of the structure of sea urchin embryo and sperm chromatin using micrococcal nuclease. Biochim Biophys Acta 475: 139–151

Keichline L D, Wassarman P M (1979) Structure of chromatin in sea urchin embryos, sperm, and adult somatic cells. Biochemistry 18: 214–219

Keller C, Gundersen G, Shapiro B M (1980) Altered in vitro phosphorylation of specific protein accompanies fertilization of Strongylocentrotus purpuratus eggs. Dev Biol 74: 86–100

Keller III T C S, Rebhun L I (1982) Strongylocentrotus purpuratus spindle tubulin. I Characteristics of its polymerization and depolymerization in vitro. J Cell Biol 93: 788–796

Keller III T C S, Jemolo D K, Burgess W H, Rebhun L I (1982) Strongylocentrotus purpuratus spindle tubulin. II Characteristics of its sensitivity to Ca^{++} and the effects of calmodulin isolated from bovine brain and Strongylocentrotus purpuratus eggs. J Cell Biol 93: 797–803

Keltch A K, Strimatter C F, Walters C P, Clowes G H A (1950) Oxidative phosphorylation by a cell-free particulate system from unfertilized Arbacia eggs. J Gen Physiol 33: 547–553

Kew D, Karnowski C, Brown J (1979) Preparation and characterization of phagosome membranes from Arbacia punctulata embryo cells. Biol Bull 157: 375 (Abstr)

Kiehardt D P, Reynolds G T, Eisen A (1977) Calcium transients during early development in echinoderms and teleosts. Biol Bull 153: 432 (Abstr)

Kijima S, Wilt F H (1969) Rate of nuclear RNA turnover in sea urchin embryos. J Mol Biol 40: 235–246

Kimura-Furukawa J, Suyemitsu T, Ishihara K (1978) Induction of the acrosome reaction on the surface of de-jellied sea urchin eggs. Exp Cell Res 114: 143–151

Kinoshita S (1968) Relative deficiency of intracellular relaxing system observed in presumptive furrowing region in induced cleavage in the centrifuged sea urchin egg. Exp Cell Res 51: 395–405

Kinoshita S (1969) Periodical release of heparin-like polysaccharide within cytoplasm during cleavage of sea urchin egg. Exp Cell Res 56: 39–43

Kinoshita S (1974a) Some observations on a protein mucopolysaccharide complex found in sea urchin embryos. Exp Cell Res 85: 31–40

Kinoshita S (1974b) Synthesis of sulfate donor in developing sea urchin embryos. Exp Cell Res 87: 382–385

Kinoshita S (1976) Properties of sea urchin chromatin as revealed by means of thermal denaturation. Exp Cell Res 102: 153–162

Kinoshita S, Saiga H (1979) The role of proteoglycan in the development of sea urchin. I. Abnormal development of sea urchin embryos caused by disturbance of proteoglycan synthesis. Exp Cell Res 123: 229–236

Kinoshita S, Yazaki I (1967) The behaviour and localization of intracellular relaxing system during cleavage in the sea urchin egg. Exp Cell Res 47: 449–458

Kinoshita S, Yoshi K (1979) The role of proteoglycan synthesis in the development of sea urchins. II The effect of administration of exogenous proteoglycan. Exp Cell Res 124: 361–369

Kinoshita S, Yoshi Y (1983) Permeable embryonic cells of sea urchins as a model for studying nucleus-cytoplasm interactions. Experientia 39: 189–190

Kinsey W H, Lennarz W J (1981) Isolation of a glycopeptide fraction from the surface of the sea urchin egg that inhibits sperm-egg binding and fertilization. J Cell Biol 91: 325–331

Kinsey W H, Se Gall G K, Lennarz W J (1979) The effect of the acrosome reaction on the respiratory activity and fertilizing capacity of echinoid sperm. Dev Biol 71: 49–59

Kinsey W H, Rubin J A, Lennarz W J (1980a) Studies on the specificity of sperm binding in echinoderm fertilization. Dev Biol 74: 245–250

Kinsey W H, Decker G L, Lennarz W J (1980b) Isolation and partial characterization of the plasma membrane of the sea urchin egg. J Cell Biol 87: 248–254

Kirsten M, Couch B, Grossman A (1973) TAME hydrolase activity and activation of sea urchin eggs. Biol Bull 145: 443a

Kleene K C, Humphreys T (1977) Similarity of HnRNA sequences in blastula and pluteus stage sea urchin embryos. Cell 12; 145–155

Klein W H, Thomas T L, Loi C, Scheller R H, Britten R J, Davidson E H (1978) Characteristic of individual repetitive sequence families in the sea urchin genome studied with cloned repeats. Cell 14: 889–900

Kobayashi I, Kimura I (1976) Effects of chemical animalization and cell dissociation on RNA synthesis in sea urchin embryos. Exp Cell Res 97: 413–417

Kobayashi Y, Ogawa K, Mohri H (1978) Evidence that the Mg-ATPase in the cortical layer of sea urchin egg is dynein. Exp Cell Res 114: 285–294

Kobayashi Y, Okuno M, Mohri H (1979) "Soluble tubulin" from sea urchin sperm flagella. Sci Pap Coll Gen Educ Univ Tokyo 29: 177–192

Kondo K (1973) Cell-binding substances in sea urchin embryos. Dev Growth Differ 15: 201–216

Kondo K (1974) Demonstration of a reaggregation inhibitor in sea urchin embryos. Exp Cell Res 86: 178–181

Kondo K, Sakai H (1971) Demonstration and preliminary characterization of reaggregation-promoting substances from embryonic sea urchin cells. Dev Growth Differ 13: 1–14

Kopf G S, Garbers D L (1979) An egg-associated factor elevates cyclic AMP concentration and induces the acrosome reaction of sea urchin spermatozoa. J Cell Biol 83: 214 (Abstr)

Kopf G S, Moy G W, Vacquier V D (1982) Isolation and characterization of sea urchin egg cortical granules. J Cell Biol 95: 924–932

Kotzin B L, Baker R F (1972) Selective inhibition of genetic transcription in sea urchin embryos. Incorporation of 5-bromodeoxyuridine into low molecular weight nuclear DNA. J Cell Biol 55: 74–81

Krach S W, Green A, Nicolson G L, Oppenheimer S B (1973) Cell surface changes occurring during development. J Cell Biol 59: no 2, part 2, 176 (Abstr)

Krach S W, Green A, Nicolson G L, Oppenheimer S B (1974) Cell surface changes occurring during sea urchin embryonic development monitored by quantitative agglutination with plant lectins. Exp Cell Res 84: 191–198

Krahl M E (1956) Oxidative pathway for glucose in eggs of the sea urchin. Biochim Biophys Acta 20: 27–32

Krahl M E, Jandorf B J, Glowes G H A (1942) Studies on cell metabolism and cell division. VII Observations on the amount and possible function of diphosphothiamine (cocarboxylase) in eggs of *Arbacia punctulata*. J Gen Physiol 25: 733

Krahl M E, Keltch A K, Walters C P, Clowes G H A (1954) Hexokinase activity from eggs of the sea urchin *Arbacia punctulata*. J Gen Physiol 38: 31–40

Krahl M E, Keltch A K, Clowes G H A (1955) Glucose-6 phosphate and 6-phosphogluconate dehydrogenases from eggs of the sea urchin *Arbacia punctulata*. J Gen Physiol 38: 431–440

Krane S M, Crane R K (1960) Changes in levels of TPN in marine eggs subsequent to fertilization. Biochim Biophys Acta 43: 369–373

Kronenberg L H, Humphreys T (1972) Double-stranded RNA in sea urchin embryos. Biochemistry 11: 2020–2026

Krystal G W, Poccia D (1979) Control of chromosome condensation in the sea urchin egg. Exp Cell Res 123: 207–219

Kuhn O, Wilt F H (1980) Double labelling of chromatin proteins in vivo and in vitro, and their two-dimensional electrophoretic resolution. Anal Biochem 105: 274–280

Kuhn O, Wilt F H (1981) Chromatin proteins of sea urchin embryos: Dual origin from an oogenetic reservoir and new synthesis. Dev Biol 85: 416–424

Kung K S (1974) On the size relationship between nuclear and cytoplasmic RNA in sea urchin embryos. Dev Biol 36: 343–356

Kunkel N S, Weinberg E S (1978) Histone gene transcripts in the cleavage and mesenchyme blastula embryo of the sea urchin *S. purpuratus*. Cell 14: 313–326

Kunkel N S, Hemminiki K, Weinberg E S (1978) Size of histone gene transcripts in different embryonic stages of the sea urchin *Strongylocentrotus purpuratus*. Biochemistry 17: 2591–2597

Kunkle M (1980) Structural components of the sea urchin sperm nucleus. J Cell Biol 87: 147 (Abstr)

Kunkle M, Longo F J (1975) Analysis of sea urchin egg cytosol proteins that bind to sperm. J Cell Biol 67: 228 (Abstr)

Kurek M P, Billig D, Stambrook P (1979) Size classes of replication units in DNA from sea urchin embryos. J Cell Biol 81: 698–703

Kuriyama R, Borisy G G (1983) Cytasters induced within unfertilized sea-urchin eggs. J Cell Sci 61: 175–190

Kusunoki S, Yasumasu I (1976) Cyclic change in polyamine concentrations in sea urchin eggs related with cleavage cycle. Biochem Biophys Res Commun 68: 881–885

Kusunoki S, Yasumasu I (1978a) Heat-labile activating substance for ornithine decarboxylase activity in sea urchin eggs. Gam Res 1: 129–136

Kusunoki S, Yasumasu I (1978b) Inhibitory effect of α-hydrazinoornithine on egg cleavage in sea urchin eggs. Dev Biol 67: 336–345

Kusunoki S, Yasumasu I (1980) Inhibition of dihydropholate reductase by palmytol-CoA and the reversal of the inhibition by spermine and spermidine in the eggs of the sea urchin *Hemicentrotus pulcherrimus*. Dev Growth Differ 22: 299–304

Lallier R (1955) Effects des ions zinc et cadmium sur le développement de l'oeuf de l'Oursin *Paracentrotus lividus*. Arch Biol 66: 75–102

Lallier R (1964) Biochemical aspects of animalization and vegetalization in the sea urchin embryos. In: Abercrombie M, Brachet J (eds) Advances in morphogenesis, vol III. Academic Press, London New York, pp 147–196

Lallier R (1970) Formation of fertilization membrane in sea urchin eggs. Exp Cell Res 63: 460–462

Lallier R (1963) Effets de l'actinomycine D sur le development de l'ouef de l'oursin *Paracentrotus lividus*. Experientia 19: 572–573

Lallier R (1971) Effects of various inhibitors of protein cross-linking on the formation of fertilization membrane in sea urchin egg. Experientia 27: 1323–1324

Lallier R (1972) Effects of concanavalin A on the development of sea urchin egg. Exp Cell Res 72: 157–163

Lallier R (1974a) Induction of fertilization membrane by ionophore A 23187 in sea urchin egg *Paracentrotus lividus*. Exp Cell Res 89: 425–426

Lallier R (1974b) Recherches sur le rôle des groupements chimiques de la surface des gamètes dans la fécondation de l'oeuf de l'oursin, *Paracentrotus lividus*. Acta Embryol Morphol Exp 3: 255–265

Lallier R (1978) Recherches sur la polarite de l'oeuf de l'oursin *Paracentrotus lividus*. Acta Embryol Morphol Exp 47–58

Lallier R (1980) A study of the role of sulfate ions on the development of the sea urchin egg. Acta Embryol Morphol Exp NS 1: 3–4

Lasky L A, Lev Z, Xin J K, Britten R J, Davidson E (1980) Messenger RNA prevalence in sea urchin embryos measured with cloned cDNAs. Proc Natl Acad Sci USA 77: 5317–5321

Lau J T Y, Lennarz W J (1983) Regulation of sea urchin glycoprotein mRNAs during embryonic development. Proc Natl Acad Sci USA 80: 1028–1032

Leader D P, Thomas A, Voorma H O (1981) The protein synthetic activity in vitro of ribosomes differing in the extent of phosphorylation of their ribosomal protein. Biochim Biophys Acta 656: 69–75

Lee A S, Britten R J, Davidson E H (1977a) Short-period repetitive-sequence interspersion in cloned fragments of sea urchin DNA. Cold Spring Harbor Symp Quant Biol 42: 1065–1076

Lee A S, Britten R J, Davidson E H (1977b) Interspersion of short repetitive sequences studied in cloned sea urchin DNA fragments. Science 196: 189–193

Lee A S, Thomas T L, Lev Z, Britten R J, Davidson E H (1980) Four sizes of transcripts produced by a single sea urchin gene expressed in early embryos. Proc Natl Acad Sci USA 77: 3259–3263

Lee H C, Epel D (1983) Changes in intracellular acidic compartments in sea urchin eggs after activation. Dev Biol 98: 446–454

Lee H C, Johnson C, Epel D (1983) Changes in internal pH associated with initiation of motility and acrosome reaction of sea urchin sperm. Dev Biol 95: 31–45

Lee Y R, Whiteley A H (1982) Transcription of the two genomes in developmentally arrested sea urchin hybrid embryos. Cell Differ 11: 311–314

Lev Z, Thomas T L, Lee A S, Angerer R G, Britten R J, Davidson E H (1980) Developmental expression of two cloned sequences coding for rare sea urchin embryo messages. Dev Biol 76: 322–340

Levine A E, Walsh K A (1979) Involvement of an acrosin-like enzyme in the acrosome reaction of the sea urchin sperm. Dev Biol 72: 126–137

Levine A E, Walsh K A (1980) Purification of an acrosine-like enzyme from sea urchin sperm. J Biol Chem 255: 4814–4820

Levine A E, Walsh K A, Fodor E J B (1978) Evidence of an acrosin-like enzyme in sea urchin sperm. Dev Biol 63: 299–306

Levy L, Moav B (1982) Developmental patterns in histones and histone synthesis during early embryogenesis of the sea urchin *Paracentrotus lividus*. Differentiation 23: 17–24

Levy S, Wood P, Grunstein M, Kedes L (1975) Individual histone mRNAs. Identification by template activity. Cell 4: 239–248

Levy S, Sures I, Kedes L H (1979) Sequence of the 5′-end of Strongylocentrotus purpuratus H2b histone mRNA and its location within histone DNA. Nature (London) 279: 737–739

Lewin R (1982a) Surprising discovery with a small RNA. Nature (London) 218: 777–778

Lewin R (1982b) Can genes jump between eukaryotic species? Science 217: 42–43

Liebermann D, Hoffman-Lieberbann B, Weinthal J, Childs G, Maxson R, Mauron A, Cohen S, Kedes L (1983) An unusual trasposon with long terminal inverted repeats in the sea urchin *Strongylocentrotus purpuratus*. Nature (London) 306: 342–347

Lifton R P, Kedes L H (1976) Size and sequence homology of masked maternal and embryonic messenger RNAs. Dev Biol 48: 47–55

Lillie F R (1914) Studies of fertilization. VI. The mechanism of fertilization in *Arbacia*. J Exp Zool 16: 523–590

Lillie F R (1919) Problem of fertilization. Chicago Univ Press, Chicago

Linck R W (1976) Flagellar doublet microtubules: fractionation of minor components and α-tubulin from specific regions of the A-tubule. J Cell Sci 20: 405–439

Lindahl P E (1933) Ueber „animalisierte" und „vegetativisierte" Seeigellarven. Wilhelm Roux' Arch Entwicklungsmech Org 128: 661–664

Lindahl P E (1939) Zur Kenntnis der Entwicklungsphysiologie des Seeigeleies. Hoppe-Seyler's Vergl Physiol 27: 233–239

Lindberg O (1943) Studien über das Problem des Kohlehydratabbaus und der Säurebildung bei der Befruchtung des Seeigels. Ark Kemi Mineral Geol 16A: no 15

Lindberg O (1945) On the metabolism of glycogen in the fertilization of the sea urchin egg. Ark Kemi Mineral Geol 20B: no 1

Liquori A M, Monroy A, Parisi E, Tripiciano A (1981) A theoretical equation for diauxic growth and its application to the kinetics of the early development of the sea urchin embryos. Differentiation 20: 174–175

Lichtfield J B, Whiteley A H (1959) Studies on the mechanism of phosphate accumulation by sea urchin embryos. Biol Bull 117: 133

Loeb J, King W O R, Moore A R (1910) Über Dominanzerscheinungen bei den hybriden Pluteus des Seeigels. Archiv f Entw-Mech 29: 354–362

Loeb L A, Fansler B (1970) Intracellular migration of DNA polymerase in early developing sea urchin embryos. Biochim Biophys Acta 217: 50–55

Loeb L A, Mazia D, Ruby A D (1967) Priming of DNA polymerase in nuclei of sea urchin embryos by native DNA. Proc Natl Acad Sci USA 57: 841–848

Loeb L A, Fansler B, Williams R, Mazia D (1969) Sea urchin nuclear DNA polymerase. I. Localization in nuclei during rapid DNA synthesis. Exp Cell Res 57: 298–304

Lois A F, Neill D J, Carroll Jr E J (1983) Acid release, cytoplasmic alkalinization, and chromosome condensation during sea urchin fertilization and amine-induced parthenogenesis. Gam Res 7: 123–132

Longo F J (1973) Origin of the male pronuclear envelope in fertilized sea urchin egg. J Cell Biol 59: no 2, part 2, 201 (Abstr)

Longo F (1977) Cytochalasin B inhibits sperm incorporation without inhibiting sperm-induced egg activation. J Cell Biol 75: 44 (Abstr)

Longo F J (1978) Effects of cytochalasin B on sperm-egg interactions. Dev Biol 67: 249–265

Longo F J (1980) Organization of microfilaments in sea urchin (*Arbacia punctulata*) eggs at fertilization. Effects of cytochalasin B. Dev Biol 74: 422–433

Longo F J (1981a) Morphological features of the surface of the sea urchin (*Arbacia punctulata*) egg: oolemma-cortical granule association. Dev Biol 84: 173–182

Longo F J (1981b) Effects of concanavalin A on the fertilization of sea urchin eggs. Dev. Biol 82: 197–202

Longo F J (1982) Integration of sperm and egg plasma membrane components at fertilization. Dev Biol 89: 409–416

Longo F J, Anderson E (1968) The fine structure of pronuclear development and fusion in the sea urchin, *Arbacia punctulata*. J Cell Biol 39: 339–368

Longo F J, Anderson E (1970a) The effects of nicotine on fertilization in the sea urchin, *Arbacia punctulata*. J Cell Biol 46: 308–326

Longo F J, Anderson E (1970b) A cytological study of the relation of the cortical reaction to subsequent events of fertilization in urethane-treated eggs of the sea urchin Arbacia punctulata. J Cell Biol 47: 646–665

Longo F J, Kunkle M (1978) Transformations of sperm nuclei upon inseminations. Curr Top Dev Biol 13: 149–185

Longo F J, Plunkett W (1973) The onset of DNA synthesis and its relation to morphogenetic events of the pronuclei in activated eggs of the sea urchin *Arbacia punctulata*. Dev Biol 30: 56–67

Longo F J, Schuel H (1973) An ultrastructural examination of polyspermy induced by soybean trypsin inhibitor in the sea urchin *Arbacia punctulata*. Dev Biol 34: 187–199

Longo F J, Schuel H, Wilson W L (1974) Mechanism of soybean trypsin inhibitor-induced polyspermy as determined by an analysis of refertilized sea urchin (*Arbacia punctulata*) eggs. Dev Biol 41: 193–201

Lönning S (1967a) Studies of the ultrastructure of sea urchin eggs and the changes induced at insemination. Sarsia 30: 31–48

Lönning S (1967b) Experimental and electron microscopic studies of sea urchin oocytes and eggs and the changes following insemination. Acta Univ Bergensis Ser Mat Res Nat Norw Univ Press Oslo 8: 1–20

Lönning S (1967c) Electron microscopic studies of the block to polyspermy. The influence of trypsin, soy bean trypsin inhibitor and chloral hydrate. Sarsia 30: 107–116

Lopo A, Vacquier (1977) The rise and fall of intracellular pH of sea urchin eggs after fertilization. Nature (London) 269: 590–592

Lopo A, Vacquier V (1979) Identification of sea urchin sperm membrane protein involved in the acrosome reaction. J Cell Biol 83: 201 (Abstr)

Lopo A C, Vacquier V D (1980a) Radioiodination and characterization of the plasma membrane of sea urchin sperm. Dev Biol 76: 15–25

Lopo A C, Vacquier V D (1980b) Antibody to a sperm surface glycoprotein inhibits the jelly-induced acrosome reaction of sea urchin sperm. Dev Biol 79: 325–333

Lorenzi M, Hedrick J L (1973) On the macromolecular composition of the jelly coat from *Strongylocentrotus purpuratus* eggs. Exp Cell Res 79: 417–422

Løvtrup A, Iverson R M (1969) Respiratory phases during early sea urchin development measured with the automatic diver balance. Exp Cell Res 55: 25–32

Løvtrup-Rein H, Løvtrup S (1980) Energy sources in the sea urchin embryo. A critic of current views. Acta Embryol Morphol Exp 1: no 3, 93–101

Lu A L, Stafford D W (1982) The sea urchin 5S ribosomal RNA gene. In: Busch H, Rothblum L (eds) The cell nucleus vol II 45–62 Academic Press, London New York

Lu A L, Steege D A, Stafford D W (1980) Nucleotide sequence of a 5S ribosomal RNA gene in the sea urchin *Lytechinus variegatus*. Nucleic Acid Res 8: 1839–1853

Lu A L, Blin N, Stafford D W (1981) Cloning and organization of genes for ribosomal RNA in the sea urchin *Lytechinus variegatus*. Gene 14: 51–62

Luduena R F, Woodward D O (1973) Isolation and partial characterization of α- and β-tubulin from outer doublets of sea urchin sperm and microtubules of chick-embryo brain. Proc Natl Acad Sci USA 70: 3594–3598

Lundblad G (1954) Proteolytic activity in sea urchin gametes. Almqvist and Wiksells, Uppsala

Lundblad G, Falksveden L G (1964) Purification of cathepsin B from oocytes and mature eggs of the sea urchin *Briossipsis lyrifera* by means of gel filtration. Acta Chem Scand 18: 2044–2050

Lundblad G, Runnström J (1962) Distribution of proteolytic enzymes in protein fractions from non-fertilized eggs of the sea urchin *Paracontrotus lividus*. Exp Cell Res 27: 328–331

Lundblad G, Schilling W (1968) Changes in cathepsin D activity and concomitant studies of macromolecular components in early sea urchin development. Ark Kemi 29: 367–377

Lundblad G, Lundblad M, Immers J, Schilling W (1966) Chromatographic analysis of the proteolytic enzymes of the sea urchin egg. II Gel filtration of extracts from unfertilized and fertilized eggs and investigation of the cath eptic activity. Ark Kemi 26: 395–409

Lundgren B (1973) Surface coating of the sea urchin larva as revealed by Ruthenium red. J Submicrosc Cytol 5: 61–70

Lundgren B, Westin M (1974) Intracellular antigen of sea urchin eggs and embryos studied on ultrathin sections. J Ultrastruct Res 46: 230–238

Lynn D A, Angerer L M, Bruskin A M, Klein W H, Angerer R C (1983) Localization of a family of mRNAs in a single cell type and its precursors in sea urchin embryos. Proc Natl Acad Sci USA 80: 2656–2660

Mabuchi I (1973) ATPase in the cortical layer of sea urchin egg: its properties and interaction with cortex protein Biochim Biophys Acta 297: 317–332

Mackintosh F R, Bell E (1967) Stimulation of protein synthesis in unfertilized sea urchin eggs by prior metabolic inhibition. Biochem Biophys Res Commun 27: 425–430

Mackintosh F R, Bell E (1969) Regulation of protein synthesis in sea urchin eggs. J Mol Biol 41: 365–380

Maggio R (1957) Mitochondrial and cytoplasmic protease activity in sea urchin eggs. J Cell Comp Physiol 50: 135–143

Maggio R (1969) Cytochrome oxidase activity in the mitochondria of unfertilized and fertilized sea urchin eggs. Exp Cell Res 16: 272–278

Maggio R, Catalano G (1963) Activation of amino acids during sea urchin development. Arch Biochem Biophys 103: 164–167

Maggio R, Ghiretti-Magaldi A (1958) The cytochrome system in mitochondria of unfertilized sea urchin eggs. Exp Cell Res 15: 95–102

Maggio R, Monroy A (1959) An inhibitor of cytochrome oxidase activity in the sea urchin egg. Nature (London) 184: 68–70

Maggio R, Aiello F, Monroy A (1960) Inhibitor of the cytochrome oxidase of unfertilized sea urchin eggs. Nature (London) 188: 1195–1196

Maggio R, Vittorelli M L, Rinaldi A M, Monroy A (1964) In vitro incorporation of amino acids into proteins stimulated by RNA from unfertilized eggs. Biochem Biophys Res Commun 15: 436–441

Maggio R, Vittorelli M L, Caffarelli-Mormino I, Monroy A (1968) Dissociation of ribosomes of unfertilized eggs and embryos of sea urchin. J Mol Biol 31: 631–636

Maitra U S, Metz C B (1974) Purification and properties of an Arbacia sperm fertilization antigen. Biol Bull 174: 490 (Abstr)

Manahan D T, Davis J P, Stephens G C (1983) Science 220: 204–206

Mangia F, Nicotra A, Manelli H (1973) Polysome pattern during the first mitotic cycle of the sea urchin eggs. Acta Embryol Exp 2: 199–204

Mano Y (1966) Role of a trypsin-like protease in "informosomes" in a trigger mechanism of activation of protein synthesis by fertilization in sea urchin eggs. Biochem Biophys Res Commun 25: 216–221

Mano Y (1968) Regulation system of protein synthesis in early embryogenesis in the sea urchin. Biochem Biophys Res Commun 33: 877–883

Mano Y (1969) Factors involved in cyclic protein synthesis in sea urchin cells during early embryogenesis. J Biochem (Tokyo) 65: 483–487

Mano Y (1970) Cytoplasmic regulation and cyclic variation in protein synthesis in the early cleavage stage of the sea urchin embryo. Dev Biol 32: 433–460

Mano Y (1971a) Mechanism of increase in the basal rate of protein synthesis in the early cleavage stage of the sea urchin. J Biochem (Tokyo) 69: 1–10

Mano Y (1971b) Cell-free cyclic variation of protein synthesis associated with the cell cycle in sea urchin embryos. J Biochem (Tokyo) 69: 11–25

Mano Y (1971c) Participation of the sulphydryl groups of a protein in the cyclic variation in the rate of protein synthesis in a cell-free system from sea urchin cells. Arch Biochem Biophys 146: 237–248

Mano Y (1977) Interaction between glutathione and the endoplasmic reticulum in cyclic protein syntheses in sea urchin embryos. Dev Biol 61: 273–286

Mano Y, Kano K (1977a) Sites of action of the KCl-soluble protein stimulating protein synthesis in sea urchin systems J Biochem (Tokyo) 81: 757–769

Mano Y, Kano K (1977b) Relation between the KCl-soluble protein-stimulating amino acid incorporation and protein from the ribosomal wash fraction in sea urchin systems. J Biochem (Tokyo) 81: 1791–1801

Mar H (1980) Radial cortical fibers and pronuclear migration in fertilized and artificially activated eggs of *Lytechinus pictus*. Dev Biol 78: 1–13

Marco R, Flecha M, Kedes L H (1977) Studies on the organization and purification of sea urchin (*L. pictus*) histone genes. Biol Bull 151: 420 (Abstr)

Marcus N H (1981) Effect of pronase on the development of reversed symmetry and duplications in the sea urchin, *Arbacia punctulata*. Dev Biol 83: 387–390

Maroun L E (1973) Cytoplasmic localization of DNA-dependent RNA polymerase in unfertilized sea urchin eggs. Exp Cell Res 79: 459–461

Martin K A, Miller O L Jr (1983) Polysome structure in sea urchin eggs and embryos: an electron microscopic analysis. Dev Biol 98: 338–348

Mathews A P (1901) The so-called cross fertilization of Asterias by Arbacia Am J Physiol 6: 216–228

Mathews C K (1975) Giant pools of DNA precursors in sea urchin eggs. Exp Cell Res 93: 47–56

Matranga V, Giarrusso G, Vasile V, Vittorelli M L (1978) Trypsin treatment which elicits DNA synthesis, removes a high molecular weight glycoprotein from the plasma membrane of sea urchin embryonic cells. Cell Biol Int Rep 2: 147–154

Matsui K (1924) Studies on the hybridization among echinoderms with special reference to the behavior of chromosomes. J Coll Agr Imp Univ Tokyo, Japan 7: 211–236

Matsumoto L, Kasamatsu H, Pikò L (1973) Mitochondrial DNA synthesis in sea urchin oocytes. J Cell Biol 59: no 2, part 2, 221 (Abstr)

Matsumoto L, Kasamatsu H, Pikò L, Vinograd J (1974) Mitochondrial DNA replication in sea urchin oocytes. J Cell Biol 63: 146–159

Mauron A, Kedes L, Hough-Evans B R, Davidson E H (1982) Accumulation of individual histone mRNAs during embryogenesis of the sea urchin *Strongylocentrotus purpuratus*. Dev Biol 94: 425–434

Maxon R E Jr, Egrie J C (1980) Expression of maternal and paternal histone genes during early cleavage stages of the echinoderm hybrid *Strongylocentrotus purpuratus* X *Lytechinus pictus*. Dev Biol 74: 335–342

Maxon R E Jr, Wilt F H (1981) The rate of synthesis of histone mRNA during the development of sea urchin embryos (*Strongylocentrotus purpuratus*). Dev Biol 83: 380–386

Maxon R E Jr, Wilt F H (1982) Accumulation of the early histone messenger RNAs during the development of *Strongylocentrotus purpuratus*. Dev Biol 94: 435–440

Maxon R, Mohun T, Gormezano G, Childs G, Kedes L (1983) Distinct organization and patterns of expression of early and late histone gene sets in the sea urchin. Nature (London) 301: 120–124

Mazia D (1958) The production of twin embryos in Dendraster by means of mercaptoethanol. Biol Bull 114: 247–254

Mazia D (1974) Chromosome cycles turned on in unfertilized sea urchin eggs exposed to NH$_4$OH. Proc Natl Acad Sci USA 71: 690–693

Mazia D, Dan K (1952) The isolation and biochemical characterization of the mitotic apparatus of dividing cells. Proc Natl Acad Sci USA 38: 826

Mazia D, Gontcharoff M (1964) The mitotic behavior of chromosome in echinoderm eggs following incorporation of bromodeoxyuridine. Exp Cell Res 35: 14–25

Mazia D, Hinegardner R T (1963) Enzymes of DNA synthesis in nuclei of sea urchin embryos. Proc Natl Acad Sci USA 50: 148–156

Mazia D, Ruby A (1974) DNA synthesis turned on in unfertilized sea urchin eggs by treatment with NH$_4$OH. Exp Cell Res 85: 167–172

Mazia D, Blumenthal G, Benson E (1948) The activity and distribution of DNase and phosphatases in the early development of *Arbacia punctulata*. Biol Bull 95: 250–251

Mazia D, Schatten G, Steinhardt R (1975) Turning on of activities in unfertilized sea urchin eggs: Correlation with changes of the surface. Proc Natl Acad Sci USA 72: 4469–4473

Mazia D, Pawelets N, Sluder G, Finze E M (1981) Cooperation of kinetochores and pole in the establishment of monopolar mitotic apparatus. Proc Natl Acad Sci USA 78: 377–381

McBlain P J, Carroll B J Jr (1977) Hyaline layer formation in sea urchin eggs does not require cortical granule exocytosis. J Cell Biol 75: 169 (Abstr)

McBlain P J, Carroll E J Jr (1978) Sea urchin egg hyaline layer. Evidence for localization of hyaline on the unfertilized egg surface. Dev Biol 75: 137–147

McCarthy R A, Brown J C (1978) Preparation of cell surface membranes from reaggregating sea urchin cells. Biol Bull 155: 456 (Abstr)

McCarthy R A, Spiegel M (1983) Serum effects on the in vitro differentiation of sea urchin micromeres. Exp Cell Res 149: 433–441

McClay D R (1979) Surface antigens involved in interactions of embryonic sea urchin cells. Curr Top Dev Biol 13: 199–214

McClay D R (1982) Cell recognition during gastrulation in the sea urchin. Cell Differ 11: 341–344

McClay D R, Chambers A F (1978) Identification of four classes of cell surface antigens appearing at gastrulation in sea urchin embryos. Dev Biol 63: 179–186

McClay D R, Hausman R E (1975) Specificity of cell adhesion: Differences between normal and hybrid sea urchin cells. Dev Biol 47: 454–460

McClay D R, Marchase R B (1979) Separation of ectoderm and endoderm from sea urchin pluteus larvae and demonstration of germ layer specific antigens. Dev Biol 71: 289–296

McClay D R, Chambers A F, Warren R H (1977) Specificity of cell-cell interactions in sea urchin embryos. Appearance of new cell-surface determinants at gastrulation. Dev Biol 56: 343–355

McClure M E, Hnilica L S (1972) Nuclear acidic protein biosynthesis during early development. J Cell Biol 55: 169 (Abstr)

McColl R S, Aronson A I (1974) Transcription from unique and redundant DNA sequences in sea urchin embryos. Biochem Biophys Res Commun 56: 47–51

McColl R S, Aronson A I (1978) Changes in transcription patterns during early development of the sea urchin. Dev Biol 65: 126–138

McGwin N F, Morton R W, Nishioka D (1983) Increased uptake of thymidine in the activation of the sea urchin eggs. Exp Cell Res 145: 115–126

Meeker G L, Inverson R M (1971) Tubulin synthesis in fertilized sea urchin eggs. Exp Cell Res 64: 129–132

Mellado R P, Delius H, Klein B, Murray K (1981) Transcription of sea urchin histone genes in *Escherichia coli*. Nucleic Acid Res 9: 3889–3906

Melli M, Spinelli G, Wysoling H, Arnold B (1977) Presence of histone mRNA sequences in high molecular weight RNA of HeLa cells. Cell 11: 651–661

Merlino G T, Water R D, Chamberlain J P, Jackson D A, ElGewely M R, Kleinsmith L J (1980) Cloning of sea urchin gene sequences for use in studying the regulation of actin gene transcription. Proc Natl Acad Sci USA 77: 765–769

Merlino G T, Water R D, Moore G P, Kleinsmith L J (1981) Change in expression of the actin gene family during early sea urchin development. Dev Biol 85: 505–508

Mescher A, Humphreys T (1973) Indipendence of maternal mRNA activation from polyadenylation in *Arbacia*. Biol Bull 145: 447 (Abstr)

Mescher A, Humphreys T (1974) Activation of maternal mRNA in the absence of poly (A) formation in fertilized sea urchin eggs. Nature (London) 249: 138–139

Metafora S, Felicetti L, Gambino R (1971) The mechanism of protein synthesis activation after fertilization of sea urchin eggs. Proc Natl Acad Sci USA 68: 600–604

Metz C B (1978) Sperm and egg receptors involved in fertilization. In: Moscona A A, Monroy A (eds) Current topics in developmental biology, vol 12. Academic Press, London New York, pp 107–147

Meza I, Huang B, Bryan J (1972) Chemical heterogeneity of protofilaments forming the outer doublets from sea urchin flagella. Exp Cell Res 74: 535–540

Miki T (1964) ATPase activity of the egg cortex during the first cleavage of sea urchin eggs. Exp Cell Res 33: 575–578

Miki-Nomura T, Kamiya R (1979) Conformational change in the outer doublet microtubules from sea urchin sperm flagella. J Cell Biol 82: 355–360

Millonig G (1969) Fine structural analysis of the cortical reaction in the sea urchin egg: After normal fertilization and after electric induction. J Submicrosc Cytol 1: 69–84

Millonig G (1970) A study on the formation and structure of sea urchin spicule. J Submicrosc Cytol 2: 157–165

Minafra S, Pucci-Minafra I, Casano C, Gianguzza F (1975) Chromatographic characterization of soluble collagen in sea urchin embryos (*Paracentrotus lividus*). Boll Zool 42: 205–208

Minafra S, Galante R, D'Antoni S, Fanara M, Coppola L, Pucci-Minafra I (1980) Collagen-associated protein-polysaccharide in the *Aristotle's lanternae* of *Paracentrotus lividus*. J Submicrosc Cytol 12: 255–266

Minganti A, Vasseur E (1959) An analysis of the jelly substance of *Paracentrotus* eggs. Acta Embryol Morphol Exp 3: 195–203

Mintz G R, De Francesco S, Lennarz W J (1981) Spicule formation by cultured embryonic cells from the sea urchin. J Biol Chem 256: 13105–13111

Mishra N K (1978) Study of ribosomal deoxyribonucleic acid of sea urchin. Mol Biol Rep 4: 241–246

Mita M, Yasumasu I (1979) Reversal of palmitoyl coenzyme A-caused inhibition of glucose-6-phosphate dehydrogenase by polyamines. Biochem Biophys Res Commun 86: 961–967

Mita M, Yasumasu I (1980) Inhibition of glucose 6-phosphate dehydrogenase and 6-phosphogluconate dehydrogenase in sea urchin eggs by palmitoyl-coenzyme A and reversal by polyamines. Arch Biochem Biophys 201: 322–329

Mita M, Yasumasu I (1981) Inhibition of dihydrofolate reductase by palmitoyl coenzyme A. Int J Biochem 13: 229–232

Mita M, Yasumasu I (1982) Mechanism of coenzyme A inhibition of dihydrofolate reductase. Int J Biochem 14: 549–551

Mita M, Yasumasu I (1983) Metabolism of lipid and carbohydrate in sea urchin spermatozoa. Gamete Res 7: 133–144.

Mitchison J M, Cummins J E (1966) The uptake of valine and cytidine by sea-urchin embryos and its relation to cell surface. J Cell Sci 1: 35–47

Mitchison J M, Swann M M (1955) The mechanical properties of the cell surface. II The sea-urchin egg from fertilization to cleavage. J Exp Biol 32: 734–750

Mizoguchi H, Yasumasu I (1982a) Archenteron formation induced by ascorbate and α-ketoglutarate in sea urchin embryos kept in SO_4^{2-} — free artificial sea water. Dev Biol 93: 119–125

Mizoguchi H, Yasumasu I (1982b) Exogut formation by the treatment of sea urchin embryos with ascorbate and α-ketoglutarate. Dev Growth Differ 24: 359–368

Mizoguchi H, Yasumasu I (1983) Inhibition of archenteron formation by the inhibitors of prolyl hydroxylase in sea urchin embryos. Cell Differ 12: 225–231

Mizoguchi H, Fujiwara A, Yasumasu I (1983) Degeneration of archenteron in sea urchin embryos caused by α,α'-dipyridyl. Differentiation 25: 106–112

Mizuno S, Whiteley H R, Whiteley A H (1973) The enrichment of egg-type RNA in cleavage stage embryos of the sand dollar *Dendraster excentricus*. Differentiation 1: 339–348

Mizuno S, Lee Y R, Whiteley A H, Whiteley H R (1974) Cellular distribution of RNA populations in 16-cell stage embryos of the sand dollar. *Dendraster excentricus.* Dev Biol 37: 18–27

Moeller M N V, McBlaine P J, Carroll E J Jr (1980) Immunochemical characterization of intermediates in the assembly of the sea urchin fertilization envelope. J Cell Biol 87: 140 (Abstr)

Molinaro M, Farace M G (1972) Changes in codon recognition and chromatographic behaviour of tRNA species during embryonic development of the sea urchin *Paracentrotus lividus.* J Exp Zool 181: 223–232

Molinaro M, Mozzi R (1969) Heterogeneity of t-RNA during embryonic development of the sea urchin *Paracentrotus lividus.* Exp Cell Res 56: 163–166

Monné L, Harde S (1950) On the formation of the blastocoel and similar embryonic cavities. Ark Zool Ser 2 1: 463–469

Monroy A (1949) On the formation of the fertilization membrane in the sea urchin *Psammechinus micr.* Exp Cell Res Suppl 1: 525–528

Monroy A (1960) Incorporation of S^{35}-methionine in the microsomes and soluble proteins during the early development of the sea urchin egg. Experientia 16: 114–117

Monroy A (1967) Fertilization. In: Florkin M, Stotz E (eds) Comprehensive biochemistry, vol 28. Elsevier Amsterdam New York, pp 369–412

Monroy A, Maggio R (1963) Biochemical studies on the early development of the sea urchin. Adv Morphol 3: 95–145

Monroy A, Nakano E (1959) Evaluation of the methods for the incorporation of radioactive compounds in the echinoderm eggs. Pubbl Stn Zool Napoli Suppl 31: 95–99

Monroy A, Tyler A (1963) Formation of active ribosomal aggregates (polysomes) upon fertilization and development of sea urchin eggs. Arch Biochem Biophys 103: 431–435

Monroy A, Maggio R, Rinaldi A M (1965) Experimentally induced activation of the ribosomes of the unfertilized sea urchin egg. Proc Natl Acad Sci USA 54: 107–111

Moon R T (1983) Poly (A)-containing messenger ribonucleoprotein complexes from sea urchin eggs and embryos: Polypeptides associated with native and UV-cross-linked mRNPs. Differentiation 24: 13–23

Moon R T, Moe K D, Hille M B (1980) Polypeptides of non polyribosome messenger ribonucleoprotein complexes of sea urchin eggs. Biochemistry 19: 2723–2730

Moon R T, Danilchik M V, Hille M B (1982) An assessment of the masked message hypothesis: Sea urchin egg messenger ribonucleoproteins are efficient templates for in vitro protein synthesis. Dev Biol 93: 389–403

Moon R T, Nicosia R F, Olsen C, Hille M B, Jeffery W R (1983) The cytoskeletal framework of sea urchin egg and embryos: developmental changes in the association of messenger RNA. Dev Biol 95: 447–458

Moore A R, Burst A S (1939) On the locus and nature of the forces causing gastrulation in the embryos of *Dendraster excentricus.* J Exp Zool 82: 159–170

Moore G P, Scheller R H, Davidson B H, Britten R J (1978) Evolutionary change in the repetition frequency of sea urchin DNA sequences. Cell 15: 649–660

Moore G P, Costantini F D, Posakony J W, Davidson E H, Britten R J (1980) Evolutionary conservation of repetitive sequence expression in sea urchin egg RNA's. Science 208: 1046–1048

Morgan T H (1894) Experimental studies on Echinoderm eggs Anat Anz 9: 141–152

Morisawa M, Mohri H (1972) Heavy metals and spermatozoan motility. I. Distribution of iron, zinc and copper in sea urchin spermatozoa. Exp Cell Res 70: 311–316

Morisawa M, Mohri H (1974) Heavy metals and spermatozoan motility. II. Turbidity changes induced by divalent cations and adenosinetriphosphate in sea urchin sperm flagella. Exp Cell Res 83: 87–94

Morris G F, Marzluff W E (1983) A factor in sea urchin eggs inhibits transcription in isolated nuclei by sea urchin RNA polymerase III. Biochemistry 22: 645–652

Morris P W, Rutter W J (1976) Nucleic acid polymerizing enzymes in developing *Strongylocentrotus franciscanus* embryos. Biochemistry 15: 3106–3113

Mortensen Th (1921) Studies of the development of echinoderms. G E C GAD, Copenhagen

Morton R W, Nishioka D (1983) Effects of cytochalasin B on the cortex of the unfertilized sea urchin egg. Cell Biol Int. Rep 7: 835–842

Moser F (1939) Studies on a cortical layer response to stimulating agents in the *Arbacia* egg. I Response to insemination. J Exp Zool 80: 423–445

Motomura I (1949) Artificial alteration of the embryonic axis in the centrifuged eggs of sea urchins. Sci Rep Tohoku Univ 4th Ser 18: 117–125

Motomura I (1957) On the nature and localization of the third factor for the thoughening of the fertilization membrane of the sea urchin egg. Sci Rep Tohoku Imp Univ Biol 23: 167–181

Moy G W, Friend D S, Vacquier V D (1977) Immunoperoxidase localization of sea urchin binding during sperm and egg adhesion. J Cell Biol 75: 61 (Abstr)

Müller W E G, Breter H J, Zahn R K (1974) Morphological and biochemical characterization of the developmental stages of fertilized eggs in *Sphaerechinus granularis Lam*. II. DNA content, DNA polymerase activity. Wilhelm Roux' Arch Entwicklungsmech Org 174: 117–133

Murakami K, Mano Y (1973) Stimulation of sea urchin DNA polymerase by protein factors. Biochem Biophys Res Commun 55: 1125–1133

Murakami-Murofushi K, Mano Y (1977) Stimulation of sea urchin DNA polymerase by protein factors. II Formation of active complex in DNA polymerase reaction. Biochim Biophys Acta 475: 251–266

Murray A, Sosnowski R (1980) The status of mRNA in eggs and embryos. Biol Bull 159: 476 (Abstr)

Nagai Y, Hoshi M (1975) Sialosphingolipids of sea urchin eggs and spermatozoa showing a characteristic composition for species and gamete. Biochim Biophys Acta 388: 146–151

Nakamura M, Yasumasu I (1974) Change in the calcium binding capacity of sea urchin egg homogenate caused by treatment with ATP and magnesium ions. Exp Cell Res 88: 121–126

Nakano E, Iwata M (1982) Collagene-binding proteins in sea urchin egg and embryos. Cell Differ 11: 339–340

Nakano E, Monroy A (1957) A method for incorporation of radioactive isotopes in the sea urchin egg. Experientia 13: 416–417

Nakano E, Monroy A (1958a) Incorporation of S^{35} methionine in the cell fractions of sea urchin eggs and embryos. Exp Cell Res 14: 236–244

Nakano E, Monroy A (1958b) Some observations on the metabolism of S^{35}-methionine during development of the sea urchin eggs. Experientia 14: 367–371

Nakano E, Ohashi S (1954) On the carbohydrate component of the jelly coat and related substances of eggs from Japanese sea urchins. Embryologia 2: 81–86

Nakano E, Giudice G, Monroy A (1958) On the incorporation of S^{35}-methionine in artificially activated sea urchin eggs. Experientia 14: 11–13

Nakazawa T, Asami K, Shoger R, Fujwara A, Yasumasu I (1970) Ca^{2+} uptake, H^+ ejection and respiration in sea urchin eggs on fertilization. Exp Cell Res 63: 143–146

Nath J, Rebhun L I (1973) Studies on cyclic AMP levels and phosphodiesterase activity in developing sea urchin. Effects of puromycin 6-dimethylaminopurine and aminophylline. Exp Cell Res 77: 319–322

Neifakh A A (1960) A study of nuclear function in the development of sea urchin *Strongylocentrotus dröbachiensis* by radiational inactivation. Dokl Biol Sci 132: 376–379

Neifakh A A (1964) Radiation investigation of nucleocytoplasmic interrelations in morphogenesis and biochemical differentiation. Nature (London) 201: 880–884

Neifakh A A, Krigsgaber M R (1968) Protein synthesis in embryonic development of the sea urchin after inactivation of nuclei. Dokl Biol Sci 183: 639–642

Nelson L (1971) Motility control mechanisms in Arbacia sperm. Biol Bull 141: 374 (Abstr)

Nelson L (1972) Neurochemical control of *Arbacia* sperm motility. Exp Cell Res 74: 269–274

Nelson L (1975) α-Bungarotoxin blockade of sperm cell function. Biol Bull 149: 438 (Abstr)

Nelson L (1976) α-bungarotoxin binding by cell membranes. Blockage of sperm cell motility. Exp Cell Res 101: 221–224

Nelson L (1980) Receptors, drug interactions and calcium in regulation of *Arbacia* sperm function. Biol Bull 159: 469 (Abstr)

Nelson L (1982) Membrane-stabilizing and calcium-blocking agents affect *Arbacia* sperm motility. Biol Bull 163: 492–503

Nelson L, Chakraborty J, Young M, Goodwin A, Kock E, Gardner M E (1980) Control of sperm cell motility: neurochemical regulation of calcium transport. In: Steinberg A, Steinberg E (eds) Testicular Development, Structures and Function. Raven Press, New York, pp 503 to 511

Nemer M (1962) Characteristics of the utilization of nucleosides by embryos of *Paracentrotus lividus*. J Biol Chem 237: 143–149

Nemer M (1963) Old and new RNA in the embryogenesis of the purple sea urchin. Proc Natl Acad Sci USA 50: 230–235

Nemer M (1975) Developmental changes in the synthesis of sea urchin embryo mRNA containing and lacking polyadenylic acid. Cell 6: 559–570

Nemer M, Bard S G (1963) Polypeptide synthesis in sea urchin embryogenesis: An examination with synthetic polyribonucleotides. Science 140: 664–666

Nemer M, Infante A A (1968) Heterogeneous ribonucleoprotein particles in the cytoplasm of sea urchin embryos. J Mol Biol 32: 543–565

Nemer M, Lindsay D T (1969) Evidence that the s-polysomes of early sea urchin embryos may be responsible for the synthesis of chromosomal histones. Biochem Biophys Res Commun 35: 156–160

Nemer M, Surrey S (1975) Messenger RNA synthesis as a function of sea urchin development: mRNAs containing and lacking poly (A). J Cell Biol 67: 305 (Abstr)

Nemer M, Graham M, Dubroff L M (1974) Co-existence of nonhistone mRNA species lacking and containing polyadenylic acid in sea urchin embryos. J Mol Biol 89: 435–454

Nemer M, Dubroff L M, Graham M (1975) Properties of sea urchin embryo messenger RNA containing and lacking poly (A). Cell 6: 171–178

Nemer M, Surrey S, Ginzburg I, Echols M M (1979) The 5' terminal capping of heterogeneous nuclear RNA at different embryonic stages of the sea urchin. Nucleic Acid Res 6: 2307–2326

Nemer M, Travaglini E C, Moody J (1983) Metallothionein gene expression inductively and developmentally regulated in the sea urchin embryo. J Cell Biol 97: 328 (Abstr)

Neri A, Roberson M, Connolly D T, Oppenheimer J B (1975) Quantitative evaluation of concanavalin A receptor site distributions on the surfaces of specific populations of embryonic cells. Nature (London) 258: 343–344

Newrock K M, Alfageme C R, Nardi R V, Cohen L H (1978a) Histone changes during chromatin remodeling in embryogenesis. Cold Spring Harbor Symp Quant Biol 42: 421–431

Newrock K M, Cohen L H, Hendricks M B, Donnelly R J, Weinberg E S (1978b) Stage-specific mRNAs coding for subtypes of H2A and H2B histones in sea urchins. Cell 14: 327–336

Newrock K M, Freedman N, Alfageme C R, Cohen L H (1982) Isolation of sea urchin embryo histone H2Aα1 and immunological identification of other stage specific H2A proteins. Dev Biol 89: 248–253

Nijhawan P, Marzluff W F (1979) Metabolism of low molecular weight ribonucleic acids in early sea urchin embryos. Biochemistry 18: 1353–1360

Nishioka D, Epel D (1977) Intracellular pH and activation of sea urchin eggs at fertilization. J Cell Biol 75: 40 (Abstr)

Nishioka D, Magagna L S (1981) Increased uptake of thymidine in the activation of the sea urchin eggs. Exp Cell Res 133: 363–372

Nishioka D, Mazia D (1977) The phosphorylation of thymidine and the synthesis of histones in ammonia-treated eggs and egg fragments of sea urchin. Cell Biol Int Rep 1: 23–30

Nishioka D, McGwin N F (1980) The involvement of the eggg cortex in the uptake and phosphorylation of thymidine in fertilized sea urchin eggs. J Cell Biol 87: 136 (Abstr)

Noll H, Matranga V, Cascino D, Vittorelli M L (1979) Reconstitution of membranes and embryonic development in dissociated blastula cells of sea urchin by reinsertion of aggregation promoting membrane proteins extracted with butanol. Proc Natl Acad Sci USA 76: 288–292

Noll H, Matranga V, Palma P, Cutrono F, Vittorelli L (1981) Species-specific dissociation into single cells of live sea urchin embryos by Fab against membrane components of *Paracentrotus lividus* and *Arbacia lixula*. Dev Biol 87: 229–241

Nonaka M, Terayama H (1975) Thymidine kinase activation by unfertilized sea urchin eggs by homogenization and fertilization. Dev. Biol 43: 322–333

Nonaka M, Terayama H (1977) Compartmentation of thymidine kinase in unfertilized sea urchin eggs and possible release of thymidine kinase from particulates in activated eggs. Dev Biol 56: 68–75

Noronha J M, Sheys G H, Buchanan J M (1972) Induction of a reductive pathway for deoxyribonucleotide synthesis during early embryogenesis of the sea urchin. Proc Natl Acad Sci USA 69: 2006–2010

Ogawa K, Gibbons I R (1976) Dynein 2: A new adenosine-triphosphatase from sea urchin sperm flagella. J Biol Chem 251: 5793–5801

Ogawa K, Mohri H (1972) Studies on flagellar ATPase from sea urchin spermatozoa I Purification and some properties of the enzyme. Biochim Biophys Acta 256: 142–155

Ogawa K, Mohri H (1975) Preparation of antiserum againsta a tryptic fragment (Fragment A) of dynein and an immunological approach to the subunit composition of dynein. J Biol Chem 250: 6476–6483

Ogawa K, Asai D J, Brokaw C J (1977a) Properties of an antiserum against native dynein from sea urchin sperm flagella. J Cell Biol 73: 183–192

Ogawa K, Mohri T, Mohri H (1977b) Identification of dynein as the outer arms of sea urchin sperm axonemes. Proc Natl Acad Sci USA 74: 5006–5010

Ogawa K, Negishi S, Obika M (1982) Immunological dissimilarity in protein component (dynein[1]) between outer and inner arms within sea urchin sperm axonemes. J Cell Biol 92: 706–713

Ohama T, Hori H, Osawa S (1983) The nucleotide sequences of 5s rRNAs from a sea-urchin. Nucleic Acid Res 11: 5181–5184

Ohsawa T, Nagai Y (1975) Immunological evidence for the localization of sialoglycosphingolipids at the cell surface of sea urchin spermatozoa. Biochim Biophys Acta 389: 69–83

Okazaki K (1960) Skeleton formation of sea urchin larvae. II Organic matrix of the spicule. Embryologia 5: 283–293

Okazaki K (1975a) Spicule formation by micromeres isolated from embryos of *Arbacia punctulata*. Biol. Bull 149: 439–440 (Abstr)

Okazaki K (1975b) Spicule formation by isolated micromeres of the sea urchin embryo. Am Zool 15: 567–581

Okazaki K, Inoué S (1976) Crystal property of the larval sea urchin spicule. Dev Growth Differ 18: 413–434

Okuno M (1980) Inhibition and relaxation of sea urchin sperm flagella by vanadate. J Cell Biol 85: 712–725

Okuno M, Brokaw C J (1979) Inhibition of movement of Triton-demembranated sea urchin sperm flagella by Mg^{2+}, ATP^{4-} and Pi. J Cell Sci 38: 105–123

O'Melia A F (1979a) The synthesis of 5S RNA and its regulation during early sea urchin development. Dev Growth Differ 21: 99–108

O'Melia A F (1979b) Quantitative measurement of rates of 5S RNA and transfer RNA synthesis in sea urchin embryos. Differentiation 15: 97–105

O'Melia A F (1984) The synthesis of 5S RNA and transfer RNA in sea urchin embryos animalized by Evans blue. Cell Biol Int Rep 8: 33–40

O'Melia A F, Villee C A (1972a) Animalizing ability of Evans blue in embryos of *Arbacia punctulata* effect on ribosomal RNA synthesis. Exp Cell Res 72: 276–284

O'Melia A F, Villee C A (1972b) De novo synthesis of transfer and 5S RNA in cleaving sea urchin embryos. Nature (London) New Biol 239: 51–53

Onitake K, Tsuzuki H, Aketa K (1972) Immunochemical study on the species specificity of sperm binding protein from the surface of the sea urchin egg. Dev Growth Differ 14: 207–215

Oppenheimer S B, Meyer J T (1982a) Isolation of species-specific and stage-specific adhesion promoting component by disaggregation of intact sea urchin embryo cells. Exp Cell Res 137: 472–476

Oppenheimer S B, Meyer J T (1982b) Carbohydrate specificity of sea urchin blastula adhesion component. Exp Cell Res 139: 451–455

Oppenheimer S B, Potter R L, Barber M L (1973) Alteration of sea urchin embryo cell surface properties by micostatin a sterol binding antibiotic. Dev Biol 33: 218–223

Ord M G, Stocken L A (1974) Thymidine uptake by *Paracentrotus* eggs during the first cell cycle after fertilization. Exp Cell Res 83: 411–414

Ord M G, Stocken L A (1978) Histones in the first cleavage cycle of fertilized sea urchin eggs. Cell Differ 7: 271–281

Orengo A, Hnilica L S (1970) In vivo incorporation of labeled aminoacids into nuclear proteins of sea urchin embryos. Exp Cell Res 62: 331–337

Örström A, Lindberg O (1940) Über den Kohlenhydratstoffwechsel bei der Befruchtung des Seeigeleies. Enzymologia 8: 367–384

Ortolani G, Tosi L, Branno M, Patricolo E (1982) Development of animal halves of sea urchin eggs after trypsin treatment. Acta Embryol Morphol Exp no 3 XVII–XVIII (Abstr)

Osanai K (1960) Development of the sea urchin egg with the inhibited breakdown of the cortical granules. Sci Rep Tohoku Univ 36: 77–87

Osanai K (1970) On the mitotic activity of the nuclei in the centrifuged sea urchin eggs, with special reference to the formation of the giant nucleus. Annu Rep Fac Educ Univ Iwate 30: 85–92

Osanai K (1972) Hybridization of sea urchins, *Hemicentrotus* and *Glytocidaris crenularis.* Annu Rep Fac Educ Univ Iwate 32: 1–11

Osanai K (1974) Interspecific hybridization of sea urchins, *Strongylocentrotus nudus* and *Strongylocentrotus intermedius.* Bull Mar Biol St Asamushi Tohoku Univ 15: 37–47

Osanai K, Kyozuka K I (1979) Cross fertilization in sea urchin eggs treated with egg water. Bull Mar Biol Stn Asamushi 16: 189–198

Oshima N (1983) Mechanical properties of perivitelline fibers of sea urchin eggs as studied by application of centrifugal and electrophoretic forces. Biol Bull 164: 483–492

Overbeek P A, Merlino G T, Peters N K, Cohn V H, Moore G P, Kleinsmith L J (1981) Characterization of five members of the actin gene family in the sea urchin. Biochim Biophys Acta 656: 195–205

Overton G C, Weinberg H S (1978) Length and sequence heterogeneity of the histone gene repeat unit of the sea urchin *Strongylocentrotus purpuratus.* Cell 14: 247–258

Ozaki H (1974) Localization and multiple forms of acetylcholinesterase in sea urchin embryos. Dev Growth Differ 16: 267–279

Ozaki H (1975) Regulation of isozymes in interspecies sea urchin hybrid embryos. In: Isozymes, vol III. Developmental Biology. Academic Press, London New York, pp 543–558

Ozaki H (1976) Molecular properties and differentiation of acetylcholinesterase in sea urchin embryos. Dev Growth Differ 18: 245–257

Ozaki H (1980) Yolk proteins of the sand dollar *Dendraster excentricus.* Dev Growth Differ 22: 365–372

Ozaki H (1982) Vitellogenesis in the sand dollar *Dendraster excentricus.* Cell Differ 11: 315–317

Ozaki H, Whiteley A H (1970) 1-Malate dehydrogenase in the development of the sea urchin *Strongylocentrotus purpuratus.* Dev Biol 21: 196–215

Parisi E, De Petrocellis B (1972) Properties of a deoxyribonuclease from a nuclear extract of *Paracentrotus lividus* embryos. Biochem Biophys Res Commun 49: 706–712

Parisi E, De Petrocellis B (1976) Sea urchin thymidylate synthetase. Changes of activity during embryonic development. Exp Cell Res 101: 59–62

Parisi E, Filosa S, De Petrocellis B, Monroy A (1978) The pattern of cell division in the early development of the sea urchin, *Paracentrotus lividus.* Dev Biol 65: 43–49

Parisi E, Filosa S, Monroy A (1979) Actinomyc in D disruption of the mitotic gradient in the cleavage stages of the sea urchin embryo. Dev Biol 72: 167–174

Passananti C, Felsani A, Giordano R, Metafora S, Spadafora C (1983) Cloning and characterization of the ribosomal genes of the sea-urchin *Paracentrotus lividus.* Heterogeneity of the multigene family. Eur J Biochem 137: 233–239

Pasternak C A (1973) Phospholipid synthesis in cleaving sea urchin eggs: Model for specific membrane assembly. Dev Biol 30: 403–410

Patterson J B, Stafford D W (1970) Sea urchin satellite deoxyribonocleic acid. Its large-scale isolation and hybridization with homologous ribosomal ribonucleic acid. Biochemistry 5: 1278–1283

Patterson J B, Stafford D W (1971) Characterization of sea urchin ribosomal satellite DNA. Biochemistry 10: 2775–2779

Paul M, Epel D (1975) Formation of fertilization acid by sea urchin eggs does not require specific cations. Exp Cell Res 94: 1–6

Paul M Johnston R N (1978) Absence of a Ca response following ammonia activation of sea urchin eggs. Dev Biol 67: 330–335

Pawlawski P P, Rodriguez L V (1974) Timing of initiation of poly-adenylation of preformed RNA in sea urchin and *Urechis caupo* zygotes. Dev Biol 40: 71–77

Payan P, Girard J P, Sardet C (1981) Na^+ movements and their oscillations during fertilization and the cell cycle in sea urchin eggs. Exp Cell Res 134: 339–344

Payan P, Girard J P, Ciapa B (1983) Mechanisms regulating intracellular pH in sea urchin eggs. Dev Biol 100: 29–38

Pearson W R, Morrow J F (1981) Discrete-length repeated sequences in eukaryotic genomes. Proc Natl Acad Sci USA 78: 4016–4020

Peltz R, Giudice G (1967) The control of skeleton differentiation in sea urchin embryos. A molecular approach. Biol Bull 133: 479 (Abstr)

Penningroth S M, Witman G B (1978) Effects of adenylyl imidodiphosphate, a monohydrolyzable adenosine triphosphate analog on reactivated and rigor wave sea urchin sperm. J Cell Biol 70: 827–833

Pennigroth S M, Cheung A, Olehnik K, Koslosky R (1982) Mechanochemical coupling in the relaxation of rigor-wave sea urchin sperm flagella. J Cell Biol 92: 733–741

Pepe G, Greco M, Cantatore P, Nicotra A (1976) Isolation of ribosomal and characterization of RNA species from mitochondria of *Paracentrotus lividus*. Bull Mol Biol Med 1: 71–79

Perry G (1977) Echinochrome oxidation in *Arbacia punctulata* embryos. Biol Bull 151: 423 (Abstr)

Perry G, Epel D (1975) Calcium control of a cyanide insensitive respiration in *Arbacia punctulata* eggs, and its relation to echinocrome. Biol Bull 149: 441 (Abstr)

Perry G, Epel D (1977) Calcium stimulation of a lipoxygenase activity accounts for the respiratory burst at fertilization of the sea urchin egg. J Cell Biol 75: 40 (Abstr)

Perry G, Epel D (1981) Ca^{2+}-stimulated production of H_2O_2 from naphtoquinone oxidation in *Arbacia* eggs. Exp Cell Res 134: 65–72

Peters C, Jeffery W R (1978) Postfertilization poly (A)-protein complex formation on sea urchin maternal messenger RNA Differentiation 2: 91–97

Peters R, Richter H P (1981) Translational diffusion in the plasma membrane of sea urchin eggs. Dev Biol 86: 285–293

Petzelt C (1976) NH_3-treatment of unfertilized sea urchin eggs turns on the Ca^{2+}-ATPase cycle. Exp Cell Res 102: 200–205

Petzelt C (1979) Biochemistry of the mitotic spindle. Intern Rev Cytol 60: 53–120

Pfohl R J (1965) Changes in alkaline phosphatase during the early development of the sea urchin, *Arbacia punctulata*. Exp Cell Res 39: 496–503

Pfohl R J (1975) Alkaline phosphatase of sea urchin embryos: Chromatographic and electrophoretic characterization. Dev Biol 44: 333–345

Pfohl R J, Giudice G (1967) The role of cell interactions in the control of enzyme activity during embryogenesis. Biochim Biophys Acta 142: 263–266

Piatigorsky J (1968) RNAse and trypsin tretment of ribosomes and polyribosomes from sea urchin eggs. Biochim Biophys Acta 166: 142–155

Piatigorsky J, Tyler A (1968) Displacement of valine from intact sea-urchin eggs by exogenous aminoacids. J Cell Sci 3: 515–527

Piatigorsky J, Whiteley A H (1965) A change in permeability and uptake of C^{14} uridine in response to fertilization in *Strongylocentrotus purpuratus* eggs. Biochim Biophys Acta 108: 404–418

Piatigorsky J, Ozaki H, Tyler A (1967) RNA and protein-synthesing capacity of isolated oocytes of the sea urchin *Lytechinus pictus*. Dev Biol 15: 1–22

Pikò L (1969) Absence of synthesis of mitochondrial DNA during early development in sea urchins. Am Zool 9: 1118 (Abstr)

Pikò L, Tyler A, Vinograd J (1967) Amount location, priming capacity and other properties of cytoplasmic DNA in sea urchin eggs. Biol Bull 132: 68–90

Pikò L, Blair O G, Tyler A, Vinograd J (1968) Cytoplasmic DNA in the unfertilized sea urchin egg: Physical properties of circular mitochondrial DNA and the occurrence of catenated forms. Proc Natl Acad Sci USA 59: 838–845

Piperno G, Fleming J (1983) Dynein arm component from axonemes of sea urchin sperm. Purification and use as antigens for monoclonal antibody production. J Submicrosc Cytol 15: 203–208

Pirrone A M, Sconzo G, Mutolo V, Giudice G (1970) Effect of chemical animalization and vegetalization on the synthesis of ribosomal RNA in sea urchin embryos. Wilhelm Roux' Arch Entwicklungsmech Org 164: 222–225

Pirrone A M, Roccheri M C, Giudice G (1973) Analyses of exogenous P^{32} incorporation into the nucleotide pool of sea urchin embryos. Wilhelm Roux' Arch Entwicklungsmech Org 172: 80–82

Pirrone A M, Roccheri M C, Bellanca V, Acierno P, Giudice G (1976) Studies on the regulation of ribosomal RNA synthesis in sea urchin development. Dev Biol 49: 311–320

Pirrone A M, Spinelli G, Acierno P, Errera M, Giudice G (1977) The RNA of unfertilized sea urchin eggs is "capped". Cell Differ 5: 335–343

Pirrone A M, Gambino R, Oddo F, Faraci M T, Luparello G, Giudice G (1979) Sea urchin embryos do not synthesize diadenosinetetraphosphate. Exp Cell Res 122: 419–422

Pirrone A M, Gambino R, Acuto S A, Butticè G, Bellavia P, Giudice G (1983) Ornithine decarboxylase activity correlates with the synthesis of DNA and not of rRNA in developing sea urchins. Cell Biol Int Rep 7: 377–381

Platz R D, Hnilica L S (1973a) Characterization of phosphorylated nuclear proteins in developing sea urchin embryos. J Cell Biol no 3, part 2, 267 (Abstr)

Platz R D, Hnilica L S (1973b) Phosphorylation of nonhistone chromatin proteins during sea urchin development. Biochem Biophys Res Commun 54: 222–227

Poccia D L, Hinegardner R T (1975) Developmental changes in chromatin proteins of the sea urchin from blastula to mature larva. Dev Biol 45: 81–89

Poccia D L, Le Vine D, Wang J C (1978) Activity of a DNA topoisomerase (nickingclosing enzyme) during sea urchin development and the cell cycle. Dev Biol 64: 273–283

Poccia D L, Salik J, Krystal G (1981) Transition in histone variants of the male pronucleus following fertilization and evidence for a maternal store of cleavage-stage histones in the sea urchin egg. Dev Biol 82: 287–296

Poenie M F, Fromson D (1979) Developmental changes in actin during fertilization and early stages of sea urchin embryos. J Cell Biol 83: 208 (Abstr)

Pollock J M Jr, Swihart M, Taylor J H (1979) Methylation of DNA in early development: 5-methyl cytosine content of DNA in sea urchin sperm and embryos. Nucleic Acid Res 5: 4855–4863

Popa G T (1927) The distribution of substances in spermatozoon (*Arbacia* and *Nereis*). Biol Bull 52: 238–245

Portman R, Schaffner W, Birnstiel M (1976) Partial denaturation mapping of cloned histone DNA from the sea urchin *Psammechinu miliaris*. Nature (London) 264: 31–33

Poznanovic G, Sevaljevic L (1980) The isolation and characterization of the nuclear matrix from sea urchin embryos. Cell Biol Int Rep 4: 701–709

Pratt M M (1980) The identification of a dynein ATPase in unfertilized sea urchin eggs. Dev Biol 74: 364–378

Pratt M M, Otter T, Salmon E D (1980) Dynein-like Mg 2+ — ATPase in mitotic spindles isolated from sea urchin embryos (*Strongylocentrotus droebachiensis*). J Cell Biol 86: 738–745

Probst E, Kressmann A, Birnstiel M L (1979) Expression of sea urchin histone genes in the oocyte of *Xenopus laevis*. J Mol Biol 135: 709–732

Prothero J W, Tamarin A (1977) The blastomere pattern in echinoderms: cleavages one to four. J Embryol Exp Morphol 40: 23–34

Pucci-Minafra I, Bosco M, Giambertone L (1968) Preliminary observations on the isolated micromeres from sea urchin embryos. Exp Cell Res 53: 177–183

Pucci-Minafra I, Casano C, La Rosa C (1972) Collagen synthesis and spicule formation in sea urchin embryos. Cell Differ 1: 157–165

Pucci-Minafra I, Minafra S, Gianguzza F, Casano C (1975) Amino acid composition of collagen extracted from the spicules of sea urchin embryos. (*Paracentrotus lividus*). Boll Zool 42: 201–204

Pucci-Minafra I, Galante R, Minafra S (1978) Identification of collagen in the Aristotle's lanternae of *Paracentrotus lividus*. J Submicrosc Cytol 10: 53–63

Pucci-Minafra I, Fanara M, Minafra S (1980) Chemical and physical changes in the organic matrix of mineralized tissues from embryo to adult of *Paracentrotus lividus*. J. Submicrosc Cytol 12: 267–273

Puigdomenech P, Palau J, Crane-Robinson C (1980) The structure of sea urchin-sperm histone 1 (H1) in chromatin and in free solution. Trypsin digestion and spectroscopic studies. Eur J Biochem 104: 263–270

Raff R A (1975) Regulation of microtubule synthesis and utilization during early embryonic development of the sea urchin. Am Zool 15: 661–678

Raff R A, Greenhouse G, Gross K W, Gross P R (1971) Synthesis and storage of microtubule proteins by sea urchin embryos. J Cell Biol 50: 516–527

Raff R A, Colot H V, Selving S E, Gross P R (1972) Oogenetic origin of mRNA for embryonic synthesis of microtubule proteins. Nature (London) 235: 211–214

Raff R A, Brandis J W, Huffman C J, Koch A L, Leister D E (1981) Protein synthesis as an early response to fertilization of the sea urchin egg: a model. Dev Biol 86: 265–271

Randany E W, Gerzer R, Garbers D L (1983) Purification and characterization of particulate guanylate cyclase from sea urchin spermatozoa. J Biol Chem 258: 8346–8351

Rappaport R (1961) Experiments concerning the cleavage stimulus in sand dollar eggs. J Exp Zool 148: 81–89

Rappaport R (1964) Geometrical relations of the cleavage stimulus in constricted sand dollar eggs. J Exp Zool 155: 225–230

Rappaport R (1965) Geometrical relations of the cleavage stimulus in invertebrate eggs. J Theor Biol 9: 51–56

Rappaport R (1967) Geometrical analysis of establishment of the cleavage furrow in a sea urchin egg. J Cell Biol 35: 109 (Abstr)

Rappaport R (1968) Geometrical relations of the cleavage stimulus in flattened, perforated sea urchin eggs. Embryologia 10: 115–130

Rappaport R (1969a) Division of isolated furrows and furrow fragments in invertebrate eggs. Exp Cell Res 56: 87–91

Rappaport R (1969b) Reversal of chemical cleavage inhibition in echinoderm eggs. J Cell Biol 43: 111 (Abstr)

Rappaport R (1975) In: Inoue's, Stephens RE (eds) Molecular and cell movement. Raven Press, New York, pp 287–303

Rappaport R (1982) Cleavage of geometrically altered cells. Cell Differ 11: 359–361

Rappaport R, Ebstein R P (1965) Duration of stimulus and latent periods preceding furrow formation in sand dollar eggs. J Exp Zool 158: 373–382

Rappaport R, Ratner J H (1967) Cleavage of sand dollar eggs with altered pattern of new surface formation. J Exp Zool 165: 89–100

Rapraeger A C, Epel D (1980) The appearance of an extracellular sulfatase during sea urchin embryogenesis. J Cell Biol 87: 135 (Abstr)

Rapraeger A C, Epel D (1981) The appearance of an extracellular arylsulfatase during morphogenesis of the sea urchin *Strongylocentrotus purpuratus*. Dev Biol 88: 269–278

Rebhun, L I, Suprenant K, Keller T C, Folley L (1982) Spindle and cytoplasmic tubulins from marine eggs. Cell Differ 11: 367–371

Regier J C, Kafatos F C (1977) Absolute rates of protein synthesis in sea urchins with specific activity measurements of radioactive leucine and leucyl-tRNA. Dev Biol 57: 270–283

Renaud F, Parisi E, Capasso A, De Prisco P (1983) On the role of serotonin and 5-methoxy-tryptamine in the regulation of cell division in sea urchin eggs. Dev Biol 98: 37–46

Repaske D R, Garbers D L (1983) A hydrogen flux mediates stimulation of respiratory activity by speract in sea urchin spermatozoa. J Biol Chem 238: 6025–6029

Ribot H, Decker S J, Kinsey W H (1983) Preparation of plasma membranes from fertilized sea urchin eggs. Dev Biol 97: 494–499

Riederer-Henderson M A, Rosenbaum J L (1975) Increase in ciliary length of *Arbacia* blastulae by concanavalin A or trypsin treatment. Biol Bull 149: 444 (Abstr)

Riederer-Henderson M A, Rosenbaum J L (1979) Ciliary elongation in blastulae of *Arbacia punctulata* induced by trypsin. Dev Biol 70: 500–509

Rikmenspoel R (1978) Movement of sea urchin sperm flagella. J Cell Biol 76: 310–322

Rinaldi A M, Monroy A (1969) Polyribosome formation and RNA synthesis in the early post-fertilization stages of the sea urchin egg. Dev Biol 19: 73–86

Rinaldi A M, Parente A (1976) Rate of protein synthesis in oocytes of *Paracentrotus lividus*. Dev Biol 49: 260–267

Rinaldi A M, Sconzo G, Albanese I, Ramirez F, Bavister B D, Giudice G (1974) Cytoplasmic giant RNA in sea urchin embryos. III Polysomal localization. Cell Differ 3: 305–312

Rinaldi A M, Storace A, Arzone A, Mutolo V (1977) Cell nucleus negatively controls mitochondrial RNA synthesis in early sea urchin development. Cell Biol Int Rep 1: 249–254

Rinaldi A M, Salcher-Cillari I, Mutolo V (1979a) Mitochondrial division in non-nucleated sea urchin eggs. Cell Biol Int Rep 3: 179–182

Rinaldi A M, De Leo G, Arzone A, Salcher I, Storace A, Mutolo V (1979b) Biochemical and electron microscopic evidence that cell nucleus negatively controls mitochondrial genomic activity in early sea urchin development. Proc Natl Acad Sci HSA 76: 1916–1920

Rinaldi A M, Salcher-Cillari I, Valenti A M (1981) Mitochondrial DNA synthesis in *Xenopus laevis* enucleated eggs. Cell Biol Int Rep 5: 987–990

Rinaldi A M, Carra E, Salcher-Cillari I, Oliva O A (1983a) The nucleus negatively controls the synthesis of mitochondrial proteins in the sea urchin egg. Cell Biol Int Rep 7: 211–218

Rinaldi A M, Salcher-Cillari I, Comito L, Carra E (1983b) A method for obtaining pure mitochondrial preparation from sea urchin sperm. Cell Biol Int Rep 7: 915–921

Roberson M, Oppenheimer S B (1975) Quantitative agglutination of specific populations of sea urchin embryo cells with concanavalin A. Exp Cell Res 91: 263–268

Roberson M, Neri A, Oppenheimer S B (1975) Distribution of concanavalin A receptor sites on specific populations of embryonic cells. Science 189: 639–640

Roberts J W, Grula J W, Posakony J W, Hudspeth R, Davidson E H, Britten R J (1983) Comparison of sea urchin and human mtDNA: Evolutionary rearrangement. Proc Natl Acad Sci USA 80: 4614–4618

Robinson K R (1976) Potassium is not compartimentalized within the unfertilized sea urchin egg. Dev Biol 48: 466–470

Roccheri M C, Di Bernardo M G, Giudice G (1979) Archenteron cells are responsible for the increase in ribosomal RNA synthesis in sea urchin gastrulae. Cell Biol Int Rep 3: 733–737

Roccheri M C, Sconzo G, Di Bernardo M G, Albanese I, Di Carlo M, Giudice G (1981a) Heat shock proteins in sea urchin embryos. Territorial and intracellular location. Acta Embryol Morphol Exp 2: 91–100

Roccheri M C, Di Bernardo M G, Giudice G (1981b) Synthesis of heat-shock proteins in developing sea urchins. Dev Biol 83: 173–177

Roccheri M C, Sconzo G, Di Carlo M, Di Bernardo M G, Pirrone A, Gambino R, Giudice G (1982a) Heat-shock proteins in sea urchin embryos. Transcriptional and posttranscriptional regulation. Differentiation 22: 175–178

Roccheri M C, Spinelli G, Sconzo G, Casano C, Giudice G (1982b) Some effects of heat shock on the regulation of macromolecular syntheses in sea urchin embryos. In: Embryonic development, part B. Cellular aspects. Liss, New York, pp 31–43

Rodgers W H, Gross P R (1978) Inhomogeneous distribution of egg RNA sequences in the early embryo. Cell 14: 279–288

Roeder R G, Rutter W J (1969) Multiple forms of DNA dependent RNA polymerase in eukaryotic organisms. Nature (London) 224: 234–237

Roeder R G, Rutter W J (1970a) Specific nucleolar and nucleoplasmic RNA polymerases. Proc Natl Acad Sci USA 65: 675–682

Roeder R G, Rutter W J (1970b) Multiple RNA polymerases and ribonucleic acid synthesis during sea urchin development. Biochemistry 9: 2543–2553

Rosenberg P A, Wallace R A (1973) Autoradiographic localization of sulfated macromolecules during development of *Arbacia punctulata*; normal and sulfate deficient media. Biol Bull 145: 439 (Abstr)

Rosenspire A J, Brenner T A, Pogell B M (1977) Expression of differentiated function in cultures of isolated sea urchin micromeres. Biol Bull 151: 427 (Abstr)

Rothfeld J M (1980) The effect of depolarization on sea urchin bindin's ability to agglutinate eggs. J Cell Biol 87: 130 (Abstr)

Rothschild Lord (1949) The metabolism of fertilized and unfertilized sea-urchin eggs. J Exp Zool 26: 100–111

Rothschild Lord, Swann M M (1952) The fertilization reaction in the sea urchin. The block to polyspermy. J Exp Biol 29: 469–483

Ruderman J V, Alexandraki D (1983) Organization and expression of the tubullin gene families in the sea urchin J Submicrosc Cytol 15: 349–352

Rudensey L M, Infante A A (1979) Translational efficiency of cytoplasmic nonpolysomal messenger ribonucleic acid from sea urchin embryos. Biochemistry 18: 3056–3074

Ruderman J V, Gross P R (1973) Histones and histone mRNA in sea urchin embryogenesis. J Cell Biol 50: no 2, part 2, 236 (Abstr)

Ruderman J V, Gross P R (1974) Histones and histone synthesis in sea urchin development. Dev Biol 36: 286–298

Ruderman J V, Pardue M L (1977) Cell-free translation analysis of messenger RNA in echinoderm and amphibian early development. Dev Biol 60: 48–68

Ruderman J V, Schmidt M R (1981) RNA transcription and translation in sea urchin oocytes and eggs. Dev Biol 81: 220–228

Ruderman J V, Baglioni C, Gross P R (1974) Histone mRNA and histone synthesis during embryogenesis. Nature (London) 247: 36–38

Ruiz-Carrillo A, Palace J (1973) Histones from embryos of the sea urchin *Arbacia lixula*. Dev Biol 35: 115–124

Runnström J (1928a) Plasmabau und Determination bei dem Ei von *Paracentrotus lividus* LK. Wilhelm Roux' Arch Entwicklungsmech Org 113: 556–581

Runnström J (1928b) Zur experimentellen Analyse der Wirkung des Lithium auf den Seeigelkeim. Acta Zool (Stockholm) 9: 365–423

Runnström J (1966) The vitelline membrane and cortical particles in sea urchin eggs and their function in maturation and fertilization. In: Adv Morphog 5: 221–325

Runnström J (1967) The animalizing action of pretreatment of sea urchin eggs with thiocynate in calcium-free sea water and its stabilization after fertilization. Ark Zool 19: 251–263

Rustad R C (1979) Ionophore-activated pigment granule movements in unfertilized sea urchin eggs. J Cell Biol 83: 217 (Abstr)

Ruzdijic S, Glisin V (1972) Towards a total analysis of polyribosome associated ribonucleoprotein particles of sea urchin embryos. Biochim Biophys Acta 269: 441–449

Ruzdijic S, Milchev G, Bajkovic N, Glisin V (1973) Some properties of the 24S particle isolated from the cytoplasm of sea urchin eggs. Biochem Biophys Res Commun 53: 224–230

Ryberg E (1973) The localization of cholinesterases and non specific esterases in the echinopluteus. Zool Scr 2: 163–170

Ryberg E (1974) The localization of biogenic amines in the echinopluteus. Acta Zool 55: 178–189

Ryberg E (1977) The nervous system of the early echinopluteus Cell. Tissue Res 179: 157–167

Ryberg E, Lundgren B (1975) Secretory cells in the foregut of the echnipluteus. Wilhelm Roux' Arch Entwicklungsmech Org 177: 255–262

Ryberg E, Lundgren B (1979) Some aspects on pigment cell distribution and function in the developing echnipluteus of *Psammechinus miliaris*. Dev Growth Differ 21: 129–140

Sakai H (1968) Contractile properties of protein threads from sea urchin eggs in relation to cell division. Int Rev Cytol 23: 89–112

Saiga H, Kinoshita S (1976) Changes of chromatin structure induced by acid mucopolysaccharides. Exp Cell Res 102: 143–153

Sakharova A V, Manukhin B N, Markova L N (1972) The role of neurohumours in early
embryogenesis. IV. Fluorometric and histochemical study of serotonin in cleaving eggs
and larvae of sea urchins. J Embryol Expt Morphol 27: 339–351

Sale W S (1983) Low-angle rotatory shadow replication of 21S dynein from sea urchin sperm
flagella. J Submicrosc Cytol 15: 217–221

Sale W S, Gibbons I R (1979) Study of the mechanism of vanadate inhibition of dynein cross-
bridge cycle in sea archin sperm flagella. J Cell Biol 82: 291–298

Salik J, Herlands L, Hoffmann H P, Poccia D (1981) Electrophoretic analysis of the stored
histone pool in unfertilized sea urchin eggs: Quantification and identification by antibody
binding. J Cell Biol 90: 385–395

Salmon E D (1982) Calcium, spindle microtubule dynamics and chromosome movement.
Cell Differ 11: 353–355

Salmon E D, Segall R R (1980) Calcium-labile mitotic spindles isolated from sea urchin eggs
(*Lytechinus variegatus*). J Cell Biol 86: 355–365

Sano K (1977) Changes in cell surface charges during differentiation of isolated micromeres
and mesomeres from sea urchin embryos. Dev Biol 60: 404–415

Sano K, Kanatani H (1980) External calcium ions are requisite for fertilization of sea urchin
eggs by spermatozoa with reacted acrosomes. Dev Biol 78: 242–246

Sargent T D, Raff R A (1976) Protein synthesis and messenger RNA stability in activated,
enucleate sea urchin eggs are not affected by actinomycin D. Dev Biol 48: 327–335

Sasaki H, Aketa K (1981) Purification and distribution of a lectin in sea urchin (*Anthocidaris
crassispina*) egg before and after fertilization. *Exp Cell Res* 135: 15–19

Sasaki H, Epel D (1983) Cortical vesicle exocytosis in isolated cortices of sea urchin eggs:
description of a turbidometric assay and its utilization in studying effects of different
media on *discharge*. Dev Biol 98: 327–337

Satir P (1965) Studies on cilia II. Examination of the distal region of the ciliary shaft and the
role of the filaments in motility. J Cell Biol 26: 805–834

Sato H (1983) Role of spindle microtubules for the anaphase chromosome movements in
fertilized sea urchin eggs. Cell Differ 11: 345–348

Sato H, Owaribe K, Miki-nomura T (1973) Existence of surface fibers in perivitelline space.
Proc 44th Annu Meet Zool Soc J, Tokyo, 239 (Abstr)

Sato H, Kato T, Takahashi T C, Ito T (1982) Analysis of D$_2$O effects on in vivo and in vitro
tubulin polymerization and depolymerization. In: Sakai H, Mohri H, Borisy G G (eds)
Biological functions of microtubules and related structures. Academic Press, London
New York, 211–226

Savic A, Richman P, Williamson P, Poccia D (1981) Alterations in chromatin structure during
early sea urchin embryogenesis. Proc Natl Acad Sci USA 78: 3706–3710

Scarano E (1969) Enzymatic modifications of DNA and embryonic differentiation. Ann
Embryol Morphol Suppl 1: 7–15

Scarano E (1971) In: Clementi F, Ceccarelli B (eds) 1st Int Symp Cell Biol Cytopharmacol.
Raven Press, New York (Venice 1969) pp 13–24

Scarano E, Augusti-Tocco A (1967) Biochemical pathways in embryos. In: Florkin M, Stotz E H
(eds) Comprehensive biochemistry vol 28, Elsevier, Amsterdam New York, pp 55–111

Scarano E, Maggio R (1957) An exchange between ^{32}P-labeled pyrophosphate and ATP
catalyzed by aminoacids in unfertilized sea urchin eggs. Exp Cell Res 12: 403–405

Scarano E, Maggio R (1959) Attivazione enzimatica dell'acetato nelle uova di riccio di mare.
G Biochim 8: 98–102

Scarano E, De Petrocellis B, Augusti-Tocco G (1964) Deoxycytidylate aminohydrolase content
in disaggregated cells from sea urchin embryos. Exp Cell Res 36: 211–213

Scarano E, Iaccarino M, Grippo P, Winckelmans O (1965) On methylation of DNA during
development of sea urchin embryos. J Mol Biol 14: 603–607

Scarano E, Iaccarino M, Grippo P, Parisi E (1967) The heterogeneity of thymine methyl group
origin in DNA pyrimidine isostichs of developing sea urchin embryos. Proc Natl Acad
Sci USA 57: 1394–1400

Schackmann R W, Shapiro B M (1981) A partial sequence of ionic changes associated with the acrosome reaction of *Strongylocentrotus purpuratus.* Dev Biol 81: 145–154

Schackmann R W, Eddy E M, Shapiro B M (1977) Ion movements during the acrosome reaction of *Strongylocentrotus purpuratus.* J Cell Biol 75: 415 (Abstr)

Schackmann R W, Eddy E M, Shapiro B M (1978) The acrosome reaction of *Strongylocentrotus purpuratus* sperm. Dev Biol 65: 483–495

Schackmann R W, Christen R, Shapiro B (1981) Membrane potential depolarization and increased intracellular pH accompany the acrosome reaction of sea urchin sperm. Proc Natl Acad Sci USA 78: 6066–6070

Schaffner W, Gross K, Telford J, Birnstiel M (1976) Molecular analysis of the histone gene cluster of *Psammechinus miliaris*: II The arrangement of the five histone-coding and spacer sequences. Cell 8: 471–478

Schaffner W, Kunz G, Daetwyler H, Telford J, Smith H O, Birnstiel M (1978) Genes and spacers of cloned sea urchin histone DNA analyzed by sequencing. Cell 14: 655–671

Schatten G (1977) The late block to polyspermy in the sea urchin. J Cell Biol 75: 35 (Abstr)

Schatten G (1979) Pronuclear movement and fusion at fertilization. J Cell Biol 83: 198 (Abstr)

Schatten G (1981) Sperm incorporation, the pronuclear migrations, and their relation to the establishment of the first embryonic axis: Time-lapse video microscopy of the movements during fertilization of the sea urchin *Lytechinus variegatus.* Dev Biol 86: 426–437

Schatten G, Hemmer M (1979) Localization of sequestered calcium in unfertilized sea urchin eggs. J Cell Biol 83: 199 (Abstr)

Schatten G, Hülser D (1983) Timing the early events during sea urchin fertilization. Dev Biol 100: 244–248

Schatten G, Mazia D (1976) The penetration of the spermatozoa through the sea urchin egg surface at fertilization. Observations from the outside on whole eggs and from the inside on isolated surfaces. Exp Cell Res 98: 325–337

Schatten G, Schatten H (1981) Effects of motility inhibitors during sea urchin fertilization. Microfilament inhibitors prevent sperm incorporation and restructuring of fertilized egg cortex whereas microtubule inhibitors prevent the pronuclear migrations. Exp Cell Res 135: 311–330

Schatten G, Schatten H, Simerly C (1982) Detection of sequestered calcium during mitosis in mammalian cell cultures and in mitotic apparatus isolated from sea urchin zygotes Cell Biol Int Rep 6: 717–724

Schatten H, Schatten G (1979) Cytochalasin-sensitive microvilli engulf entire sperm in eggs denuded of vitelline layers. J Cell Biol 83: 199 (Abstr)

Schatten H, Schatten G (1980) Surface activity at the egg plasma membrane during sperm incorporation and its cytochalasin B sensitivity scanning electron microscopy and time-lapse video microscopy during fertilization of the sea urchin Lytechinus variegatus. Dev. Biol 78: 435–449

Scheller R H, Thomas T L, Lee A S, Klein W H, Niles W D, Britten R J, Davidson E H (1977) Clones of individual repetitive sequences from sea urchin DNA constructed with synthetic ECO RI sites. Science 196: 197–200

Scheller R H, Costantini F D, Kozlowski M R, Britten R J, Davidson E H (1978) Specific representation of cloned repetitive DNA sequences in sea urchin RNAs. Cell 15: 189–203

Scheller R H, Anderson D M, Posakony I W, McAllister L B, Britten R J, Davidson E H (1981) Repetitive sequences of the sea urchin genome. Subfamily structure and evolutionary conservation. J Mol Biol 149: 15–39

Schmell E, Lennarz W J (1974) Phospholipid metabolism in eggs and embryos of the sea urchin *Arbacia punctulata.* Biochemistry 13: 4114–4121

Schmell E, Earles B J, Breaux C, Lennarz W J (1977) Identification of a sperm receptor on the surface of the eggs of the sea urchin *Arbacia punctulata.* J Cell Biol 72: 35–46

Schmidt T, Epel D (1983) High hydrostatic pressure and the dissection of the fertilization responses. I. The relationship between cortical granule exocytosis and proton efflux during fertilization of the sea urchin egg. Exp Cell Res 146: 235–248

Schmidt T, Patton C, Epel D (1982) Is there a role for the Ca^{2+} influx during fertilization of the sea urchin egg? Dev Biol 90: 284–290

Schneider R G, Lennarz W J (1976) Glycosyl transferases of eggs and embryos of *Arbacia punctulata*. Dev Biol 53: 10–20

Scholander P F, Leivestad H, Sundnes G (1958) Cycling in the oxygen consumption of cleaving eggs. Exp Cell Res 15: 505–511

Schreuer M, Czihak G (1978) Effect of 5-bromodeoxyuridine on differentiation. I Probability distribution of BUdR containing DNA-strands in subsequent divisions. Differentiation 11: 89–101

Schroeder T E (1972) The contractile ring. II Determining its brief existence, volumetric changes, and vital role in cleaving *Arbacia* eggs. J Cell Biol 53: 419–434

Schroeder T E (1973) Actin in dividing cells: contractile ring filaments bind heavy meromyosin. Proc Natl Acad Sci USA 70: 1688–1692

Schroeder T E (1978) Microvilli on sea urchin eggs: A second burst of elongation. Dev Biol 64: 342–346

Schroeder T E (1979) Surface area change at fertilization: Resorption of the mosaic membrane. Dev Biol 70: 306–326

Schroeder T E (1980a) Espressions of the prefertilization polar axis in sea urchin egg. Dev Biol 79: 428–443

Schroeder T E (1980b) The jelly canal marker of polarity for sea urchin oocytes, eggs and embryos. Exp Cell Res 128: 490–494

Schroeder T E (1981a) Development of a "primitive" sea urchin (*Eucidaris tribuloides*): Irregularities in the hyaline layer, micromeres, and primary mesenchyme. Biol Bull 161: 141–151

Schroeder T E (1981b) The origin of cleavage forces in dividing eggs. A mechanism in two steps. Exp Cell Res 134: 231–240

Schroeder T E (1982) Distinctive features of the cortex and cell surface of micromeres: Observations and cautions. Cell Differ 11: 289–290

Schuel H, Kesner H (1977) Mechanism of fertilization envelope elevation in sea urchin eggs. J Cell Biol 75: 166 (Abstr)

Schuel H, Schuel R (1981) A rapid sodium-dependent block to polyspermy in sea urchin eggs. Dev Biol 87: 249–258

Schuel H, Lorand L, Chen K, Wilson W L (1972a) A trypsin-like enzyme in cortical granules of sea urchin eggs, and its role in fertilization. Biol Bull 143: 476 (Abstr)

Schuel H, Wilson W L Bressler R S, Kelly J W, Wilson J R (1972b) Purification of cortical granules from unfertilized sea urchin egg homogenates by zonal centrifugation. Dev Biol 29: 307–320

Schuel H, Wilson W D, Chen K, Lorand L (1973) A trypsin-like proteinase localized in cortical granules isolated from unfertilized sea urchin eggs by zonal centrifugation. Role of the enzyme in fertilization. Dev Biol 34: 175–186

Schuel H, Kelly J W, Berger B R, Wilson W L (1974) Sulfated acid mucopolysaccharides in the cortical granules of eggs. Effects of quaternary ammonium salts on fertilization. Exp Cell Res 88: 24–30

Schuel H, Cardasis C, Herman L, Wilson W L (1975a) Ultrastructural localization of calcium in the early stages of fertilization in *Arbacia* eggs. Biol Bull 149: 446 (Abstr)

Schuel H, Wilson W L, Wilson J R, Bressler R S (1975b) Heterogenous distribution of "lysosomal" hydrolases in yolk platelets isolated from unfertilized sea urchin eggs by zonal centrifugation. Dev Biol 46: 404–412

Schuel H, Longo F J, Wilson W L, Troll W (1976a) Polyspermic fertilization of sea urchin eggs treated with protease inhibitors: Localization of sperm receptor sites at the egg surface. Dev Biol 49: 178–184

Schuel H, Troll W, Lorand L (1976b) Physiological responses of sea urchin eggs to stimulation by calcium ionophore A 23187 analyzed with protease inhibitors. Exp Cell Res 103: 442–447

Schuler M A, Keller E B (1981) The chromosomal arrangement of two linked actin genes in the sea urchin *Strongylocentrotus purpuratus*. Nucleic Acid Res 9: 591–604

Schuler M A, Osker P M, Keller E B (1983) DNA sequence of two linked actin genes of sea urchins. Mol Cell Biol 3: 448–456

Sconzo G, Giudice G (1971) Synthesis of ribosomal RNA in sea urchin embryos. V. Further evidence for an activation following the hatching blastula stage. Biochim Biophys Acta 254: 447–451

Sconzo G, Giudice G (1976) A study on the effect of the inhibition of polyadenylylation on the production of giant cytoplasmic RNA in sea urchin embryos. Wilhelm Roux' Arch Dev Biol 179: 163–169

Sconzo G, Pirrone A M, Mutolo V, Giudice G (1970a) Synthesis of ribosomal RNA during sea urchin development. III Evidence for an activation of transcription. Biochim Biophys Acta 199: 435–440

Sconzo G, Pirrone A M, Mutolo V, Giudice G (1970b) Synthesis of ribosomal RNA in disaggregated cells of sea urchin embryos. Biochim Biophys Acta 199: 441–446

Sconzo G, Vitrano E, Bono A, Di Giovanni L, Mutolo V, Giudice G (1971) Synthesis of rRNA in sea urchin embryos. IV. Maturation of rRNA precursor. Biochim Biophys Acta 232: 132–139

Sconzo G, Bono A, Albanese I, Giudice G (1972) Studies on sea urchin oocytes: II Synthesis of RNA during oogenesis. Exp Cell Res 72: 95–100

Sconzo G, Albanese I, Rinaldi A M, Lo Presti G F, Giudice G (1974) Cytoplasmic giant RNA in sea urchin embryos. II. Physicochemical characterization. Cell Differ 3: 297–304

Sconzo G, Roccheri M C, Di Liberto M, Giudice G (1977) Studies on the structure and possible function of the RNA "cap" in developing sea urchins. Cell Differ 5: 323–332

Sconzo G, Mutolo A M, Giallongo A, Di Bernardo M G, Giudice G (1978) DNA amplification in sea urchin oocytes. Wilhelm Roux' Arch Dev Biol 184: 351–353

Sconzo G, Roccheri M C, Di Carlo M, Di Bernardo M G, Giudice G (1983) Synthesis of heat shock proteins in dissociated embryonic sea urchin cells. Cell Differ 12: 317–320

Seale R L, Aronson A I (1973a) Chromatin-associated proteins of the developing sea urchin embryo. I Kinetics of synthesis and characterization of non-histone proteins. J Mol Biol 75: 633–646

Seale R L, Aronson A I (1973b) Chromatin-associated proteins of the developing sea urchin embryo. II Acid-soluble proteins. J Mol Biol 75: 617–658

SeGall G K, Lennarz W J (1978) The role of jelly coat in the induction of the acrosomal reaction in sperm. J Cell Biol 79: 170 (Abstr)

SeGall G K, Lennarz W J (1979) Chemical characterization of the component of the jelly coat from sea urchin eggs responsible for induction of the acrosome reaction. Dev Biol 71: 33–48

SeGall G K, Lennarz W (1981) Jelly coat induction of the acrosome reaction in echinoid sperm. Dev Biol 86: 87–93

Selvig S E, Gross P R, Hunter A D (1970) Cytoplasmic synthesis of RNA in the sea urchin embryo. Dev Biol 23: 343–365

Selvig S E, Greenhouse G A, Gross P R (1972) Cytoplasmic synthesis of RNA in the sea urchin embryo. II. Mitochondrial transcription. Cell Differ 1: 5–14

Senger D R, Gross P R (1978) Macromolecule synthesis and determination in sea urchin blastomeres at the sixteen-cell stage. Dev Biol 65: 404–415

Senger D R, Arceci R J, Gross P R (1978) Histones of sea urchin embryos. Transients in transcription, translation, and the composition of chromatin. Dev Biol 65: 416–425

Sevaljevic L (1974a) Developmental changes of sea urchin histones. Roux' Arch Dev Biol 174: 210–214

Sevaljevic L (1974b) Developmental changes of chromatin nonhistone proteins of sea urchin embryos. Wilhelm Roux'Arch Dev Biol 174: 215–221

Sevaljevic L, Koviljka K (1975) Buoyant density centrifugation of sea urchin embryo chromatin on sucrose-glucose gradient. Mol Biol Rep 2: 27–34

Shapiro B M (1975) Limited proteolysis of egg surface components is an early event following fertilization of the sea urchin, *Strongylocentrotus purpuratus*. Dev Biol 46: 88–102

Shapiro B M, Schackmann R W, Gabel C A (1981) Molecular approaches to the study of fertilization. Annu Rev Biochem 50: 815–844

Shapiro B, Schackmann R W, Christen R, Deits T, Kay E, Wiedman M (1982) Triggering of the acrosome reaction and assembly of the fertilization membrane in the sea urchin. Cell Differ 11: 263–265

Shaw B R, Cognetti G, Sholes W M, Richards R G (1981) Shift in nucleosome populations during embryogenesis: Microheterogeneity in nucleosomes during development of the sea urchin embryo. Biochemistry 20: 4971–4978

Shearer C, De Morgan W, Fuchs H M (1913) On the experimental hybridization of echinoids. Philosophical Trans B 204 255–362

Shen S S, Steinhardt R A (1978) Direct measurement of intracellular pH during metabolic derepression in the sea urchin egg. Nature (London) 272: 253–254

Shen S S, Steinhardt R A (1979) Intracellular pH and the sodium requirement at fertilization. Nature (London) 282: 87–89

Shen S S, Steinhardt R A (1980) Intracellular pH controls the development of new potassium conductance after fertilization of sea urchin egg. Exp Cell Res 125: 55–61

Shepherd G W, Nemer M (1980) Developmental shifts in frequency distribution of polysomal RNA and their post transcriptional regulation in the sea urchin embryo. Proc Natl Acad Sci USA 77: 4653–4656

Shepherd G W, Rondinelli E, Nemer M (1983) Differences in abundance of individual RNAs in normal and animalized sea urchin embryos. Dev Biol 96: 520–528

Shimada H (1983) Cellular factor stimulating DNA synthesis in nuclei isolated from sea urchin embryos. Dev Biol 97: 454–459

Shimada H, Terayama H (1976) Discontinuous DNA replication in developing sea urchin embryos. Dev Biol 54: 151–156

Shimada H, Terayama H, Fujiwara A, Yasumasu I (1982) Melittin, a component of bee venom, activates unfertilized sea urchin eggs. Dev Growth Differ 24: 7–16

Shimada H, Haraguchi T, Nagano H, Fujiwarawa A, Yasumasu I (1983) Inhibition of DNA polymerase of sea urchin by palmitoyl coenzyme A. Biochem Biophys Res Commun 110: 902–907

Shioda M, Nagano H (1983) Localization of DNA polymerase α on the nuclear membrane in sea urchin embryos. Exp Cell Res 146: 349–360

Shioda M, Nagano H, Mano Y (1977) Cytoplasmic localization of DNA polymerase-α and -β of sea urchin eggs. Biochem Biophys Res Commun 78: 1362–1368

Shioda M, Nagano H, Mano Y (1980) Association of DNA polymerase α and β with rough endoplasmic reticulum in sea urchin eggs and changes in subcellular distribution during early embryogenesis. Eur J Biochem 108: 345–355

Shioda M, Nagano H, Mano Y (1982) Transition of DNA polymerase-α and endoplasmic reticulum during gastrulation of the sea urchin. Dev Biol 91: 111–120

Shoger R L, Asami K, Yasumasu I, Fujiwara A (1973) Activation of phophorylase in sea urchin eggs by Ca^{2+} and cyclic 3'-5'-AMP. A possible mechanism of the regulation of its activity at fertilization. Exp Cell Res 62: 375–382

Shott R J, Lee J J, Britten R J, Davidson E H (1984) Differential expression of the actin gene family of *Strongylocentrotus purpuratus*. Dev Biol 101: 295–306

Showman R M, Foerder C A (1979) Removal of the fertilization membrane of sea urchin embryos employing aminotriazole. Exp Cell Res 120: 253–255

Showman R M, Wells D E, Anstrom J, Hursh D H, Raff R A (1982) Message-specific sequestration of maternal histone mRNA in the sea urchin egg. Proc Natl Acad Sci USA 79: 5944–5947

Showman R M, Wells D E, Anstrom J A, Hursh D A, Leaf D S, Raff R A (1983) Subcellular localization of maternal histone mRNAs and the control of histone synthesis in the sea urchin embryo. In: Malacinski G M, Klein W H (eds) Molecular aspects of early development. Plenum, New York, pp 109–130

Shutt R H, Kedes L H (1974) Synthesis of histone mRNA sequences in isolated nuclei of cleavage stage sea urchin embryos. Cell 3: 283–290

Simmel E B, Karnofsky D A (1961) Observations on the uptake of tritiated thymidine in the pronuclei of fertilized sand dollar embryos. J Biophys Biochem Cytol 10: 59–65

Simpson R T (1981) Modulation of nucleosome structure by histone subtypes in sea urchin embryos. Proc Natl Acad Sci USA 78: 6803–6807

Simpson R T, Safford D W (1983) Structural features of a phased nucleosome core particle. Proc Natl Acad Sci USA 80: 51–55

Sittman D B (1978) Ph D thesis, Univ North Carolina

Siu C H, Sedensky M, Crippa M (1973) RNA-directed DNA polymerase activity in sea urchin development. Biol Bull 143: 478 (Abstr)

Skoultchi A, Gross P R (1973) Maternal histone messenger RNA: Detection by molecular hybridization. Proc Natl Acad Sci USA 70: 2840–2844

Slater D W, Spiegelman S (1966) An estimation of genetic messages in the unfertilized echinoid egg. Proc Natl Acad Sci USA 56: 164–170

Slater D W, Slater I, Gillespie D (1972) Post-fertilization synthesis of polyadenylic acid in sea urchin embryos. Nature (London) 240: 333–337

Slater I, Slater D W (1972) DNA polymerase potentials of sea urchin embryos. Nature (London) New Biol 237: 81–85

Slater I, Slater D W (1974) Polyadenylylation and transcription following fertilization. Proc Natl Acad Sci USA 71: 1103–1107

Slater D W, Spiegelman S (1968) Template capabilities and size distribution of echinoid RNA during early development. Biochim Biophys Acta 166: 82–93

Slater I, Gillespie D, Slater D W (1973) Cytoplasmic adenylylation and processing of maternal RNA. Proc Natl Acad Sci USA 70: 406–411

Smith M J, Hough B R, Chamberlin M E, Davidson E M (1974) Repetitive and non-repetitive sequence in sea urchin heterogeneous nuclear RNA. J Mol Biol 85: 103–126

Smith J, Serunian L, Phillips W, Murray A, Horowitch S, Rubin G (1979) Repeated genomic sequences cloned from the sea urchin *Lytechinus pictus*. Biol Bull 157: 395 (Abstr)

Sobieski D A, Eden F C (1981) Clustering and methylation of repeated DNA: persistance in avian development and evolution. Nucleic Acid Res 22: 6001–6015

Solursh M, Katow H (1982) Initial characterization of sulfated macromolecules in the blastocoels of mesenchyme blastulae of *Strongylocentrotus purpuratus* and *Lytechinus pictus*. Dev Biol 94: 326–336

Spadafora C, Geraci G (1975) The subunit structure of sea urchin sperm chromatin: A kinetic approach. FEBS Lett 57: 79–82

Spadafora C, Geraci G (1976) A site of discontinuity in the interaction between DNA and histones in nucleosomes of sea urchin embryo chromatin. Biochem Biophys Res Commun 69: 291–295

Spadafora C, Igo-Kemenes T, Zachau H G (1973) Changes in tRNAs and aminoacyl tRNA synthetases during sea urchin development. Biochim Biophys Acta 312: 674–684

Spadafora C, Noviello L, Geraci G (1976a) Chromatin organization in nuclei of sea urchin embryos. Comparison with the chromatin organization of the sperm. Cell Differ 5: 225–232

Spadafora C, Bellard M, Compton J, Lee, Chambon P (1976) The DNA repeat lengths in chromatin from sea urchin sperm and gastrula cells are markedly different. FEBS Lett 69: 281–285

Spiegel E, Howard L (1983) Development of cell junctions in sea urchin embryos. J Cell Sci 62: 27–48

Spiegel E, Spiegel M (1977a) A scanning electron microscope study of early sea urchin reaggregation. Exp Cell Res 108: 412–420

Spiegel E, Spiegel M (1977b) Microvilli in sea urchin eggs. Difference in their formation and type. Exp Cell Res 109: 462–465

Spiegel E, Spiegel M (1979) The hyaline layer is a collagen-containing extracellular matrix in sea urchin embryos and reaggregating cells. Exp Cell Res 123: 434–441

Spiegel E, Burger M, Spiegel M (1980) Fibronectin in the developing sea urchin embryo. J Cell Biol 87: 309–313

Spiegel E, Burger M M, Spiegel M (1983) Fibronectin and laminin in the extracellular matrix and basement membrane of sea urchin embryos. Exp Cell Res 144: 47–55

Spiegel M, Burger M M (1982) Cell adhesion during gastrulation. A new approach. Exp Cell Res 139: 377–382

Spiegel M, Rubinstein N A (1972) Synthesis of RNA by dissociated cells of the sea urchin embryo. Exp Cell Res 70: 423–430

Spiegel M, Spiegel E (1975) The reaggregation of dissociated embryonic sea urchin cells. Am Zool 15: 583–606

Spiegel M, Spiegel E (1978a) The morphology and specificity of cell adhesion of echinoderms embryonic cells. Exp Cell Res 117: 261–268

Spiegel M, Spiegel E (1978b) Sorting out of sea urchin embryonic cells according to cell type. Exp Cell Res 117: 269–271

Spiegel M, Tyler A (1966) Protein synthesis in micromeres of the sea urchin egg. Science 151: 1233–1234

Spiegel M, Spiegel E S, Meltzer P S (1970) Qualitative changes in the basic protein fraction of developing embryos. Dev Biol 21: 73–86

Spieth J, Whiteley A H (1980) Effect of 3'-deoxyadenosine (Cordycepin) on early development of the sand dollar, Dendraster excentricus. Dev Biol 79: 95–106

Spinelli G, Gianguzza F, Casano C, Acierno P, Burckhardt J (1979) Evidences of two different set of histone genes active during embryogenesis of the sea urchin Paracentrotus lividus. Nucleic Acid Res 6: 545–560

Spinelli G, Melli M, Arnold E, Casano C, Gianguzza F, Ciaccio M (1980) High molecular weight RNA containing histone messenger in the sea urchin Paracentrotus lividus. J Mol Biol 139: 111–122

Spinelli G, Casano C, Gianguzza F, Ciaccio M, Palla F (1982a) Transcription of sea urchin mesenchyme blastula histone gene after heat shock. Eur J Biochem 128: 509–513

Spinelli G, Albanese I, Anello L, Ciaccio M, Di Liegro I (1982b) Chromatin structure of histone genes in sea urchin sperms and embryos. Nucleic Acid Res 10: 7977–7992

Spirin A S (1966) On "masked" forms of messenger RNA in early embryogenesis and in other differentiating systems. Curr Top Dev Biol 1: 2–38

Spirin A S (1979) Messenger ribonucleoproteins (informosomcs) and RNA-binding proteins. Mol Biol Rep 5: 55–58

Spirin A S, Nemer M (1965) Messenger RNA in early sea urchin embryos: Cytoplasmic particles. Science 150: 214–217

Spirin A S, Belitsina N V, Ajtkhozin M A (1964) Synthesis of messenger RNA (mRNA) and relation of mRNA to protein synthesizing structures of the cytoplasm were studied at different stages of early development of loach (Misgurnus fossilis L.) Zh Obshch Biol 25: 321–335

Spudich A, Greenberg Gifford R, Speedick J A (1982) Molecular aspects of cortical actin filament formation upon fertilization. Cell Differ 11: 281–284

Spudich J A, Amos L A (1979) Structure of actin filament bundles from microvilli of sea urchin eggs. J Mol Biol 129: 319–331

Spudich A, Spudich J A (1979) Actin in Triton-treated cortical preparation of unfertilized and fertilized sea urchin eggs. J Cell Biol 82: 212–226

Stafford D W, Guild W R (1969) Satellite DNA from sea urchin sperm. Exp Cell Res 55: 347–350

Stafford D W, Sofer W H, Iverson R M (1964) Demonstration of polyribosomes after fertilization of the sea urchin egg. Proc Natl Acad Sci USA 53: 313–316

Stavy L, Gross P R (1967) The protein synthetic lesion in unfertilized eggs. Proc Natl Acad Sci USA 57: 735–742

Stavy L, Gross P R (1969) Protein synthesis in vitro with fractions of sea urchin eggs and embryos. Biochim Biophys Acta 182: 193–202

Stearns L W (1974) Sea urchin development: Cellular and molecular aspects. Dowden, Hutchinson and Ross, Pennsylvania

Steinbruck H (1902) Über Bastardbildung zwischen Strongylocentrotus und Sphaerechinus. Arch f Entw-Mech 14: 1–48

Steinhardt R A, Epel D (1974) Activation of sea urchin eggs by a calcium ionophore. Proc Natl Acad Sci USA 71: 1915–1919

Steinhardt R A, Mazina D (1973) Development of K^+-conductance and membrane potentials in unfertilized sea urchin eggs after exposure to NH_4OH. Nature (London) 241: 400–401

Steinhardt R A, Lundin L, Mazia D (1971) Bioelectric responses of the echinoderm egg to fertilization. Proc Natl Acad Sci USA 68: 2426–2430

Steinhardt R A, Shen S, Mazia D (1972) Membrane potential, membrane resistance, and an energy requirement for the development of potassium conductance in the fertilization reaction of echinoderm eggs. Exp Cell Res 72: 195–203

Steinhardt R A, Epel D, Garroll E J Jr, Yanagimachi R (1974) Is calcium ionophore a universal activator for unfertilized eggs? Nature (London) 252: 41–43

Steinhardt R A, Zucker R, Schatten G (1977) Intracellular calcium release at fertilization in the sea urchin egg. Dev Biol 58: 185–196

Stephens R E (1970a) Thermal fractionation of outer fiber double microtubules into A and B-subfiber components: A- and B-tubulin. J Mol Biol 47: 353–363

Stephens R E (1970b) Isolation of nexin-the linkage protein responsible for mantainance of the nine-fold configuration of flagellar axonemes. Biol Bull 139: 438 (Abstr)

Stephens R E (1972a) Studies on the development of the sea urchin *Strongylocentrotus droebachensis*. I. Ecology and normal development. Biol Bull 142: 132–144

Stephens R E (1972b) Studies on the development of *Strongylocentrotus droebachiensis*. III Embryonic synthesis of ciliary proteins. Biol Bull 142: 489–504

Stephens R E (1977) Differential protein synthesis and utilization during cilia formation in sea urchin embryos. Dev Biol 61: 311–329

Stephens R E (1978) Primary structural differencs among tubulin subunits from flagella, cilia, and the cytoplasm. Biochemistry 17: 2882–2891

Strathmann R D (1975) Larval feeding in echinoderms. Am Zool 15: 717–730

Strickland W N, Strickland M, De Groot P C, Von Holt C, Wittmann-Liebold B (1980a) The primary structure of histone H1 from sperm of the sea urchin *Parechinus angulosus*. 1. Chemical and enzymatic fragmentation of the protein and sequence of aminoacids in the four N-terminal cyanogen bromide peptides. Eur J Biochem 104: 559–566

Strickland W N, Strickland M, Brandt W F, Von Holt C, Lehmann A, Wittmann-Liebold B (1980b) The primary structure of histone H1 from sperm of the sea urchin *Parechinus angulosus* 2. Sequence of the C-terminal CNBr peptide and the entire primary structure. Eur J Biochem 104: 567–578

Strickland W N, Strickland M, De Groot P C, Von Holt C (1980c) The primary structure of histone H2A from the sperm cell of the sea urchin *Parechinus angulosus*. Eur J Biochem 109: 151–158

Stunnenberg H G, Birnstiel M L (1982) Bioassay for components regulating eukaryotic gene expression: A chromosomal factor involved in the generation of histone mRNA 3′ termini. Proc Natl Acad Sci USA 79: 6201–6204

Subirana J A (1970) Nuclear proteins from a somatic and a germinal tissue of the echinoderm *Holothuria tubulosa*. Exp Cell Res 63: 253–260

Subirana J A, Palau J, Cozcolluela C, Ruiz-Carrillo A (1970) Very lysine rich histones of echinoderms and molluscs. Nature (London) 228: 992–993

Summers R G, Hylander B L (1974) An ultrastructural analysis of early fertilization in the sand dollar, *Echinaracnius parma*. Cell Tissue Res 150: 343–368

Summers R G, Hylander B L (1976) Primary gamete binding. Quantitative determination of its specificity in echinoid fertilization. Exp Cell Res 100: 190–194

Summers R G, Colwin L H, Colwin A L, Turner R (1971) Fine structure of the acrosomal regio in spermatozoa of two echinoderms, *Ctenodiscus* (starfish) and *Thyone* (holoturian). Biol Bull 141: 404 (Abstr)

Summers R G, Hylander B L, Colwin L H, Colwin A L (1975) The functional anatomy of the echinoderm spermatozoon and its interaction with the egg at fertilization. Am Zool 15: 523–551

Suprenant K A, Rebhun L I (1983) Assembly of unfertilized sea urchin egg tubulin at physiological temperatures. J Biol Chem 258: 4518–4525

Sures I, Maxam A, Cohn R H, Kedes L H (1976) Identification and location of the histone H2A and H3 genes by sequence analysis of sea urchin (*S. purpuratus*) DNA cloned in *E. coli*. Cell 9: 495–502

Sures I, Lowry J, Kedes L H (1978) The DNA sequence of sea urchin (*S. purpuratus*) H2A, H2B and H3 histone coding and spacer regions. Cell 15: 1033–1044

Sures I, Levy S, Kedes L H (1980) Leader sequences of *Strongylocentrotus purpuratus* histone mRNAs start a unique heptanucleotide common to all five histone genes. Proc Natl Acad Sci USA 77: 1265–1269

Surrey S, Nemer M (1976) Methylated blocked 5′ terminal sequences of sea urchin embryos messenger RNA classes containing and lacking poly (A). Cell 9: 589–596

Surrey S, Ginzburg I, Nemer M (1979) Ribosomal RNA synthesis in pre- and post-gastrula-stage sea urchin embryos. Dev Biol 71: 83–99

Suzuki N, Nomura K, Ohtake H, Isaka S (1981) Purification and the primary structure of sperm-activating peptides from the jelly coat of sea urchin eggs. Biochem Biophys Res Commun 99: 1238–1244

Suzuki-Hori C, Nagano H, Mano Y (1977) DNA polymerase-B from the nuclear fraction of sea urchin embryos: Characterization of the purified enzyme. J Biochem (Tokyo) 82: 1613–1621

Swan E (1953) The Strongylocentrotudae evolution 7: 269–273

Swann M M, Mitchison J M (1958) The mechanism of cleavage in animal cells. Biol Rev 33: 103–135

Sy J, McCarty K S (1970) Characterization of 5.8S RNA from a complex with 26–S ribosomal RNA from *Arbacia punctulata*. Biochim Biophys Acta 199: 86–94

Sy J, McCarty K S (1971) Formation in vitro of a 5.8-S–26 S sea urchin rRNA complex. Biochim Biophys Acta 228: 517–525

Taglietti V (1979) Early electrical responses to fertilization in sea urchin eggs. Exp Cell Res 120: 448–451

Takahashi K, Kamimura S (1983) Dynamic aspects of microtubule sliding in sperm flagella. J Submicrosc Cytol 15: 1–3

Takahashi T C, Sato H (1982) Thermodynamic analysis of the effect of (D_2O) on mitotic spindles in developing sea urchin eggs. Cell Struct Funct 7: 349–357

Takahashi Y M, Sugiyama M (1973) Relation between the acrosome reaction and fertilization in the sea urchin I. Fertilization in Ca-free sea water with egg-water-treated spermatozoa. Dev Growth Differ 15: 261–267

Takashima Y (1960) Studies on the ultrastructure of the cortical granules in sea urchin eggs. Tokushima J Exp Med 6: 341–349

Takeshima K, Nakano E (1982a) Ribosomal proteins of sea urchin eggs. I. Characterization of cytoplasmic ribosomal proteins. Cell Differ 11: 223–234

Takeshima K, Nakano E (1982b) Ribosomal proteins of sea urchin eggs. II. Taxonomical differences. Mol Genet 186: 566–568

Takeshima K, Nakano E (1983) Modification of ribosomal proteins in sea urchin eggs following fertilization. Eur J Biochem 137: 437–443

Tamini E (1943) Ricerche sulla vegetativizzazione nello sviluppo dei ricci di mare. Rendiconti 76: 363

Taylor B A, Burdick C J (1975) Histone acetylation during the early stages of sea urchin (*Arbacia punctulata*) development. Biol Bull 149: 447–448 (Abstr)

Tegner M J (1972) Sea urchin sperm egg interactions studied with the scanning electron microscopc. J Cell Biol 55: 258 (Abstr)

Tegner M J (1974) Fertilization in sea urchin eggs without vitelline layers: A scanning electron microscope and experimental study. J Cell Biol 63: 344 (Abstr)

Tegner M J, Epel D (1973) Sea urchin sperm-egg interaction studied with the electron microscope. Science 179: 685–688

Tencer R, Brachet J (1973) Studies on the effects of bromodeoxyuridine (BUdR) on differentiation. Differentiation 1: 51–64

Tennent D H (1910) Echinoderm hybridization Pubbl Carnegie Inst Wash 132: 117–151

Tennent D H (1912a) The correlation between chromosomes and particular characters in hybrid echinoid larvae. Amer Nat: 46 68–75

Tennent D H (1912b) The behavior of the chromosomes in cross-fertilized Echinoid eggs. J Morph 23 17–29

Tennent D H (1922) Studies of the hybridization of echinoids, Cidaris tribuloides Pubbl Carnegie Ins-Wash 312 1–42

Tennent D H (1929) Activation of the eggs of Echinometra mathaei by sperms of the Crinoids Comatula pectinata and Comatula purpurea. Pubbl Carnegie Inst Wash 391: 105–114

Tenner A J, Humphreys T (1973) Activation of ribosomal RNA accumulation in sea urchin plutei by insulin. Biol Bull 145: 457 (Abstr)

Terman S A (1970) Relative effect of transcription-level and translation-level control of protein synthesis during early development of the sea urchin. Proc Natl Acad Sci USA 65: 985–992

Terman S A, Gross P R (1965) Translation-level control of protein synthesis during early development. Biochem Biophys Res Commun 21: 595–600

Thaler M M, Cox M C L, Villee C A (1970) Histones in early embryogenesis. Developmental aspects of composition and synthesis. J Biol Chem 345: 1479–1483

Thomas T L, Britten R J, Davidson E H (1982) An interspersed region of the sea urchin genome represented in both maternal poly (A) RNA and embryo nuclear RNA. Dev Biol 94: 230–239

Tilney L G (1976a) The polymerization of actin II. How nonfilamentous actin becomes nonrandomsly distributed in sperm: evidence for the association of this actin with membranes. J Cell Biol 69: 51–72

Tilney L G (1976b) The polymerization of actin III. Aggregates of nonfilamentous actin and its associated proteins: a storage form of actin. J Cell Biol 69: 73–89

Tilney L G (1977) How H^+, and Ca^{++}, and nucleation from a newly described organelle control actin polymerization during the acrosomal reaction. J Cell Biol 75: 249 (Abstr)

Tilney L G (1978) Polymerization of actin. V. A new organelle, the actomere, that initiates the assembly of actin filaments in Thyone sperms. J Cell Biol 77: 551–564

Tilney L G (1979) Actin, motility, and membranes. In: Cone R A, Dawling J E (eds) Membrane traduction mechanisms. Raven Press, New York, pp 163–186

Tilney L G, Inoué S (1982) Acrosomal reaction of Thyone sperm. II. The kinectics and possible mechanism of acrosomal process elongation. J Cell Biol 93: 820–827

Tilney L G, Jaffe L A (1980) Actin, microvilli and fertilization cone of sea urchin eggs. J Cell Biol 87: 771–782

Tilney L G, Kallenbach N (1979) Polymerization of actin VI. The polarity of the actin filaments in the acrosomal process and how it might be determined. J Cell Biol 81: 608–623

Tilney L G, Hatano S, Ishikawa H, Mooseker M S (1973) The polymerization of actin its role in the generation of the acrosomal process of certain Echinoderm sperm. J Cell Biol 59: 109–126

Tilney L G, Bryan J, Bush D J, Fujiwara K, Mooseker M S, Murphy D B, Snyder D H (1973) Microtubules: Evidence for 13 protofilaments. J Cell Biol 59: 267–275

Tilney L G, Hiehart D P, Sardet C, Tilney M (1978) Polymerization of actin. IV. Role of Ca^{++} and H^+ in the assembly of actin and in membrane fusion in the acrosomal reaction of echinoderm sperm. J Cell Biol 77: 536–550

Timourian H (1966) Protein synthesis during first cleavage of sea urchin embryos. Science 154: 1055

Timourian H, Watchmaker G (1970) Protein synthesis in sea urchin eggs. II. Changes in amino acid uptake and incorporation at fertilization. Dev Biol 23: 487–491

Timourian H, Watchmaker G (1971) Bipolar and tetrapolar cleavage time in sea urchin eggs. Exp Cell Res 68: 428–430

Timourian H, Watchmaker G (1975) The sea urchin blastula: extent of cellular determination. Am Zool 15: 607–627

Timourian H, Clothier G, Watchmaker G (1973) Reaggregation of sea urchin blastula cells. I Intrinsic differences in the blastula cells. Dev Biol 31: 252–263

Tonegawa Y (1973) Isolation and characterization of a particulate cell-aggregation factor from sea urchin embryos. Dev Growth Differ 14: 337–352

Tonegawa Y (1982) Cell aggregation factor and endogenous lectin in sea urchin embryos. Cell Differ 11: 335–337

Tonelli A, Hunt T (1977) Studies on messenger RNA template activity during early development of Lytechinus pictus. Biol Bull 153: 447–448 (Abstr)

Tosi L, Scarano E (1973) Effect of trypsin on DNA methylation in isolated nuclei from developing sea urchin embryos. Biochem Biophys Res Commun 55: 470–476

Tosi L, Granieri A, Scarano E (1972) Enzymatic DNA modifications in isolated nuclei from developing sea urchin embryos Exp Cell Res 72: 257–264

Treigyte G, Ginetis A (1979) Specific changes in the biosynthesis and acetylation of nucleosomal histones in the early stages of embryogenesis of sea urchin. Exp Cell Res 121: 127–134

Troll W, Schuel H, Wilson W L (1974) Induction of polyspermic fertilization of Arbacia eggs by specific protease inhibitors leupeptin and antipapin. Biol Bull 147: 502 (Abstr)

Troll W, Schuel H, Lorand L (1975) Calcium ionophore A23187 induced block to sperm penetration of *Arbacia* eggs is prevented by protease inhibitors. Biol Bull 149: 449 (Abstr)

Tsai Y H, Ansevin A T, Hnilica L S (1975) Association of tissue-specific histones with DNA. Thermal denaturation of native, partially dehistonized, and reconstituted chromatins. Biochemistry 14: 1257–1265

Tsukahara J, Sugiyama M (1969) Ultrastructural changes in the surface of the oocyte during oogenesis of the sea urchin *Hemicentrotus pulcherrimus*. *Embryologia* 10: 343–355

Tsuzuki H, Yoshida M, Onitake K, Aketa K (1977) Purification of the sperm-binding factor from the egg of the sea urchin, *Hemicentrotus pulcherrimus*. Biochem Biophys Res Commun 76: 502–511

Tufaro F, Brandhorst B P (1979) Similarity of proteins synthesized by isolated blastomeres of early sea urchin embryos. Dev Biol 72: 390–397

Tufaro F, Brandhorst B P (1982) Restricted expression of paternal genes in sea urchin interspecific hybrids. Dev Biol 92: 209–220

Tupper J T (1973) Potassium exchangeability potassium permeability and membrane potential: Some observations in relation to protein synthesis in the early echinoderm embryo. Dev Biol 32: 140–154

Tupper J T (1974) Inhibition of increased potassium permeability following fertilization of the echinoderm embryo: Its relationship to the initiation of protein synthesis and potassium exchangeability. Dev Biol 38: 332–345

Turner R S Jr (1980) Evidence for two types of cell adhesions in cleavage stage sea urchin embryos. J Cell Biol 87: 133 (Abstr)

Turner R S Jr, Watanabe M, Schnitman J R (1977) Cell adhesive changes during sea urchin development. J Cell Biol 75: 67 (Abstr)

Tyler A (1949) Properties of fertilizin and related substances of eggs and sperm of marine animals. Am Nat 83: 195–219

Tyler A (1956) Physico-chemical properties of the fertilizins of the sea urchin *Arbacia punctulata* and the sand dollar *Echinaracnius parma*. Exp Cell Res 10: 377–386

Tyler A (1958) Changes in efflux and influx of potassium upon fertilization in eggs of *Arbacia punctulata* measured by use of K^{42}. Biol Bull 115: 339–340

Tyler A (1966) Incorporation of amino acids into protein by artificially activated non-nucleate fragments of sea urchin eggs. Biol Bull 130: 450–461

Tyler A, Metz C B (1955) Effects of fertilizin treatment of sperm and trypsin treatment of eggs on homologous and cross-fertilization in sea urchins. Publ Stn Zool Napoli 27: 128–145

Tyler A, Monroy A (1956) Change in the rate of release of K^{42} upon fertilization of eggs of *Arbacia punctulata* Biol Bull 111: 296a

Tyler A, Monroy A (1959) Change in rate of transfer of potassium across the membrane upon fertilization of eggs of *Arbacia punctulata*. J Exp Zool 142: 675–690

Tyler A, Monroy A, Kao C Y, Grundfest H (1956) Membrane potential and resistance of the starfish egg before and after fertilization. Biol Bull 111: 153–177

Tyler A, Piatigorsky J, Ozaki H (1966) Influence of individual amino acids on uptake and incorporation of valine, glutamic acid and arginine by unfertilized and fertilized sea urchin eggs. Biol Bull 131: 204–217

Tyler A, Tyler B S (1966) The gametes; some procedures and properties. In: Boolotian R A (ed) Physiology of echinodermata. Wiley, New York, pp 639–682

Tyler A, Tyler B S, Piatigorsky J (1968) Protein synthesis by unfertilized eggs of sea urchins. Biol Bull 134: 209–219

Uehara T (1971) Experimental studies on the site of propagation of the fertilization wave in the sea urchin egg. Dev Growth Differ 13: 165–172

Uehara T, Katou K (1983) Changes of the membrane potential at the time of fertilization in the sea urchin egg with special reference to the fertilization wave. Dev Growth Differ 14: 175–184

Ullu E, Esposito V, Melli M L (1982) Evolutionary conservation of the human 7S RNA sequences. J Mol Biol 161: 195–201

Uno Y, Hoshi M (1978) Separation of the sperm agglutinin and the acrosoma reaction-inducing substance in egg jelly of starfish. Science 200: 58–59

Usui N, Yoneda M (1982) Ultrastructural basis of the tension increase in sea-urchin eggs prior to cytokinesis. Dev Growth Differ 24: 479–490

Vacquier V D (1969) The isolation and preliminary analysis of the hyaline layer of sea urchin eggs. Exp Cell Res 54: 140–142

Vacquier V D (1974) Cortical granules of sea urchin eggs: Isolation of the intact granule layer and dehiscence of granules by Ca^{2+}. J Cell Biol 63: 355 (Abstr)

Vacquier V D (1975a) Calcium activation of esteroproteolytic activity obtained from sea urchin egg cortical granules. Exp Cell Res 90: 454–456

Vacquier V D (1975b) The isolation of intact cortical granules from sea urchin eggs. Calcium ions trigger granule discharge. Dev Biol 43: 62–74

Vacquier V D (1975c) The appearance of β-1,3-glucanase in hatching embryos of the sea urchin, *Echinometra vanbrunti*. Exp Cell Res 93: 202–206

Vacquier V D (1981) Dynamic changes of the egg cortex. Dev Biol 84: 1–26

Vacquier V D, Brandriff B (1975) DNA synthesis in unfertilized sea urchin eggs can be turned on and off by the addition and removal of procaine hydrochloride. Dev Biol 47: 12–31

Vacquier V D, Mazia D (1968a) Twinning of sand dollar embryos by means of dithiotreitol. The structural basis of blastomere interactions. Exp Cell Res 52: 209–221

Vacquier V D, Mazia D (1968b) Twinning of sea urchin embryos by treatment with dithiothreitol. Roles of cell surface interactions and of the hyaline layer. Exp Cell Res 52: 459–468

Vacquier V D, Moy G W (1977) Isolation of bindin, the protein responsible for adhesion of sperm to sea urchin eggs. Proc Natl Acad Sci USA 74: 2456–2460

Vacquier V D, Moy G W (1980) The cytolytic isolation of the cortex of the sea urchin egg. Dev Biol 77: 178–190

Vacquier V D, O'Dell D S (1975) Concanavalin A inhibits the dispersion of the cortical granule contents of sand dollar eggs. Exp Cell Res 90: 465–468

Vacquier V D, Payne J E (1973) Methods for quantitating sea urchin sperm-egg binding. Exp Cell Res 83: 227–235

Vacquier V D, Epel D, Douglas L A (1972a) Sea urchin eggs release protease activity at fertilization. Nature (London) 237: 34–36

Vacquier V D, Legner M J, Epel D (1972b) Protease activity establishes the block against polyspermy in sea urchin eggs. Nature 240: 352–353

Vallee R B, Bloom G S (1983) Isolation of sea urchin egg microtubules with taxol and identification of mitotic spindle microtubule-associated proteins with monoclonal antibodies. Proc Natl Acad Sci USA 80: 6259–6263

Vanyushin B F, Mazin A L, Vasilyev U K, Belozersky A N (1973) The content of 5-methylcytosine in animal DNA: The species and tissue specificity. Biochim Biophys Acta 299: 397–403

Vasseur E (1952) The chemistry and physiology of the jelly coat of the sea urchin eggs. Diss, Stockholm, pp 1–32

Vasseur E (1952) Geographic variation in the Norwegian sea urchins, *Strongylocentrotus droebachiensis* and *S. palladius*. Evolution 6: 87–100

Venezky D L, Angerer L M, Angerer R C (1981) Accumulation of histone repeat transcripts in the sea urchin egg pronucleus. Cell 24: 385–391

Vernon H M (1900) Cross fertilization among echinoids Arch f Entw-Mech 9: 464–478

Veron M, Foerder C, Eddy E M, Shapiro B M (1977) Sequential biochemical and morphological events during assembly of the fertilization membrane of the sea urchin. Cell 10: 321–328

Villacorta-Moeller M N, Carroll E J Jr (1982) Sea urchin embryo fertilization envelope: immunological evidence that soluble envelope proteins are derived from cortical granule secretion. Dev Biol 94: 415–424

Vitelli L, Weinberg E S (1983) An inverted sea urchin histone gene sequence with break points between TATA boxes and mRNA cap sites. Nucleic Acid Res 11: 2135–2154

Vittorelli M L, Caffarelli-Mormino I, Monroy A (1969) Poli U stimulation of single ribosomes and of ribosomes engaged in polysomes of sea urchin eggs and embryos. Biochim Biophys Acta 186: 408–411

Vittorelli M L, Cannizzaro G, Giudice G (1973) Trypsin treatment of cells dissociated from sea urchin embryos elicits DNA synthesis. Cell Differ 2: 279–284

Vittorelli M L, Matranga V, Feo S, Giudice G, Noll H (1980) Inverse effects of thymidine incorporation in dissociated blastula cells of the sea urchin *Paracentrotus lividus* induced by butanol treatment and Feb addition. Cell Differ 9: 63–70

Von Ledebur-Villiger M (1975) Thymidine uptake by developing sea urchin embryos. Exp Cell Res 96: 344–350

Voroboyev V I (1969) The effects of histone on RNA and protein synthesis in sea urchin embryos at early stages of development. Exp Cell Res 55: 168–170

Voroboyev V I, Gineitis A A, Vinogradova I A (1969a) Histones in early embryogenesis. Exp Cell Res 57: 1–7

Voroboyev V I, Gineitis A A, Kostyleva E I, Smirnova T A (1969b) The effect of histones on the early embryogenesis of sea urchins. Exp Cell Res 55: 171–175

Wadsworth P, Sloboda R D (1983) Microinjection of fluorescent tubulin into dividing sea urchin cells. J Cell Biol 97: 1249–1254

Wagenaar E B (1983a) The timing of synthesis of proteins required for mitosis in the cell cycle of the sea urchin embryo. Exp Cell Res 144: 393–403

Wagenaar E B (1983b) Increased free Ca^{2+} levels delay the onset of mitosis in fertilized and artificially activated eggs of the sea urchin. Exp Cell Res 148: 73–82

Warburg O (1908) Beobachtungen über die Oxydationsprozesse im Seeigelei. Hoppe-Seyler's Z Physiol-Chem 57: 1–16

Ward G E, Vacquier V D (1983) Dephosphorylation of a major sperm membrane protein is induced by egg jelly during sea urchin fertilization. Proc Natl Acad Sci 80: 5578–5582

Ward G E, Vacquier V D, Michel S (1983) The increased phosphorylation of ribosomal protein S6 in *Arbacia punctulata* is not a universal event in the activation of the sea urchin eggs. Dev Biol 95: 360–371

Watanabe M, Thomas W C, Turner R S Jr (1980) Developmental study of adhesions between cells in sea urchin embryos. J Cell Biol 87: 141 (Abstr)

Watanabe M, Bertolini D R, Schnitman J R, Turner R S Jr (1982) Reconstitution of embryo-like structures from sea urchin embryo cells: distinction between cell-cell and cell-substrate association. Differentiation 21: 79–85

Weinberg E S, Birnstiel M L, Purdom I F, Williamson R (1972) Genes coding for polysomal 9S RNA of sea urchins: Conservation and divergence. Nature (London) 240: 225–228

Weinberg E S, Overton G C, Shutt R H, Reeder R H (1975) Histone gene arrangement in the sea urchin, *Strongylocentrotus purpuratus*. Proc Natl Acad Sci USA 72: 4815–4819

Weinberg E S, Overton G C, Hendricks M B, Newrock K M, Cohen L H (1978) Histone gene heterogeneity in the sea urchin *Strongylocentrotus purpuratus*. Cold Spring Harbor Symp Quant Biol 49: 1093–1100

Weinberg E S, Hendricks M B, Hemminki K, Kuwabara P E, Farrelly L A (1983) Timing and rates of synthesis of early histone mRNA in the embryo of *Strongylocentrotus purpuratus*. Dev Biol 98: 117–129

Weinblum D, Güngerich H, Geisert M, Zahn R K (1973) Occurrence of repetitive sequences in the DNA of some marine invertebrates. Biochim Biophys Acta 299: 231–240

Wells D E, Showman R M, Klein W H, Raff R A (1981) Delayed recruitment of maternal histone H3 mRNA in sea urchin embryos. Nature (London) 292: 477–478

Westin M (1976) Immunological studies on the incorporation of radioactive leucine into sea urchin embryonic proteins: extractability and turn-over of newly synthesized proteins. J Embryol Exp Morphol 35: 507–519

Westin M, Perlman P (1972a) Stage specific changes during early development of sea urchin embryos studied by means of two-dimensional electrophoresis. Acta Embryol Morphol Exp Suppl 471–481

Westin M, Perlman P (1972b) Immunofluorescence studies of developmental changes in sea urchin eggs and embryos. Exp Cell Res 72: 233–239

Whitaker M J, Steinhardt R A (1981) The relation between the increase in reduced nicotinamide nucleotide and the initiation of DNA synthesis in sea urchin eggs. Cell 25: 95–103

Whitaker M J, Steinhardt R A (1983) Evidence in support of the hypothesis of an electrically mediated fast block to polyspermy in sea urchin eggs. Dev Biol 95: 244–248

Whiteley A H, Baltzer F (1958) Development respiratory rate and content of deoxyribonucleic acid in the hybrid *Paracentrotus* ♀ × *Arbacia* ♂. Pubbl Stn Zool Napoli 30: 402–457

Whiteley A H, Chambers E L (1960) The differentiation of a phosphate transport mechanism in the fertilized egg of the sea urchin. In: Ranzi S (ed) Symp Germ Cells Dev, Pallanza, pp 387–401

Whiteley A H, Chambers E L (1966) Phosphate transport in fertilized sea urchin eggs. II Effects of metabolic inhibitors and studies on differentiation. J Cell Physiol 68: 309–323

Whiteley A H, Mizuno S (1981) Electron microscope visualization of giant polysomes in sea urchin embryos. Wilhelm Roux's Arch Entwicklungsmech Org 190: 73–82

Whiteley A H, McCarthy B J, Whiteley H R (1966) Changing populations of messenger RNA during sea urchin development. Proc Natl Acad Sci USA 65: 519–525

Whiteley H R, Whiteley A H (1975) Changing populations of reiterated DNA transcripts during early echinoderm development. Curr Top Dev Biol 9: 39–88

Whiteley H R, McCarthy B J, Whiteley A H (1970) Conservativism of base sequences in RNA for early development of echinoderms. Dev Biol 21: 216–242

Wiley H S, Johnston R N, Beach D, Epel D (1977) Studies of the change in NAD kinase activity during the activation of eggs of *Arbacia punctulata*. Biol Bull 153: 449 (Abstr)

Wilson F E, Blin N, Stafford D (1976) A denaturation map of sea urchin ribosomal DNA. Chromosoma 58: 247–253

Wilson F E, Blin N, Stafford D W (1977) Structure of the ribosomal RNA gene region in *Lytechinus variegatus*. J Cell Biol 75: 343 (Abstr)

Wilt F H (1963) The synthesis of RNA in sea urchin embryos. *Strongylocentrotus purpuratus*. Biochem Biophys Res Commun 11: 447–451

Wilt F H (1973) Polyadenylation of maternal RNA of sea urchin eggs after fertilization. Proc Natl Acad Sci USA 70: 2345–2349

Wilt F H (1977) The dynamics of maternal poly (A) containing mRNA in fertilized sea urchin eggs. Cell 11: 673–681

Wilt F H, Mazia D (1974) The stimulation of cytoplasmic polyadenylylation in sea urchin eggs by ammonia. Dev Biol 37: 422–424

Winkler M M, Crainger J L (1978) Mechanism of action of NH_4Cl and other weak bases in the activation of sea urchin eggs. Nature (London) 273: 536–538

Winkler M M, Steinhardt R A (1981) Activation of protein synthesis in a sea urchin cell-free system. Dev Biol 84: 432–439

Winkler M M, Baker E, Hunt T (1979) Protein synthesis in cell-free extracts of *Lytechinus pictus* eggs. Biol Bull 157: 402 (Abstr)

Winkler M M, Steinhardt R A, Grainger L J, Minning L (1980) Dual ionic controls for the activation of protein synthesis at fertilization. Nature (London) 287: 558–560

Winkler M M, Matson G B, Hershey J B, Bradbury E M (1982) ^{31}P-NMR study of the activation of the sea urchin egg. Exp Cell Res 139: 217–222

Winkler M M, Bruening G, Hershey J W B (1983) An absolute requirement for the 5' cap structure for mRNA translation in sea urchin eggs. Eur J Biochem 137: 227–232

Wolcott O L (1981) Effect of potassium and lithium ions on protein synthesis in the sea urchin embryo. Exp Cell Res 132: 464–468

Wolcott D L (1982) Does protein synthesis decline in lithium treated sea urchin embryos because RNA synthesis is inhibited? Exp Cell Res 137: 427–431

Wold B J, Klein W H, Hough-Evans B R, Britten R J, Davidson E H (1978) Sea urchin embryo mRNA sequences expressed in the nuclear RNA of adult tissues. Cell 14: 941–950

Wolf D E, Edidin M, Kinsey W, Lennarz W J (1979) Changes in the diffusion probes on sea urchin plasma membrane. J Cell Biol 83: 205 (Abstr)

Wolf D E, Kinsey W, Lennarz W, Edidin M (1981) Changes in the organization of the sea urchin egg plasma membrane upon fertilization: Indication from the lateral diffusion rates of lipid-soluble fluorescent dyes. Dev Biol 81: 133–138

Wolpert L, Gustafson T (1961a) Studies on the cellular basis of morphogenesis of the sea urchin embryo. Exp Cell Res 25: 311–325

Wolpert L, Gustafson T (1961b) Studies on the cellular basis of morphogenesis of the sea urchin embryo. Exp Cell Res 25: 374–382

Wolsky A, de Issekutz Wolsky M (1961) The effect of actinomycin D on the development of *Arbacia* eggs. Biol Bull 121: 414–425

Wong L J C (1980) Effect of sea urchin sperm chromatin on histone acetylation. Biochem Biophys Res Commun 97: 1362–1369

Woodland H R, Wilt F H (1980a) The functional stability of sea urchin histone mRNA injected into oocytes of *Xenopus laevis*. Dev Biol 75: 199–213

Woodland H R, Wilt F H (1980b) The stability and translation of sea urchin histone messenger RNA molecules injected into *Xenopus laevis* eggs and developing embryos. Dev Biol 75: 214–221

Woods D E, Fitschen W (1978) The mobilization of maternal histone messenger RNA after fertilization of the sea urchin egg. Cell Differ 7: 103–114

Wortzman M S, Baker R F (1980) Specific sequences within single-stranded regions in sea urchin embryo genome. Biochim Biophys Acta 609: 84–96

Wortzman M S, Baker R F (1981) Two classes of single-stranded regions in DNA from sea urchin embryos. Science 211: 588–590

Wu R S, Wilt F H (1973) Poly A metabolism in sea urchin embryos. Biochem Biophys Res Commun 54: 704–714

Wu R S, Wilt F H (1974) The synthesis and degradation of RNA containing polyriboadenylate during sea urchin embryogeny. Dev Biol 41: 352–370

Wu R S, Nishioka D, Bonner W M (1982) Differential conservations of histone 2A variants between mammals and sea urchins. J Cell Biol 93: 426–431

Wu M, Holmes D S, Davidson N, Cohn R H, Kedes L H (1976) The relative positions of sea urchin histone genes on the chimeric plasmids pSp2 and pSp17 as studied by electron microscopy. Cell 9: 163–169

Xin J H, Brandhorst B D P, Britten R J, Davidson E H (1982) Cloned embryo mRNAs not detectably expressed in adult sea urchin coelomocytes. Dev Biol 89: 527–531

Yablonka-Reuveni Z, Hille M B (1983) Isolation and distribution of elongation factor 2 in eggs and embryos of sea urchins. Biochemistry 22: 5205–5212

Yamada Y, Aketa K (1981) Vitelline layer lytic activity in sperm extracts of sea urchin, *Hemicentrotus pulcherrimus*. Gam Res 4: 193–202

Yamada Y, Aketa K (1982) Purification and characterization of trypsin-like enzyme of sperm of the sea urchin, *Hemicentrotus pulcherrimus*. Dev Growth Differ 24: 125–134

Yamada Y, Matsui T, Aketa K (1982) Characterization of a chrymotrypsin-like enzyme from sperm of the sea urchin, *Hemicentrotus pulcherrimus*. Eur J Biochem 122: 57–62

Yanagisawa T (1968) Studies on echinoderm phosphagens. IV Changes in the content of arginine phosphate in sea urchin egg after fertilization and the effect of some metabolic inhibitors. Exp Cell Res 53: 525–536

Yanagisawa T (1975a) Respiration and energy metabolism. In: Czihak G (ed) The sea urchin embryo. Springer, Berlin Heidelberg New York, pp 510–549

Yanagisawa T (1975b) Carbohydrate metabolism and related enzymes. In: Czihak G (ed) The sea urchin embryo. Springer, Berlin Heidelberg New York

Yanagisawa T, Amemiya S (1982) Genomic complexities of sea urchins *Asthenosoma yimai* and *Araeosoma owstoni* (Echinothurbida). Cell Differ 11: 309–310

Yanagisawa T, Yazaki I, Mohri H (1973) An immunological study on flagellar tubulin. Sci Pap Coll Gen Educ Univ Tokyo 23: 155–158

Yanagisawa T, Isono N (1966) Acid soluble nucleotides in the sea urchin egg. I. Ion exchange chromatographic separation and characterization. Embryologia 9: 170–183

Yang S S, Comb D G (1968) Distribution of multiple forms of lysyl tRNA during early embryogenesis of the sea urchin *Lytechinus variegatus*. J Mol Biol 31: 139–142

Yano Y, Miki-Nomura T (1980) Sliding velocity between outer doublet microtubules of sea-urchin sperm axonemes. J Cell Sci 44: 169–186

Yano Y, Miki-Nomura T (1981) Recovery of sliding ability in arm-depleted flagellar axonemes after recombination with extracted dynein I. J Cell Sci 48: 223–231

Yasumasu I (1976) Change in the activity of pyruvate dehydrogenase complex in sea urchin eggs following fertilization. Dev Growth Differ 18: 123–131

Yasumasu I, Koshihara H (1963) Aminoacyl RNA and transfer enzyme in sea urchin eggs. Zool Mag 72: 259–262

Yasumasu I, Asami K, Shoger R L, Fujiwara A (1973) Glycolysis of sea urchin eggs. Rate limiting steps and activation at fertilization. Exp Cell Res 30: 361–371

Yasumasu I, Fujivara A, Shoger R L, Asami K (1975) Distribution of some enzymes concerning carbohydrate metabolism in sea urchin eggs. Exp Cell Res 92: 444–450

Yasumasu I, Saitoh M, Fujmoto N, Kusunoki S (1979) Changes in activities of thymidilate synthase and dihydrofolate reductase in sea urchin eggs after fertilization. Dev Growth Differ 21: 237–243

Yasumasu I, Fujiwara A, Hino A (1980) Inhibition of respiration in sea urchin spermatozoa following interaction with fixed unfertilized eggs. II Capacity of the glutaraldehyde-fixed unfertilized eggs for the inhibition of the sperm respiration. Dev Growth Differ 22: 429–436

Yasumasu I, Mita M, Fujimoto N, Fujiwara A (1982) Changes in the activities of dihydrofolate reductase and thioredoxin reductase in sea urchin eggs following fertilization. Cell Differ 11: 297–298

Ycas M (1950) Studies on the respiratory enzymes of sea urchin eggs. Ph D dissertation, Cal Inst Tech

Ycas M (1954) The respiratory and glycolytic enzymes of sea urchin eggs. J Exp Biol 31: 208–217

Yoshida M, Aketa K (1978) Localization of species-specific sperm-binding factor in sea urchin eggs with immunofluorescent probe. Acta Embryol Exp 269–278

Yokota Y, Nakano E (1979) Multiple molecular forms of acid nitrophenylphosphatase in sea urchin eggs and embryos. Gamete Res 2: 177–185

Yoshida M, Aketa K (1979) Effect of papain digested, univalent antibody against sperm-binding factor on the fertilizability of sea urchin eggs. Dev Growth Differ 21: 431–436

Yoshida M, Aketa K (1982) Partial purification of the sperm-binding factor from the egg of the sea urchin *Anthocidaris crassispina* followed by an immunological method. Dev Growth Differ 24: 55–63

Yoshida M, Aketa K (1983) A 225 kD glycoprotein is the active core structure of the sperm-binding factor of the sea urchin, *Anthocidaris crassispina*. Exp Cell Res 148: 243–248

Yoshimi T, Yasumasu I (1978) Vegetalization of sea urchin larvae induced with cAMP phosphodiesterase inhibitors. Dev Growth Differ 20: 213–218

Yoshimi T, Yasumasu I (1979) Prevention by hydroxyurea of vegetalization of sea urchin larvae induced by cAMP phosphodiesterase inhibitors. Dev Growth Differ 21: 271–280

Yoshioka T, Inoue H (1981) Activation of sea urchin eggs by a channel-forming ionophore amphotericin B. Exp Cell Res 132: 461–464

Young E M, Raff R A (1979) Messenger ribonucleoprotein particles in developing sea urchin embryos. Dev. Biol 72: 24–40

Yukawa O, Koshihara H (1973) Acetylation of proteins in the course of sea urchin development. Dev Biol 33: 477–481

Zdunsky D, Hamkalo B, Kedes L (1977) Actively transcribing chromatin of sea urchin (*Arbacia punctulata*) visualized by electron microscopy. Biol Bull 151: 435 (Abstr)

Zeikus J G, Taylor M W, Buck C A (1969) Transfer RNA changes associated with early development and differentiation of the sea urchin *Strongylocentrotus purpuratus*. Exp Cell Res 57: 74–78

Zeitz L, Ferguson R, Garfinkel E (1968) Radiation-induced effects on DNA synthesis in developing sea urchin eggs. Radiat 34: 200

Zeuthen E (1960) Cycling on oxygen consumption in cleaving eggs. Exp Cell Res 19: 1—6

Zimmerman A M, Silberman L (1964) Further studies on incorporation of H^3 thymidine in *Arbacia* eggs under hydrostatic pressure. Biol Bull 137: 365

Zimmerman A M, Zimmerman S (1967) Action of colcemid in sea urchin eggs. J Gell Biol 34: 483–488

Zucker R S, Steinhardt R A, Winkler M M (1978) Intracellular calcium release and the mechanisms of pathenogenetic activation of the sea urchin egg. Dev Biol 65: 285–295

References of the Addendum

Allemand D, De Renzis G, Ciapa B, Girard P, Payan P (1984) Characterization of valine transport in sea urchin eggs. Biochim Biophys Acta 772: 337–343

Alliegro M C, Schuel H (1983) Is there specificity in the induction of polyspermy in sea urchins by protease inhibitors? Biol Bull 165: 512 (Abstr)

Alliegro M, Schuel H (1984) Kinetic and electrophoretic analysis of sea urchin trypsin-like protease. J Cell Biol 99: 85 (Abstr)

Anstrom J A, Parks A L, Raff R A (1984) Intracellular localization and surface expression of a set of cell-lineage specific antigens in sea urchin embryos. J Cell Biol 99: 125 (Abstr)

Benson S, Benson M (1984) Biochemical and morphological characterization of the organic matrix of sea urchin embryo spicules. J Cell Biol 99: 162 (Abstr)

Benson S, Jones E M, Crise-Benson N, Wilt F (1983) Morphology of the organic matrix of the spicule of the sea urchin larva. Exp Cell Res 148: 249–253

Bibring Th, Baxandall J, Harter C C (1984) Sodium-Dependent pH regulation in active sea urchin sperm. Dev Biol 101: 425–435

Blaukenship J, Benson (1984) Collagen metabolism and spicule formation in sea urchin micromeres. Exp Cell Res 152: 98–104

Blum F, Nachtigall M, Troll W (1983) Superoxide dismutase biomimetic compounds prevent fertilization in *Arbacia punctulata* eggs. Biol Bull 165: 513 (Abstr)

Branning G B, Winkler M M (1984) Changes in Poly A and Poly U binding proteins following fertilization. J Cell Biol 99: 262 (Abstr)

Bray S, Hunt T (1983) Developmental studies of a major maternal mRNA in *Arbacia punctulata*. Biol Bull 165: 499–500 (Abstr)

Brookbank J W (1984) H1 histones of adult tissues of the sea urchin, *Strongylocentrotus purpuratus*. Biol Bull 166: 452–455

Burke R D (1984) Pheromonal control of metamorphosis in the Pacific sand dollar, *Dendraster excentricus*. Nature 225: 442–443

Chambers E L, Lynn J W, McCulloh D H (1984) Sperm entry is blocked by clamped membrane potentials near resting in sea urchin oocytes. J Cell Biol 99: 56 (Abstr)

Chambers S A M, Truschel M R, Stafford D, McClay D R (1984a) Isolation of the β-gluconase gene and its expression during sea urchin embryonic development. J Cell Biol 99: 124 (Abstr)

Chambers S A M, Cognetti G, Ramsay Shaw B (1984b) Diffusible factors are responsible for differences in nuclease sensitivity among chromatins originating from different cell types. Exp Cell Res 154: 213–223

Chandler D E (1984) Exocytosis in the isolated sea urchin egg cortex as viewed by rotary platinum shadowing. J Cell Biol 99: 53 (Abstr)

Christen R (1983) Acidic vesicles and the uptake of amines by sea urchin eggs. Exp Cell Res 143: 319–325

Christen R, Schackmann R W, Shapiro B M (1983) Metabolism of sea urchin sperm Interrelationships between intracellular pH, ATPase activity, and mitochondrial respiration. J Biol Chem 258: 5392–5399

Cervello M, Di Ferro D, Matranga V, La Placa G, Vittorelli M L (1984) Different aggregation properties of sea urchin embryonic cells at different developmental stages. I Stage specificity of aggregation factors solubilized by butanol. Cell Biol Int Rep 8: 787–791

Ciapa B, De Renzis G, Girard J P, Payan P (1984a) Sodium-potassium exchange in sea urchin egg. I Kinetic and biochemical characterization at fertilization. J Cell Physiol 121: 235–242

Ciapa B, Allemand D, Payan P, Girard J P (1984b) Sodium-potassium exchange in sea urchin eggs. II Ionic events stimulating the Na+ — K+ Pump activity at fertilization. J Cell Physiol 121: 243–250

Coffe C, Foucault G, Raymond M N, Pudles J (1984) Changes in the tubulin organization during the cytoplasmic cohesiveness cycles in procaine activated eggs. J Submicrosc Cytol 16: 29

Cornall R, Bornslaeger E, Hunt T (1983) What makes cyclin cycle? Biol Bull 165: 513 (Abstr)

Cox K H, DeLeon D V, Angerer L M, Angerer R C (1984) Detection of mRNAs in sea urchin embryos by in situ hybridization using asymmetric RNA probes. Dev Biol 101: 485–502

Croll J H, Jackson R C (1984) In vitro exocytosis in isolated sea urchin egg cortex: polycation inhibition. J cell Biol 99: 56 (Abstr)

Crain W R Jr, Durica D S, Van Doren (1981) Actin gene expression in developing sea urchin embryos. Mol Cell Biol 1: 711–720

Darszon A, Gould M, De La Torre L, Vargas I (1984) Response of isolated sperm plasma membranes from sea urchin to egg jelly. Eur J Biochem 144: 515–522

De Petrocellis B, Pratibha M, Maharajan V (1984) dCMP-Aminohydrolase activity during early sea urchin development. An example of negative enzyme control during embryogenesis. Exp Cell Res 152: 188–194

Dube F, Schmidt T, Johnson C H, Epel D (1984) The rise of intracellular pH is only transiently required for initiation of early development and increased protein synthesis in sea urchin embryos. J Cell Biol 99: 266 (Abstr)

Eisen A, Kiehart D P, Wieland S J, Reynolds G T (1984) Temporal sequence and spatial distribution of early events of fertilization in single sea urchin eggs. J Cell Biol 99: 1647–1654

Farach M C, Carson D D (1984) Inhibition of sea urchin spiculogenesis by monoclonal antibody that blocks calcium uptake. J Cell Biol 99: 126 (Abstr)

Farach H A, Mundy D K, Decker G, Lennarz W J, Strittmater W J (1984) A possible role for metalloendoproteases in sea urchin fertilization. J Cell Biol 99: 263 (Abstr)

Fujisawa H, Amemiya S (1982) Effects of zinc and lithium ions on the strengthening cell adhesion in sea urchin blastulae. Experientia 38: 852–853

George E L, Bray S, Rosenthal E T, Hunt T (1983) A major maternally encoded 41K protein in both Spisula and Arbacia binds to an anti-tubulin affinity column. Biol Bull 165: 515 (Abstr)

Glabe C G (1984) Sea urchin sperm bindin associates with phospholipid vesicles and induces the fusion of mixed-phase vesicles in vitro. J Cell Biol 99: 3 (Abstr)

Grant S R, Lennarz W J (1984) Identification of N-linked glycoproteins during sea urchin embryo development. J Cell Biol 99: 95 (Abstr)

Gundersen G G, Shapiro B M (1984) Sperm surface proteins persist after fertilization. J Cell Biol 99: 1343–1353

Guraya S S (1982) Recent progress in the structure, origin, Composition, and function of cortical granules in animal egg. Int Rev Cytol 78: 257–360

Harkey M A (1983) Determination and differentiation of micromeres in the sea urchin embryo. In: (ed) M M Burger and R Weber Time, Space, and Pattern in Embryonic Development. Alan R Liss, New York, pp 131–155

Hindenach B R, Stafford D W (1984) Nucleotide sequence of the 18S-26S rRNA intergene region of the sea urchin. Nucleic Acid Res 12: 1737–1747

Hinkley R E, Wright B D (1984a) Effects of Halothane on fertilization and early development in the sea urchin. J Cell Biol 99: 264 (Abstr)

Hinkley R E, Wright B D (1984b) Adaptation of detergent-ectraction methods for the isolation of sperm asters from sea urchin eggs. J Cell Biol 99: 126 (Abstr)

Hisanaga S, Pratt M M (1984) Calmodulin interaction with cytoplasmic and flagellar dynein: calcium-dependent binding and stimulation of adenosinetriphosphatase activity. Biochem 23: 3032–3037

Holt C A, Childs G (1984) A new family of tandem repetitive early histone genes in the sea urchin *Lytechinus pictus*: evidence for concerted evolution within tandem arrays. Nucleic Acid Res 12: 6455–6471

Hoshi M, Moriya T, Aoyagi T, Umezawa H, Mohri H, Nagai Y (1979) Effects of hydrolase inhibitors on fertilization of sea urchins: I Protease inhibitors. Gamete Res 2: 107–119

Hosoya H, Mabuchi I (1984) A 45,000-mol-wt protein-actin complex from unfertilized sea urchin egg affects assembly properties of actin. J Cell Biol 99: 994–1001

Howlett S, Miller J, Schultz G (1983) Induction of heat shock proteins in early embryos of *Arbacia punctulata*. Biol Bull 165: 500 (Abstr)

Ishimoda-Tagagi T, Chino I, Sato H (1984) Evidence for the involvement of muscle tropomyosin in the contractile elements of the coelom-esophagus complex in sea urchin embryos. Dev Biol 105: 365–376

Iwata M, Nakano E (1983) Characterization of sea-urchin fibronectin. Biochem J 215: 205–208

Iwata M, Nakano E (1984) Cell-to-substratum adhesion of dissociated embryonic cells of the sea urchin, *Pseudocentrotus depressus*. Roux's Arch Dev Biol 193: 71–77

Jacobs H T, Posakony J A, Grula J W, Roberts J W, Xin J H, Britten R J, Davidson E H (1983) Mitochondrial DNA sequences in the nuclear genome of *Strongylocentrotus purpuratus*. J Mol Biol 165: 609–632

Justice R W, Carroll E J Jr (1984) Effects of protease, peroxidase and glucanase inhibitors on sea urchin fertilization envelope development. J Cell Biol 99: 124 (Abstr)

Kaneko T, Terayama H (1974) Quantitative analysis of DNA in sea urchin eggs and subcellular distribution of DNA in the eggs. Anal Biochem 58: 439–448

Katow H, Solursh M (1979) Ultrastructure of blastocoel material and gastrulae of the sea urchin *Lytechinus pictus* (1). J Exp Zool 210: 561–567

Katow H, Solursh M (1980) Ultrastructure of primary mesenchyme cell ingression in the sea urchin *Lytechinus pictus*. J Exp Zool 213: 231–246

Katow H, Yamada K M, Solursh M (1982) Occurrence of fibronectin on the primary mesenchyme cell surface during migration in the sea urchin embryo. Differentiation 22: 120–124

Kay E S, Weidman P J, Shapiro B M (1984) Orientation of oveperoxidase in the sea urchin fertilization membrane determines its apparent substrate specificity. J Cell Biol 99: 263 (Abstr)

Kazazoglou T, Schackmann R W, Shapiro B M (1984) Binding of Ca^{2+} Antagonists to sperm plasma membranes. J Cell Biol 99: 258 (Abstr)

Killian C E, Nishioka D (1984) Guanosine phosphorylation and GTP pool maintenance in fertilized eggs and early embryos of sea urchins. J Cell Biol 99: 263 (Abstr)

Krieg P A, Molton D A (1984) Formation of the 3' end of histone mRNA by post-transcriptional processing. Nature 308: 203–206

Kubota L F, Carroll E J (1984) Sea urchin sperm bindin interacts with a 305kD vitelline envelope sperm receptor polypeptide. J Cell Biol 99: 396 (Abstr)

Kuwabara P E, Greer K, Maekawa S, Weinberg E S (1983) Changes in histone synthesis during Arbacia development. Biol Bull 165: 500 (Abstr)

Lee H C (1984) The Na^+/H^+ exchanger in flagella isolated from sea urchin sperm is sensitive to membrane potential. J Cell Biol 99: 187 (Abstr)

Leaf D S, Harkey M A, Hursh D A, Loo H, Mengerson A M, Raff R A (1984) Gene expression in the primary mesenchyme cells of sea urchin embryo. J Cell Biol 99: 125 (Abstr)

Longo F J (1984a) Transformations of sperm nuclei incorporated into sea urchin (*Arbacia punctulata*) embryos at different stages of the cell cycle. Dev Biol 103: 168–181

Longo F J (1984b) Integration of sperm and egg plasma membrane components following insemination. J Cell Biol 99: 263 (Abstr)

Lynn J W, Chambers E L (1984) Voltage clamp studies of fertilization in sea urchin eggs. I Effect of clamped membrane potential on sperm entry, activation, and development. Dev Biol 102: 98–109

Lynn J W, McCulloh D H, Chambers E L (1984) Membrane depolarization, sperm entry and insemination cone morphology in sea urchin oocytes. J Cell Biol 99: 57 (Abstr)

Maglott D R (1983) Heat shock response in the atlantic sea urchin Arbacia punctulata. Experientia 39: 268–270

Maglott D R (1984) Diamide reversibly induces a stress response in the sea urchin *Arbacia punctulata*. Cell Biol Int Rep 8: 747–754

Matranga V, Adragna N, Cervello M, Vittorelli M L (1984) Different aggregation properties of sea urchin embryonic cells at different developmental stages. II Stage specific response to fibronectin and collagen. Cell Biol Int Rep 8: 797–807

Maxon R, Mohun T, Gormezano G, Childs G, Kedes L (1983) Distinct organizations and patterns of expression of early and late histone gene sets in the sea urchin. Nature 301: 120–125

McCarthy R A, Spiegel M (1983a) Protein composition of the hyaline layer of sea urchin embryos and reaggregating cells. Cell Differ 13: 93–102

McCarthy R A, Spiegel M (1983b) The enhancement of reaggregation of sea urchin blastula cells by exogenous proteins. Cell Differ 13: 103–114

McCulloch D H, Chambers E L, Lynn J W (1984) Membrane currents associated with sea urchin sperm-oocyte interactions. J Cell Biol 99: 57 (Abstr)

Mita M, Hino A, Yasumasu I (1984) Effect of temperature on interaction between eggs and spermatozoa of sea urchin. Biol Bull 166: 68–77

Miyachi Y, Iwata M, Sato H, Nakano E (1984) Effect of fibronection on cultured cells derived from isolated micromeres of the sea urchin, *Hemicentrotus pulcherrimus*. Zoological Sci 1: 265–271

Moore C P (1984) Evolutionary conservation of DNA cooling for maternal RNA in sea urchins. Biochem Biophys Res Commun 123: 278–285.

Morioka M, Shimada H (1984) Synthesis of diadenosine $5',5'''$-P^1,P^4-tetraphosphate (AP_4A) in sea urchin embryos. Cell Differ 14: 53–58

Moss R, Schuel R, Schuel H (1983) FPL-55712, a leukotriene antagonist, promotes polyspermy in sea urchins. Biol Bull 165: 516 (Abstr)

Moy G W, Kopf G S, Gache C, Vacquier V (1983) Calcium-mediated release of glucanase activity from cortical granules of sea urchin eggs. Dev Biol 100: 267–274

Nakano E, Hino A, Furuse K (1984) Effects of surfactants on the fertilizing capacity and acrosome reaction of sea urchin spermatozoa. Gamete Res 9: 115–125

Nemer M, Travaglini E C, Rondinelli E, D'Alonzo J (1984) Developmental regulation, induction, and embryonic tissue specificity of sea urchin metallothionein gene expression. Dev Biol 102: 471–482

Niman H L, Hough-Evans B R, Vacquier V D, Britten R J, Lerner R A, Davidson E H (1984) Proteins of the sea urchin egg vitelline layer. Dev Biol 102: 390–401

Nishioka D, Killian C E, Chacon C T, Sgagias M K (1984) Increased uptake of thymidine in the activation of sea urchin eggs. III Effects of Aphidicolin. J Cell Physiol 118: 27–33

Nomura K, Suzuki N, Ohtake H, Isaka S (1983) Structure and action of sperm activating peptides from the egg jelly of a sea urchin, *Anthocidaris Crassispina*. Biochem Biophys Res Commun 167: 147–154

Nuccitelli R, Grey R D (1984) Controversy over the fast, partial, temporary block to polyspermy in sea urchins: a reevaluation. Dev Biol 103: 1–17

O'Brien E T, Asai D J, Jacobs R S, Wilson L (1984) Stypoldione inhibits sea urchin cell division by disrupting cytoskeleted organization. J Cell Biol 99: 182 (Abstr)

Okabayashi K, Nakano E (1983) The cytochrome system of sea urchin eggs and embryos. Arch Biochem Biophys 225: 271–278

O'Melia A F (1983) Rates of 5S RNA and tRNA synthesis in sea urchin embryos animalized by Evans Blue. Biol Bull 165: 502 (Abstr)

Parisi E, De Prisco P, Capasso A, Del Prete M (1984) Serotonin and sperm motility. Cell Biol Int Rep 8: 95

Paweltz N, Mazia D, Finze E M (1984) The centrosome cycle in the mitotic cycle of sea urchin eggs. Exp Cell Res 152: 47–65

Peters R, Sardet C, Richter H P (1984) Mobility of membrane lipids and proteins at the animal and vegetal pole of the sea urchin egg. Dev Growth Differ 26: 105–110

Poccia D, Greenough T, Green G R, Nash E, Erickson J, Gibbs M (1984) Remodeling of sperm chromatin following fertilization: nucleosome repeat length and histone variant transitions in the absence of DNA synthesis. Dev Biol 104: 274–286

Piperno G (1984) Monoclonal antibodies to dynein subunits reveal the existence of cytoplasmic antigens in sea urchin egg. J Cell Biol 94: 1842–1850

Podell S B, Vacquier V D (1984a) Wheat germ agglutinin blocks the acrosome reaction in *Strongylocentrotus purpuratus* sperm by binding a 210,000-mol-wt membrane protein. J Cell Biol 99: 1598–1604

Podell S B, Vacquier V D (1984b) Inhibition of sea urchin sperm acrosome reaction by antibodies directed against two sperm membrane proteins. Characterization and mechanism of action. Exp Cell Res 155: 467–476

Poenie M, Alderton J, Steinhardt R, Tsien R (1984) Calcium activity correlates with the activation state and specific events in the cell cycle. J Cell Biol 99: 429 (Abstr)

Posakony J W, Flytzanis C N, Britten R J, Davidson E H (1983) Interspersed sequence organization and developmental representation of cloned poly (A) RNAs from sea urchin eggs. J Mol Biol 167: 361–389

Raff R A, Anstrom J A, Huffman C J, Leaf D S, Loo J H, Showman R M, Wells D E (1984) Origin of a gene regulatory mechanism in the evolution of echinoderms Nature 310: 312–314

Ransick A J, Pollock E G (1984) Experimental analysis of plasma membrane IMP clusters in sea urchin embryos. J Cell Biol 99: 124 (Abstr)

Richards R G, Ramsay Shay B (1984) Temporal changes in nucleosome populations during sea urchin early development. Biochemistry 23: 2095–2102

Richter L, Winkler M M (1984) mRNA availability limits protein synthesis in sea urchin eggs. J Cell Biol 99: 262 (Abstr)

Rossignol D P, Scher M, Waechter C J, Lennarz W J (1984a) Metabolic interconversion of dolichol and dolichyl phosphate during development of the sea urchin embryo. J Biol Chem 258: 9122–9127

Rossignol D P, Decker G L, Lennarz W J, Tsong T Y, Teissie J (1984b) Induction of calcium-dependent, localized cortical granule breakdown in sea-urchin eggs by voltage pulsation. Biochim Biophys Acta 763: 346–355

Rossignol D P, Earles B J, Decker G L, Lennarz W J (1984c) Characterization of the sperm receptor on the surface of eggs of *Strongylocentrotus purpuratus*. Dev Biol 104: 308–321

Sale W S, Goodenough U W, Heuser J E (1984) The structure of dynein 1 of sea urchin sperm tail axonemes. J Cell Biol 99: 44 (Abstr)

Sardet C (1984) The ultrastructure of the sea urchin egg cortex isolated before and after fertilization. Dev Biol 105: 196–210

Sasaki H (1984) Modulation of calcium sensitivity by a specific cortical protein during sea urchin egg cortical vesicle exocytosis. Dev Biol 101: 125–135

Sawada H, Miura M, Yokosawa H, Ishi S (1984) Purification and characterization of trypsine-like enzyme from sea urchins eggs: substrate specificity and physiological role. Biochem Biophys Res Commun 121: 598–604

Schatten G, Schatten H, Paweletz N, Spector T, Petzelt C (1984a) Latrunculin inhibits cytokinesis and fragments stress fibers in cultured cells, and prevents fertilization and development in sea urchins. J Cell Biol 99: 181 (Abstr)

Schatten G, Maul G G, Simerly C, Balczon R, Schatten H (1984b) Nuclear lamins and perichromin-like antigen distribution during mouse and sea urchin fertilization: New appearance of lamins during pronuclear formation, in first mitotic chromosomes and exclusion in degenerating polar body nucleus. J Cell Biol 99: 125 (Abstr)

Scholey J M, Neighbors B, McIntosh J R, Salmon E D (1984) Isolation of microtubules and a dynein-like MgATPase from unfertilized sea urchin eggs. J Biol Chem 259: 6516–6525

Schuel H, Moss R, Schuel R (1984) Evidence for a leukotriene mediated block to polyspermy in sea urchins. J Cell Biol 99: 262 (Abstr)

Schwager S, Brandt W F, Von Holt C (1983) The isolation of isohistones by preparative gel electrophoresis from embryos of the sea urchin *Parenchinus angulosus*. Biochim Biophys Acta 741: 315–321

Sedlar P A, Bryan L (1984) Partial characterization of a 45kDa actin filament severing protein from sea urchin eggs. J Cell Biol 99: 308 (Abstr)

Shupe K, Weinberg E (1983) In vitro transcription of histone genes in isolated nuclei from S. purpuratus. Biol Bull 165: 518 (Abstr)

Silver R B (1983) Inhibition of mitotic anaphase and cytokinesis and reduction of spindle birefringence following microinjection of anti-calcium transport enzyme IgGs into Echinaracnius parma blastomeres. Biol Bull 165: 495 (Abstr)

Silver R B, Saft M S, Taylor A R, Cole R D (1983) Identification of nonmitochondrial creatine kinase enzymatic activity in isolated sea urchin mitotic apparatus. J Biol Chem 258: 13287–13291

Silver R B, Saft M S, Handler M A (1984) Calcium regulation in the mitotic apparatus of sea urchin: identification of the MA-Ca-Pump, & enzymological studies of the MA membranes. J Cell Biol 99: 443 (Abstr)

Sluder G, Miller F, Rieder C (1984) Experimental separation of pronuclei in fertilized sea urchin eggs: chromosomes alone are not sufficient to organize the mitotic spindle. J Cell Biol 99: 445 (Abstr)

Spiegel E, Spiegel M (1980) The internal clock of reaggregating embryonic sea urchin cells. J Exp Zool 213: 271–281

Steinert G, Felsani A, Kettmann R, Brachet J (1984) Presence of rRNA in the heavy bodies of sea urchin eggs; an in situ ibridization study with the electron microscope. Exp Cell Res 154: 203–212

Suprynowicz F A, Mazia D (1984) Fluctuation of the Ca^{2+}-sequestering activity of permeabilized sea urchin embryos during the cell cycle. J Cell Biol 99: 429 (Abstr)

Swezey R R, Epel D (1984) Ambiquitous glucose-6-phosphate dehydrogenase in sea urchin eggs. J Cell Biol 99: 54 (Abstr)

Terashita S, Kato T, Sato H (1983a) Reaction mechanism of 21S dynein ATPase from sea urchin sperm. I Kinetic properties in the steady state. J Biochem 93: 1567–1574

Terashita S, Kato T, Sato H, Tonomura Y (1983b) Reaction mechanism of 21S dynein ATPase from sea urchin sperm. II Formation of reaction intermediates. J Biochem 93: 1575–1581

Tombes R M, Schackmann R W, Shapiro B M (1984) Sea urchin sperm creatine kinase activity is essential for optimal respiration and motility. J Cell Biol 99: 258 (Abstr)

Timmer J S, Vacquier V D (1984) Monoclonal antibodies to a sea urchin sperm glycoprotein inhibit the acrosome reaction. J Cell Biol 99: 260 (Abstr)

Turner E, Shapiro B M (1984) Oxidase activity of sea urchin oveperoxidase. J Cell Biol 99: 262 (Abstr)

Turner P R, Sheetz M P, Jaffe L A (1984) Fertilization increases the polyphosphoinositide content of sea urchin eggs. Nature 310: 414–415

Uzman J A, Wilt F H (1984) Distinguishing RNA polymerase initiation and elongation in control of total RNA and histone mRNA synthesis in sea urchin embryos. Dev Biol 106: 174–180

Venkatasubramanian K, Solursh M (1984) Adhesive and migration behavior of normal and sulfate deficient sea urchin cells in vitro. Exp Cell Res 154: 421–431

Vitelli L, Rusconi S, Birnstiel M L (1984) Expression of microinjected sea urchin histone genes in developing sea urchin embryos. Atti AGI; SIBBM; ABCD Siena 1984

Wagenaar E B, Mazia D (1978) In: Dirksen E R, Prescott D M, Fox C F (eds) ICN-UCLA Symp on Molecular and Cell Biology vol XII Cell Reproduction. Academic Press, New York, pp 359–545

Wang L L, Spudich J A (1984) A 45,000-mol-wt protein from unfertilized sea urchin eggs severs actin filaments in a calcium-dependent manner and increases the steady-state concentration of nonfilamentous actin. J Cell Biol 99: 844–851

Watanabe N, Shimada H (1983) Effects of emetine on initiation of DNA synthesis in embryonic cells of sea urchin. Cell Differ 13: 239–245

Wessel G M and McClay D R (1984) Characterization of germ-layer specific molecules in the sea urchin embryo. J Cell Biol 99: 125 (Abstr)

Wessel G M, Marchase R B, McClay D R (1984) Ontogeny of the basal lamina in the sea urchin embryo. Dev Biol 103: 235–245

Yazaki I (1984) The egg originated and local distribution of the surface of sea urchin embryo cells detected by immuno-fluorescence. Acta Embryol Morphol Exp 5: 3–22

Yokota Y, Nakano E (1984) Two acid phosphatases in sea urchin eggs and embryos. Comp Biochem Physiol 79B: 17–22

Zimmerman S, Zimmerman A M, Fullerton G D, Luduena R F, Cameron I L (1984) Water ordering during the cell cycle of sea urchin eggs. J Cell Biol 99: 446 (Abstr)

Subject Index